Physics of
Submicron Devices

MICRODEVICES
Physics and Fabrication Technologies

Series Editors: Julius J. Muray† and Ivor Brodie

SRI International
Menlo Park, California

ELECTRON AND ION OPTICS
Miklos Szilagyi

GaAs DEVICES AND CIRCUITS
Michael Shur

ORIENTED CRYSTALLIZATION ON AMORPHOUS SUBSTRATES
E. I. Givargizov

PHYSICS OF SUBMICRON DEVICES
David K. Ferry and Robert O. Grondin

SEMICONDUCTOR LITHOGRAPHY
Principles, Practices, and Materials
Wayne M. Moreau

† *Deceased.*

A Continuation Order Plan is available for this series. A continuation order will bring delivery of each new volume immediately upon publication. Volumes are billed only upon actual shipment. For further information please contact the publisher.

Physics of
Submicron Devices

David K. Ferry and
Robert O. Grondin

College of Engineering and Applied Science
Center for Solid State Electronics Research
Arizona State University
Tempe, Arizona

Plenum Press • New York and London

Library of Congress Cataloging in Publication Data

Ferry, David K.
    Physics of submicron devices / David K. Ferry and Robert O. Grondin.
        p.   cm. — (Microdevices)
    Includes bibliographical references and index.
    ISBN 0-306-43843-7
    1. Semiconductors. 2. Solid state physics. 3. Electronics—Materials. 4. Electron
transport. 5. Microstructure. I. Grondin, Robert Oscar. II. Title. III. Series.
QC611.F42   1991   91-31155
537.6′22 — dc20   CIP

ISBN 0-306-43843-7

© 1991 Plenum Press, New York
A Division of Plenum Publishing Corporation
233 Spring Street, New York, N.Y. 10013

Printed in the United States of America

Preface

The purposes of this book are many. First, we must point out that it is not a device book, as a proper treatment of the range of important devices would require a much larger volume even without treating the important physics for submicron devices. Rather, the book is written principally to pull together and present in a single place, and in a (hopefully) uniform treatment, much of the understanding on relevant physics for submicron devices. Indeed, the understanding that we are trying to convey through this work has existed in the literature for quite some time, but has not been brought to the full attention of those whose business is the making of submicron devices.

It should be remarked that much of the important physics that is discussed here may not be found readily in devices at the 1.0-μm level, but will be found to be dominant at the 0.1-μm level. The range between these two is rapidly being covered as technology moves from the 256K RAM to the 16M RAM chips. Indeed, devices can be made much smaller, as Si MOSFETs have been made by IBM and MIT with effective gate lengths as short as 70 nm, while MESFETs and HEMTs have been made at Arizona State University and in Japan with gate lengths as short as 25 nm. Whether devices with these gate lengths will ever make it into production in VLSI chips is problematical. It is far more important to note that these devices will provide windows into the relevant physics that will set the ultimate limits on downscaling of semiconductor devices. As much as the minimum feature size (usually the gate length) on VLSI chips has shrunk, the *predicted* limit to down-scaling has shrunk, often at the same rate. Less than two decades ago, it was stated by one knowledgeable individual (whose identity shall remain unknown to the reader) that MOSFETs could never be made (in integrated circuits) with gate lengths less than 10 μm. We clearly do much better. We illustrate in Figure P1 how this predicted limit to down-scaling has progressed through the recent decades.

The current "projected" minimum feature size (indicated by the symbol next to the question mark in Figure P1) is based upon the observation in 20–25-μm HEMTs that tunneling through the depletion region prevents the devices from being "pinched off" and therefore greatly reduces the transconductance. It is not at all sure that this will be a limitation on downsizing, as clever engineers find

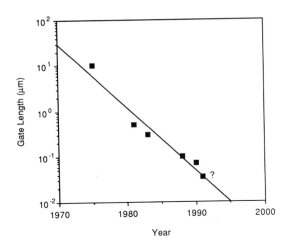

FIGURE P1. Projected minimum feature size.

other ways to introduce devices. However, the downsizing arguments point out the need to understand much physics, which has been ignored as "... second- and third-order effects, ..." in current devices. If for no other reason, it is important to understand the physics that becomes important in submicron (and ultrasubmicron) devices just as it is important to understand the circuit interaction effects.

In the process above, we try to provide a good summary of a widely scattered literature. The focus is generally on physics instead of devices, but even then the physics is not usually discussed in a conventional solid-state or semiconductor physics book. Here, rather than greatly worry over band structure and phonon spectra calculations, we instead employ the physical pictures produced by such calculations in a detailed study of electric current flow in submicron systems.

Chapter 1 reviews the material usually found in a conventional solid-state physics book, but we try to illustrate certain critical assumptions (e.g., treating scattering processes as being independent) and the weakness of certain commonly discussed ideas (e.g., tunneling times). Chapter 2, on the other hand, reviews a number of important technological processes involved in the preparation of submicron structures. Then in Chapter 3 the important properties of interfaces, whether heterojunctions between semiconductors or between an insulator and a semiconductor, are discussed. We also introduce simple models of the operation of the common devices.

Chapter 4 begins, and Chapter 5 finishes, a detailed tutorial overview of semiclassical transport theory. All commonly used, and some rarely used, schemes are discussed and their interrelationship explored. The central theme that develops is that the ensemble Monte Carlo technique is for a variety of good reasons the workhorse technique of hot-carrier transport calculations. What may not be as obvious is that many of the supposedly computationally simpler approaches may not be computationally simpler. These approaches often look simple, but when one starts asking about performing calculations in two or three spatial dimensions one finds that the equation sets grow rapidly in complexity. Such an extension, however, is basically quite trivial for Monte Carlo models. The boundary conditions also are not simplified when compared to Monte Carlo cases, and in small devices this can be quite important. Last, and often overlooked, since people usually do not publish numerical methods that fail, is the issue of numerical stability. While the Monte Carlo approach is inherently stable, the coupled sets of nonlinear partial differential equations produced in other approaches often pose formidable problems in a numerical implementation.

Chapter 5 concludes with a review of numerous attempts at experimentally seeing the effects of transient hot-carrier transport effects. The holy grail is to somehow see a velocity overshoot. We hope that this review will illustrate that while it seems clear from a variety of experiments that a velocity overshoot is occurring, a quantitative experimental observation has not yet been produced.

Chapter 6 discusses the important materials area of alloys between different semiconductors. These are both useful and interesting, but it is important to know that some "warts" appear in the simple theories that have been formulated. Then, in Chapter 7 the important electron–electron interaction and the screening effects introduced by this process are discussed. This is followed in Chapter 8 by a discussion of the relatively new area of surface superlattices.

We end the book in Chapters 9 and 10 with a discussion of quantum transport and of noise in submicron devices. These are areas in which very few experiments have been performed and relatively few calculations done. Perhaps the most important observation there is that some of the guesses that one is tempted to make about the spatial and temporal correlations seen in hot-carrier transport are not only wrong but in some cases predict trends opposite to the correct ones. We suspect that a detailed understanding of the noise limits faced in submillimeter-wave systems still lies outside the limits of our theories.

It is assumed that the reader has a basic understanding of solid state physics and electronics, at least to the level of a text such as Kittel's or Ashcroft and Mermin's. This also assumes that the reader has had an introductory course in quantum mechanics, although such important topics as tunneling and time-dependent perturbation theory are reviewed in Chapter 9, where they are important for very small devices and very high electric fields.

David K. Ferry
Robert O. Grondin

Tempe, Arizona

Contents

1

An Introductory Review

1.1. GENERAL INTRODUCTION

Semiconductor technology has long been dependent on controlling or at least using phenomena that occur on very short time and space scales. Millimeter-wave devices such as Gunn diodes or impact avalanche transit-time diodes (IMPATTs) depend critically on controlling the phase shifts between particle currents and terminal voltage waveforms in frequency ranges where times of less than a picosecond can visibly alter device performance. Monolithic silicon metal-oxide semiconductor field-effect transistor (MOSFET) technology depends on creating an inversion layer whose width is on the order of 100 Å and in which purely quantum mechanical "size" effects can be seen. The above examples are old. Arguably we have been lucky in these millimeter-wave devices in that phenomena occurring on times which seemed to lie beyond the resolution of any foreseeable measurement system actually provided the base for working devices. In the MOSFET world these quantum effects did not provide a basis for any actual devices but merely complicated the understanding of certain parameters used in the device model. Technological advances have now created a situation however in which we can fabricate semiconductor structures of submicron and even nanometer dimension in which quantum and other "novel" physical mechanisms are used in device operation and other advances have simultaneously created situations in which it is possible to perform measurements with subpicosecond resolution.

One example of a relatively new structure of nanometer dimension is the two-dimensional carrier gas[1] formed at the heterojunction interface between two semiconductors of different electronic structure. The MOS system, of course, is the oldest such example, but in more recent years the combination of hetero-structures between two lattice-matched crystalline semiconductors has formed the basis for many such systems. Of particular importance is the combination of such a heterostructure with a selective doping of the two adjoined semiconductor layers in the high-electron-mobility transistor (HEMT). Such transistors are excellent low-noise, high-frequency amplifiers, and have been used in some of the fastest logic gates ever built. There are other means of forming structures in which size quantization effects occur. For example, by combining several hetero-junctions it is possible to form quantum wells.[2] Quantum wells provide the basis for interesting optical devices. Additionally, we can make a quantum well which has thin barrier regions separating it from its contacts. Such resonant tunnel diodes[3] exhibit a negative differential conductance, which has allowed them to be used as oscillators for frequencies in the millimeter-wave range, and their nonlinearity has been used to sense radiation in the far infrared. Various proposals for forming three-terminal transistors with these structures exist as well. By stacking many layers of alternate materials together, one can form a superlattice.[4] The periodicity of the superlattice leads to the creation of minibands and minigaps in the energy bands of the system which are not seen in normal bulk semiconductors. Such structures offer interesting potential for sensors and other nonlinear devices as well.

The above structures all use nanometer epitaxial techniques for their fabrication. Electron-beam and x-ray lithography can also be used to fabricate nanometer devices. For example, nanometer lithography also can be used to define a periodic array of metallization on the surface of a sample, a structure known as a lateral surface superlattice (LSSL).[5] The LSSL has potential advantages not shared by superlattices built by epitaxial techniques. We also can imagine combinations of more than one method of obtaining a novel submicron device. One such possibility is the Bloch FET,[6] an FET device in which an LSSL is deposited inside an oxide over which a separate metal gate is then applied. Alternatively it can be formed by using an LSSL as the gate metallization of an HEMT device.

Recent years have also seen revolutionary advances in our ability to perform fast temporal measurements. These have been driven by the development of lasers which provide optical pulses which are less than 10 fs long. Such pulses can be used to excite transients in semiconductors and can also be used to probe the response of such systems.[7] Microwave engineering meanwhile has also advanced significantly both in terms of the accuracy of the measurements and in terms of the development of commercially available instrumentation for the low-millimeter-wave region. This has created a situation in which it is possible to think of quantitatively comparing theory and experiment on the subpicosecond scale.

The world seen in technology therefore is changing. Space scales and time scales which invalidate many of the common assumptions used in discussing semiconductor structures and measurements are now attainable in a variety of institutions, and ongoing experimental studies are pressing the limits of our understanding. In this book we will explore this situation. There are two main perspectives for this effort and therefore two main audiences for this book. The first is an engineering perspective in which we hope to obtain devices which perform interesting electronic functions. The other perspective is that of basic science, where we hope to deepen our understanding of the physics and chemistry of solids and interfaces.

In this chapter we will try to lay a foundation for the remainder of the book. This will primarily involve a brief review of semiconductor physics and the central problems faced in a semiconductor device model. While a single chapter obviously will not replace an entire text on semiconductor physics, we do hope to accomplish several main tasks that are not always achieved in ordinary texts. First, we hope to clearly establish the connections of semiconductor physics with Maxwell's equations and to clarify the role which, from the engineering perspective, must be played by most of the physical discussions in this book. This role is the establishment of a set of constitutive relations which when combined with Maxwell's equations provide a complete description of the physics of an individual device or measurement problem. We also will discuss another feature of the physical systems of interest here which differs from commonly seen textbook situations. There, one often discusses an infinite bulk semiconductor, while here we often have very small regions in which interfaces and boundaries play a dominant role. These interfaces and boundaries (contacts) must be properly treated. Having established these goals, we will then briefly review the main structure of semiconductor physics. As we will be dropping large amounts of computational and analytical detail, we can sharply define the main physical assumptions made in this development. Having outlined the forest, we then will introduce some of the specific problems which will be addressed in the course of the book.

1.2. CONSTITUTIVE RELATIONS

Whether we call our small, fast semiconductor structure a device or a sample, it will be embedded in an electrical or optical system. As the structures will interact with macroscopic fields, we will have to eventually solve Maxwell's equations if we are to understand their properties, and these equations serve as our starting point. They are

$$\nabla \times \mathbf{F} = -\frac{\partial \mathbf{B}}{\partial t}, \tag{1.1}$$

$$\nabla \times \mathbf{H} = \frac{\partial \mathbf{D}}{\partial t} + \mathbf{J}_{\text{cond}} = \mathbf{J}_{\text{tot}}, \tag{1.2}$$

$$\nabla \cdot \mathbf{D} = \rho, \tag{1.3}$$

$$\nabla \cdot \mathbf{B} = 0, \tag{1.4}$$

and

$$\nabla \cdot \mathbf{J}_{\text{cond}} = -\frac{\partial \rho}{\partial t}, \tag{1.5}$$

where **F** is the electric field, **D** the electric flux density, **H** is the magnetic field, **B** is the magnetic flux density, $\mathbf{J}_{cond}$ is the electrical conduction current density, and ρ is the electric charge density. Only three of the above equations, two vector and one scalar, are independent while we have five vector variables and one scalar variable. Therefore, we need three additional vector equations. These additional equations, generally referred to as constitutive equations, will differ from Maxwell's equations in that they are problem-dependent and not universal.

Our problem is illustrated in Figure 1.1. We have four interacting systems: fields, charges (both mobile and fixed), intrinsic magnetic moments, and a crystal lattice. Maxwell's equations describe one-half of the interactions, and our problem is to describe the other half. As shown, the mobile charge carriers exchange energy and momentum with both the fields and the lattices. Maxwell's equations completely describe the interaction of the fields with the currents and charges. Since the fields interact with only currents and charges, these equations can be universal. The currents and charges, however, interact with more than the fields, and additional equations are needed to describe how the interactions of currents and charges with the rest of the system constrain their interaction with the fields.

We even know what the nature of the three additional constitutive relationships must be. The fields exert a force on any charges in the system, and this force is described by the Lorentz force law

$$\mathbf{F}_L = q_t(\mathbf{F} + \mathbf{v} \times \mathbf{B}), \tag{1.6}$$

where q_t is a point charge moving with velocity **v**. Examination of (1.6) shows that only the electric field **F** can change the energy, and time-averaged momentum, of a mobile charged body. Therefore we expect to find some relationship containing dissipative effects between $\mathbf{J}_{cond}$ and **F**. While the magnetic field does not directly provide energy to the charge, it can alter the total path length followed by the particle in a two- or three-dimensional system as the particle traverses the system. This increased path length allows for greater dissipation, and therefore

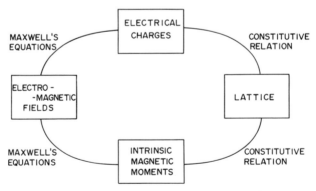

FIGURE 1.1. The interacting systems considered are shown in this figure. Maxwell's equations describe the interaction of the fields with the charges, while additional interactions can occur between the charges and the lattice and between the fields and the lattice.

we expect to see a potential magnetic-field dependence in our conductivity relations. There may be other charges in the system which, while not truly mobile, will be displaced somewhat by the force of the electric field $\mathbf{F}$, thus producing a polarization charge density. Through Gauss's law we therefore expect to see a corresponding polarization shift in the electric flux density $\mathbf{D}$. This shift is described by the second constitutive relationship, a relationship between the electric field $\mathbf{F}$ and the electric flux density $\mathbf{D}$. This relationship describes the dielectric properties of the physical system. Similarly we expect to see a last constitutive relationship between $\mathbf{B}$ and $\mathbf{H}$ as a result of the interaction between the magnetic field and any intrinsic microscopic magnetic moments.

1.3. SURFACES, INTERFACES, AND BOUNDARY CONDITIONS

The other factor needed in the determination of the performance of a submicron semiconductor structure is a set of boundary conditions. This is in fact usually the predominant factor in device design as well as modeling. It is our ability to use interfaces to control current flow that underlies the operation of the devices of interest. Beyond this practical point of view lies the fact that the physical and chemical processes that are associated with surfaces and interfaces offer a variety of puzzles and intriguing results. The study of how we can understand, design, build, and test these interfaces forms the second main focus of the book.

Ordinarily, the fact that we do not really know what is happening at the boundaries of our device and therefore cannot actually specify a correct set of boundary conditions is not viewed as a significant problem in the modeling of a semiconductor device. We argue that the errors introduced through our ignorance will be important only near the contacts. Here, however, we will be considering submicron structures where all points of interest may be near the surface at which the questionable boundary condition is applied.

The situation is complicated as even boundary conditions that appear quite natural and innocuous can introduce spurious boundary effects.[8] One simple example is a widely used boundary condition on potential at an interface across which no current flows. It is common to assume a Neumann condition ($\mathbf{n} \times \mathbf{F} = 0$) for the electric potential at this interface, and it is sometimes even argued that this is a physically correct consequence of there being no current flow across the interface. This is, of course, not true as it is the gradient of the electrochemical potential, as discussed in Chapter 4, and not the gradient of the electric potential, that determines current flow. Furthermore it is simple to show that the assumption of a Neumann condition on potential at this interface is equivalent to introducing a nonexistent contribution to either the surface charge at the interface or the field on the other side. We need only to solve the problem shown in Figure 1.2. The correct boundary conditions are

$$\mathbf{n} \times \mathbf{F}_1 = \mathbf{n} \times \mathbf{F}_2 \qquad (1.7)$$

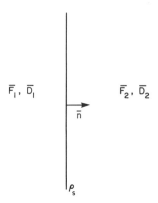

FIGURE 1.2.   The dielectric interface boundary conditions.

and

$$\mathbf{n} \cdot (\mathbf{D}_2 - \mathbf{D}_1) = \rho_s, \tag{1.8}$$

where $\mathbf{n}$ is the surface normal vector and ρ_s is the real surface charge density.

The Neumann boundary condition incorrectly asserts that the gradient of the electric potential perpendicular to the interface is zero on the side which corresponds to the semiconductor region of interest. This, of course, means that the perpendicular field component in the semiconductor goes to zero at the interface. Examination of (1.8) shows that this assumption cannot be made without the introduction of either a spurious field on the other side of the interface or a spurious contribution to the surface charge at the interface. Therefore, as we noted at the beginning of the discussion, an innocuous boundary condition is introducing artifacts into the solution. Strangely enough, this fallacious argument is only made when the insulating material which has been placed in contact with the semiconductor is air. If the insulator is silicon dioxide, not only is the field at the interface not set to zero, it is used as a parameter which controls the transport process through the inversion layer.

Surfaces which pass currents are also of great importance and also pose interesting problems. It is important to remember that for these surfaces, the contacts to our devices and samples, there exist boundary conditions on charge and potential which play an important role in connecting the seemingly inaccessible internal physics of the device with measurable external currents. The vehicle by which this occurs is the induced current or Ramo–Shockley theorem.[9,10] Consider the situation shown in Figure 1.3. We have a point charge located at x moving with velocity v between two ideal conductors. The existence of an applied potential produces an electric field in the region lying between the two conducting plates. Since these plates are ideal conductors, the electric flux vector is zero in the interior of the plates. Therefore a surface charge will be induced on them by both the applied potential and any charges such as our test charge. The contribution of the test charge, however, is controlled geometrically and

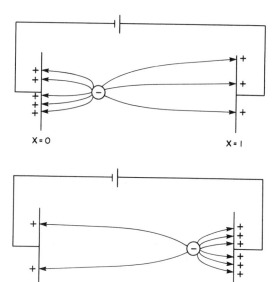

FIGURE 1.3. The simplest case of the induced-current theorem. As the charge moves between the two conductors a current must flow in the external circuit which supplies the induced surface charges on the conducting plates.

therefore as our test charge moves through the system the induced surface charge on the plates must vary in time. This in turn requires a current flow in the external circuit. Note that the induced current is determined by the necessity of maintaining the proper boundary conditions at the device-circuit interface.

The microscopic situation at an interface or surface is even more difficult. The general difficulty will be that in any quantum mechanical calculation the boundary conditions must be restrictions on the wave functions. Choosing such a set of restrictions can be difficult, especially if one wishes to model current flow (we will return to this point later in this chapter when we discuss tunneling). There are problems as well in situations where we have adopted a semiclassical, particle-based model and must decide how many particles enter the region being simulated from the exterior on a per unit time basis. This is easy for the steady-state case where electric current continuity demands particle current continuity. Then, provided one knows how many carriers are inside the device, one can simply reinject a carrier from one end whenever it leaves the other. In a transient situation, however, electric current continuity does not demand particle current continuity, and the number of carriers inside the region being simulated is in fact changing in time. Simple reinjection schemes therefore are fallacious for these transient situations.

The difficulties associated with the microscopic boundary conditions lie at the back of a very widely discussed point of interest: the alignment of energy bands at or near an interface. There is no really good method for calculating the electronic quantum states at a surface or interface. The tendency is to model these systems as an abrupt transition from the energy states of one system to the energy states of the other, with some additional surface or interface states thrown in at or near the interface. The central issue of debate concerns how we align the bands in these situations. We will return to this question at various points throughout the book.

1.4. STATES, BANDS, ZONES, AND LATTICES

One of the central topics of concern in this book will be the development of constitutive relations for use in studying submicron semiconductor devices. The additional constitutive relations which we must develop describe the interaction between the charges and the lattice, and interactions between the applied fields and the lattice. This means that we necessarily must attack the somewhat tricky problem of relating microscopic physics to macroscopic variables. To some extent therefore, our approach must differ from that commonly found in field theory texts where it is commonly assumed, for example, that the polarization charge mentioned above occurs on a space scale which is smaller than those of interest in the field solution.[11,12]

The starting point in the development of a constitutive relation from first principles is quantum theory. While we will more often start a few steps away from first principles, a quick overview is useful. We have an assembly of atoms arranged in a crystal lattice. Some of the electrons participate in bonding or are relatively mobile and contribute to charge transport. Other electrons remain in atomic core states and are closely associated with the positive charge of some nucleus, and we treat these localized atomic charges as ions located at or near lattice sites. A true first-principles calculation therefore would use a crystal Hamiltonian which would include energy contributions from all of the electrons and all of the ions and solve for a total crystal wave function which would describe all of the electrons and all of the core ions. *We will not solve this problem, and no one else does either.* Instead a series of approximations and assumptions are made that allow us to separate various manageable pieces of the big problem. There is a general hierarchy of subdivisions which is usually followed.

1. Separate the ionic motion away from the electronic motion via the Born–Oppenheimer method, in which the so-called adiabatic approximation is employed. This allows us to separate our total wave function into the product of an electronic wave function and an ionic wave function. Physically, it is related to an adiabatic approximation in which we assert that as the ionic masses are at least three, and generally four, orders of magnitude larger than the electronic mass, we can assume that the electrons will adjust essentially instantaneously to changes in the ionic positions. This divides our big problem into two separate problems: a lattice problem with a lattice Hamiltonian and an ionic wave function, and an electronic problem with an electronic Hamiltonian and an electronic wave function.

2. We now assume that the ions may move away from their equilibrium positions. A force which tends to restore the ions to the equilibrium

position is included in the lattice Hamiltonian, and we solve for the lattice vibrations, with the solution being expressed in terms of phonons and the phonon dispersion diagram. Generally we do this for the harmonic case in which the restoring force is directly proportional to the distance by which the ion is displaced. This is quite good for relatively small displacements. For larger displacements, anharmonic terms appear. These are of importance as they couple various phonon modes, thereby allowing energy to be converted from one mode to another.

3. We place the ions in their equilibrium lattice positions and then solve the electronic problem for the electronic states. This produces the band structure or E-$\mathbf{k}$ diagram.

4. We then can consider the dynamical evolution of the electrons. This is done by applying perturbation theory with the basis states being those described by the E-$\mathbf{k}$ diagram. There are two perturbing influences which must be considered: applied force fields and various disturbances in the lattice, including defects, impurities, and phonons.

As the electrons are charged there must be electron–electron interactions. In steps 2 through 4 some choices need to be made about how these latter many-body interactions are to be modeled. The basic choices are to (a) ignore them totally and do a pure single electron calculation; (b) treat the electron-electron interaction as an average background effect which may readjust some of the energy levels and states; or (c) attempt an actual many-body calculation. The electrons also interact in a nonclassical fashion through the Pauli exclusion principle, a fact which usually appears through some exchange integral.

Almost all of the discussion presented here will concern itself with crystalline semiconductor structures. The fact that we have a crystal lattice, of course, introduces a variety of useful tools, some of which we will now briefly review. We start with a lattice in real space with underlying lattice vectors $\mathbf{a}_i$, where i is 1, 2, or 3. Any two lattice points can be connected by a vector $\mathbf{R}$ which can be expressed as a linear combination of $\{\mathbf{a}_i\}$ with integer coefficients. Since we have a periodic system, it is natural to consider Fourier analysis and expect that there will be some set of harmonic spatial frequencies which are closely related to the real-space lattice constants or periods. These harmonic spatial frequencies are the reciprocal lattice axis vectors $\{\mathbf{A}_i\}$ defined by

$$\mathbf{A}_1 = 2\pi \frac{\mathbf{a}_2 \times \mathbf{a}_3}{\mathbf{a}_1 \cdot \mathbf{a}_2 \times \mathbf{a}_3}, \tag{1.9a}$$

$$\mathbf{A}_2 = 2\pi \frac{\mathbf{a}_3 \times \mathbf{a}_1}{\mathbf{a}_1 \cdot \mathbf{a}_2 \times \mathbf{a}_3}, \tag{1.9b}$$

and

$$\mathbf{A}_3 = 2\pi \frac{\mathbf{a}_1 \times \mathbf{a}_2}{\mathbf{a}_1 \cdot \mathbf{a}_2 \times \mathbf{a}_3}. \tag{1.9c}$$

They satisfy

$$\mathbf{A}_i \cdot \mathbf{a}_j = 2\pi\delta_{ij}. \tag{1.10}$$

We can express any function $h(\mathbf{r})$ whose periodicity is that of the lattice as

$$h(\mathbf{r}) = \sum_{\mathbf{G}} h_{\mathbf{G}} \exp(i\mathbf{G} \cdot \mathbf{r}), \tag{1.11}$$

where $\mathbf{G}$ is a reciprocal lattice vector formed from a linear combination of integer multiples of $\{\mathbf{A}_i\}$.

In a one-dimensional space lattice, a spatially varying phenomena like an electromagnetic wave is sampled at each of the equally spaced lattice points. The sampling theorem tells us there is a maximum spatial frequency for which we can meaningfully extract information about the "spectrum" of the signal by this sampling process. We can extend this viewpoint and view three-dimensional real-space lattice points as being locations at which we sample some three-dimensional signal. The result will be a maximum three-dimensional spatial frequency vector about which information is retained. The mathematical space in which the spatial frequency vectors are found will have a lattice structure whose lattice vectors are the reciprocal lattice vectors described above. This lattice is called the reciprocal lattice. The portion of this reciprocal lattice which contains the set of spatial frequencies adequately described by the three-dimensional spatial sampling is the first Brillouin zone.

In Figure 1.4 we illustrate the interweaved face-centered cubic (fcc) lattice structure of both the diamond lattice (Si) and the zinc-blende lattice (most III-V compounds). This lattice consists of an underlying fcc lattice in which we insert two atoms for each lattice point, one at (000) and the other at $(a/4, a/4, a/4)$. The corresponding first Brillouin zone is shown in Figure 1.5. It is a truncated body-centered cube in which a plane, perpendicular to the line from the cube center to the cube corner, bisects this line to form one face of the Brillouin zone. There are three main symmetry points: Γ which lies at the center, L which lies at the point where the line from center to corner passes through the zone boundary, and X which lies on the cube face along the line passing between the centers of two adjoining cubes. There are obviously three X points and four L points, as neighboring zones share half of each point.

Once the adiabatic approximation has been used and our lattice points identified, we then proceed to a solution of the energy bands. The first challenge will be to reduce our many-electron problem to a more tractable single-electron problem. We will briefly review this procedure. The outline we present is that followed by Madelung.[13] We start with a many-electron wave function $\Phi(\mathbf{r}_1, \mathbf{r}_2, \ldots, \mathbf{r}_N)$ which is the solution to Schrödinger's equation with a many-electron Hamiltonian of the form

$$H = \sum_{j} H_j + \sum_{j'} H_{jj'}, \tag{1.12}$$

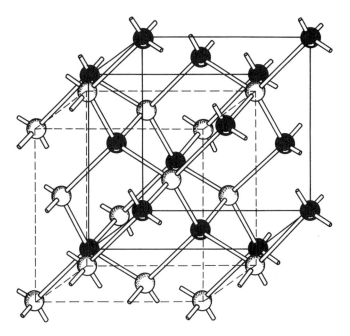

FIGURE 1.4. Two interweaving face-centered cubic lattices make up the diamond or zinc blende lattices.

where H_j is a single-particle Hamiltonian which contains the electron-ion interaction for an electron located at $\mathbf{r}_j$, and $H_{jj'}$ describes the Coulombic electron-electron interaction of electrons at $\mathbf{r}_{j'}$ and $\mathbf{r}_j$. The ansatz which we would like to make is that the many-electron wave function can be written as a product of single-electron wave functions; that is,

$$\Phi(\mathbf{r}_1, \mathbf{r}_2, \ldots, \mathbf{r}_N) = \prod_j \phi_j(\mathbf{r}_j). \tag{1.13}$$

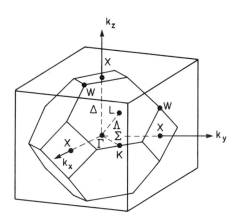

FIGURE 1.5. The first Brillouin zone of the fcc lattice. The Γ, X, and L symmetry points are shown.

The presence of the electron–electron interaction term does not allow us to do this. Instead we use a variational method and minimize the expectation value of the energy, that is, $\langle \Phi | H | \Phi \rangle$, using (1.13). A set of Lagrangian multipliers E_j are introduced, and we find that the expectation is minimized for single-electron wave functions which solve the Hartree equation

$$\left[-\frac{\hbar^2}{2m} \nabla^2 + V(\mathbf{r}) + \frac{e^2}{4\pi\varepsilon_0} \sum_{j' \neq j} \int \frac{|\phi_{j'}(\mathbf{r}_{j'})|^2}{\mathbf{r} - \mathbf{r}_{j'}} \, d\mathbf{r}_{j'} \right] \phi_j(\mathbf{r}) = E_j \phi_j(\mathbf{r}), \quad (1.14)$$

where ε_0 is the electrical permittivity of free space. The Lagrangian multipliers E_j are obviously energy eigenvalues of the Hamiltonian of (1.14). The Hartree equation describes a single electron moving under the influence of the electron-ion interaction, which is included in $V(\mathbf{r})$, and a Coulomb potential of the average spatial distribution of the other electrons.

In the above we did not include the Pauli exclusion principle. This is done by first using combinatorial arguments concerning how N electrons could be distributed at the N locations $\mathbf{r}_j$. The result is the Slater determinant representation

$$\Phi = (N!)^{-1/2} \begin{vmatrix} \phi_1(\mathbf{r}_1, \mathbf{s}_1) & \cdots & \phi_N(\mathbf{r}_1, \mathbf{s}_1) \\ \vdots & & \\ \phi_1(\mathbf{r}_N, \mathbf{s}_N) & \cdots & \phi_N(\mathbf{r}_N, \mathbf{s}_N) \end{vmatrix} \quad (1.15)$$

where $\mathbf{s}_j$ are spin coordinates. Note that if evaluated this determinant will be a complicated function of products of individual wave functions. Interchanging two electrons, however, will interchange two columns of the determinant and thereby flip the sign of Φ. This is the antisymmetry with respect to particle interchange required by the exclusion principle.

Using the Slater determinant, we would again repeat our variational calculation of the energy expectation. After some algebra we would obtain a new equation which must be solved by the functions $\phi(\mathbf{r})$ in order to minimize $\langle E \rangle$. This is the Hartree–Fock equation and consists of the Hartree equation with an additional term on the left-hand side. The additional term is the exchange interaction term

$$\frac{e^2}{4\pi\varepsilon_0} \sum_{j' \neq j} \int \frac{\phi_{j'}^*(\mathbf{r})\phi_j(\mathbf{r}')}{\mathbf{r} - \mathbf{r}'} \, d\mathbf{r}'.$$

The Lagrangian multipliers E_j can be shown to correspond to single carrier energies. The last complication which we face is that the exchange interaction term depends explicitly on ϕ_j. This complication can be dealt with by the Slater approximation,[13] in which we average the portion of this term which represents the interaction of two different electrons over j. The result of this exercise is the production of a one-electron Schrödinger equation

$$\left[-\frac{\hbar^2}{2m} \nabla^2 + V(\mathbf{r}) \right] \psi(\mathbf{r}) = E\psi(\mathbf{r}), \quad (1.16)$$

where $V(\mathbf{r})$ contains both the lattice potential and the averaged interaction terms of the Hartree–Fock approximation. Therefore even though we are performing a single-electron calculation, we can include at least some form of averaged electron–electron interactions of both Coulombic and exchange natures. These interactions have the effect of shifting the energy levels in our solution. Note that the discussion which now follows does not critically depend on their presence.

In a one-electron model we solve the Schrödinger equation for a background potential which, while unknown in detail, will have the same spatial periodicity as the crystal lattice. This allows us to use the Bloch–Floquet theorem and assert that the electronic wave functions will have the form

$$\psi_{\mathbf{k}}(\mathbf{r}) \sim \exp(i\mathbf{k} \cdot \mathbf{r})g_{\mathbf{k}}(\mathbf{r}), \tag{1.17}$$

where $g(\mathbf{r})$ is a function with the same periodicity as the lattice, and the vector $\mathbf{k}$ which appears in the exponential factor is formed from the reciprocal lattice vectors. The corresponding states and functions are Bloch states and functions.

When we actually carry out the solution for our states and wave functions, we find bands of allowed energies separated by energy gaps. These bands are periodic in k-space with the period equal to the width of the Brillouin zone. We typically therefore must supplement a description of a Bloch state by assigning a band identifier to the description; i.e.,

$$\psi_{\mathbf{k},n}(\mathbf{r}) \sim \exp(i\mathbf{k} \cdot \mathbf{r})g_{\mathbf{k},n}(\mathbf{r}), \tag{1.18}$$

where n is the band identifier.

In Figure 1.6 we show the band diagrams for Si and GaAs. The main features of the Si band structure is that it is an indirect gap semiconductor with valence band maxima located at Γ, while the conduction band minima are located approximately 80% of the way out along the [100] direction out toward the X points. The corresponding constant-energy surfaces are shown in Figure 1.7 and consist of six ellipsoids lying along the [100] axes with centers at the conduction band minima. GaAs, on the other hand, is a direct-gap semiconductor with both valence band maxima and conduction band minima lying at Γ. The corresponding constant-energy surface shown in Figure 1.7 is a sphere centered at Γ. Of great importance, however, are the satellite valleys or higher conduction band minima located at L and at X. Due to the periodicity of the bands across the Brillouin zone there will be three equivalent X valleys and four equivalent L valleys (one-half of each valley lies in an adjacent zone, as mentioned above).

The bands shown were calculated by Chelikowsky and Cohen,[14] using pseudopotentials. There are a wide variety of techniques for calculating band structures. We have already illustrated at least some of the difficulties with our discussion of the underlying approximations needed to obtain a single-electron equation from the many-electron problem. There are situations of technological importance where these many-body effects become important, the most obvious example being band-gap narrowing effects in the heavily doped and strongly-forward-biased emitter regions of bipolar junction transistors. Beyond that there

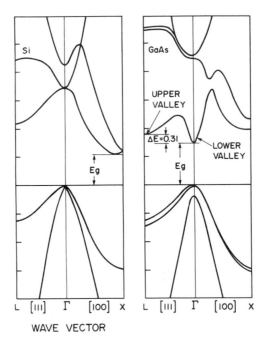

FIGURE 1.6.   The energy bands of Si and GaAs. After Chelikowsky and Cohen.[14]

are other difficulties as well. The most prominent of these is that we really do not know the details of the functional form of the underlying lattice potential. It is important to remember that because of these difficulties one encounters increasing uncertainty, both theoretical and experimental, as to the nature of the bands as one moves away from the symmetry points. It is possible to incorporate "accurate" band structures in modern device models. Appropriate caution should be taken, however, in evaluating the contribution made from the extra details of such bands. Theorists cannot even calculate band gaps correctly (values which are 25% too small are viewed as excellent fits), and data such as shown in Figure 1.7 usually have been "adjusted." These band calculations cannot currently fit

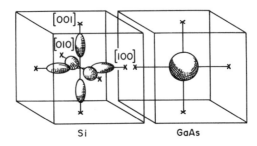

FIGURE 1.7.   The constant-energy surfaces of Si and GaAs.

both conduction band gaps and effective masses even after "adjustment." Thus, claims that significant improvements in accuracy are obtained by including detailed band structures must be viewed with some skepticism. However, genuinely self-consistent calculations may reveal hidden inconsistencies in conventional models, and with improved band structure determination may lead to better accuracy, particularly for situations in which many carriers are excited into states well away from conduction band minima and valence band maxima.

There also exists a density of states for each of these bands. Consider a macroscopic object constructed from unit cells with lattice vectors $\{a_i\}$. We construct our object such that there are $\{N_i\}$ unit cells lying along the three directions. We then apply a boundary condition on the Bloch states at all three surfaces of the macroscopic object. Generally, a periodic boundary condition is used; that is,

$$\psi_{\mathbf{k},n}(\mathbf{r}) = \psi_{\mathbf{k},n}(\mathbf{r} + N_i \mathbf{a}_i). \tag{1.19}$$

This boundary condition is obviously arbitrary and corresponds to a strange topology for our three-dimensional case. (A ring forms in one dimension, a doughnut in two.) Its use is usually justified by arguing that our macroscopic object is indeed macroscopic. The k-vectors of the Bloch functions therefore must satisfy

$$\mathbf{k} \cdot N_i \mathbf{a}_i = n2\pi, \tag{1.20}$$

where n is an integer. The volume of a state in k-space then must be $(2\pi)^3/V$, where V is the volume of the macroscopic object. Remembering that there are two spin states associated with every k, we can write a density of states per band for k-space as

$$g(\mathbf{k}) \, d\mathbf{k} = \frac{1}{4\pi^3} \, d\mathbf{k}. \tag{1.21}$$

The density of states can also be expressed as a function of energy. This is done by realizing that we wish to express the total number of states with energy of E or less, $N(E)$, in terms of an integration of an energy-dependent density-of-states function $g(E)$. This will correspond to an integration over the k-space region enclosed by the constant-energy surface with energy E. The k-space integration can be split into an integration over energy and a surface integral over constant energy surfaces (which may not necessarily be spherical). For band index n we then obtain

$$N(E) = \int g(E) \, dE = \int_E g(\mathbf{k}) \, d\mathbf{k} = \frac{1}{4\pi^3} \int_{E_{\min}}^{E} \int_{E_n \text{ const}}^{E} \frac{dE_n \, ds}{\nabla_k E_n(\mathbf{k})}. \tag{1.22}$$

The energy-dependent density-of-states function therefore is

$$g_n(E)\, dE = \left[\frac{1}{4\pi^3} \int_{E_n \text{ const}} \frac{ds}{\nabla_k E_n(\mathbf{k})} \right] dE. \qquad (1.23)$$

The density of states for a free electron is

$$g(E)\, dE = \frac{1}{2\pi^2} \left(\frac{2m}{\hbar^2} \right)^{3/2} E^{1/2}\, dE. \qquad (1.24)$$

It is common to define a density-of-states effective mass that allows us to use the form of (1.24) as the result of (1.23). In general, this effective mass will be energy-dependent and affected by the underlying deviations of the constant-energy surfaces from a spherical form

$$m^*(E)^{3/2} = \frac{\hbar^3}{2^{5/2} \pi E^{1/2}} \int_{E_n \text{ const}} \frac{ds}{\nabla_k E_n(\mathbf{k})}. \qquad (1.25)$$

For this reason, elliptical energy surfaces are often more useful. There the effective mass is split into longitudinal (m_1) and transverse (m_t) components. The overall effective mass is $(m_1 m_t^2)^{1/3}$.

1.5. BALLISTIC ELECTRON DYNAMICS IN LATTICES

We now begin our study of how the electrons respond to the perturbing influences of applied fields and lattice imperfections and vibrations. We first start by considering only an applied force field. For the present, we will follow the usual approach, which Gunn[15] aptly described as sweeping the quantum mechanics under rugs called "band structure" and "effective mass." Later, most notably in Chapter 9, we will look under these rugs. For now, though, we stay on top. The central idea is to use an effective mass to conceal the fact that our electron is not a truly free particle. We start with the energy-band diagram and use the group velocity as our carrier velocity; that is,

$$\mathbf{v}_g = \frac{1}{\hbar} \nabla_k E. \qquad (1.26)$$

If we apply a force $\mathbf{F}'$, then the electron energy must change with time in accordance with

$$\frac{dE}{dt} = \mathbf{F}' \cdot \mathbf{v}(\mathbf{k}) = \frac{\mathbf{F}'}{\hbar} \cdot \nabla_k E. \qquad (1.27)$$

However, we also note that the change in electron energy must also satisfy

$$\frac{dE}{dt} = \frac{d\mathbf{k}}{dt} \cdot \nabla_k E,$$ (1.28)

thereby leading us to conclude that

$$\mathbf{F}' = \hbar \frac{d\mathbf{k}}{dt}.$$ (1.29)

This equation, which serves as an equivalent to Newton's law, can be derived in other ways[16] more clearly related to the view of the field as a perturbation. The argument presented above essentially is an application of the correspondence principle and asserts that the motion of a packet of waves whose dispersion diagram is the E-k diagram must in the appropriate limit reproduce the trajectory of the corresponding classical particle.[17] The function $\hbar\mathbf{k}$ is called the crystal momentum and may differ greatly from the true momentum. If one recalls that the Bloch state functions are a product of an exponential of $\mathbf{k}$ and a prefactor with the same periodicity as the lattice, it is easy to show by calculating the expectation value of the true momentum that most of it is associated with the prefactor while the crystal momentum is associated with the exponential. Only the exponential term, however, is affected by an external field.

Since we really would like to predict the electronic motion in real space, we would like to write an equation of the form

$$\mathbf{F}' = \mathbf{m}^* \frac{d}{dt} \mathbf{v}_g,$$ (1.30)

where $\mathbf{m}^*$ is an effective mass tensor. By applying equations (1.27) and (1.30), it is easy to show that

$$(m^*_{ij})^{-1} = \frac{1}{\hbar^2} \frac{\partial^2 E(\mathbf{k})}{\partial k_i \, \partial k_j}$$ (1.31)

must be the elements of our effective mass tensor. We will refer to this mass as the density-of-states effective mass. A different mass is obtained if we instead choose to equate the crystal momentum to the product of the group velocity and this different effective mass; that is,

$$\hbar k = m^*_d v_g.$$ (1.32)

For a one-dimensional system it is easy to show that this second effective mass, which we will call the dynamic mass, equals the density-of-states effective mass for the case of a parabolic band. In general, however, they are not the same.

If $E(\mathbf{k})$ is a parabolic function, then we reclaim the normal relations expected for a free particle using either effective mass. However, $E(\mathbf{k})$ is not a parabolic function for electrons in a semiconductor. Our energy must in fact be a periodic function of $\mathbf{k}$, with a period set by the width of the first Brillouin zone and therefore cannot be a parabolic function of $\mathbf{k}$.

This periodicity leads to at least one counterintuitive result of importance here. Consider a simple one-dimensional sinusoidal energy band

$$E(k) = E_0 \cos(ka) \tag{1.33}$$

and a constant applied force F_A. Then k is a linear function of t with slope $F_A/\hbar$, and we have a time-varying velocity of the form

$$v_g(t) = \frac{1}{\hbar} \left(\frac{dE}{dk} \right)_t = -\frac{E_0 a}{\hbar} \sin \left\{ a \left[\frac{F_A t}{\hbar} + k(0) \right] \right\}. \tag{1.34}$$

Integrating the time-varying velocity, we find that the position of the electron is

$$x(t) = \frac{E_0}{F_A} \cos \left\{ a \left[\frac{F_A t}{\hbar} + k(0) \right] \right\} + x(0). \tag{1.35}$$

Something strange is happening here. We have applied a constant force and yet our electron is undergoing a purely periodic oscillation in real space and on a time-average basis goes nowhere. In the most simplistic picture, we envision the electron as having first traveled with the field, gaining energy from it until the electron energy rose to the band-limited maximum value. Then, for strange quantum mechanical reasons, it turned around and traveled against the field as it returned to its original location and, in the process, returned all of the previously gained energy to the field. While this result, the Zener or Bloch or Zener–Bloch oscillation, was theoretically noted over 50 years ago[18] it still remains a strange, experimentally unverified, and somewhat controversial idea that lies at the very heart of most of semiconductor physics.

1.6. TUNNELING

Tunneling is another quantum phenomena which will be important in many of our discussions. As Zener[18] noted, it sets an upper bound on the fields for which the Bloch oscillation is expected. The role of the field is usually visualized as shown in Figure 1.8. It sets a slope to the energy bands in real space. Therefore when the electron reaches the top of one of our bands during the course of its Bloch oscillation, it has the same total energy as it would have if it was in the minimum-band energy state of the next highest band and located a distance

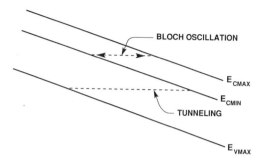

FIGURE 1.8. Simple energy bands in the presence of an applied field. The Bloch oscillation and tunneling possibilities are both illustrated.

E_G/F_A away from its present location in real space. The finite probability that the carrier will tunnel to that energy state is estimated to be[18]

$$P \sim \exp\left(\frac{-\pi^2 E_G^2 am}{eF_A \hbar^2}\right). \tag{1.36}$$

This mechanism can cause dielectric breakdown of our semiconductor as well as limiting the upper field at which Bloch oscillations are expected. However, in practical semiconductors the breakdown mechanism is usually impact ionization, a topic of later discussion in this chapter.

Tunneling through potential barriers also plays an important role in interfaces where we obtain such potential barriers. Consider a one-dimensional potential barrier $V(x)$. The transmission coefficient, which is the ratio of the transmitted electron-wave amplitude to the incident-wave amplitude, can be written as[19]

$$T(E_x) = \exp\left\{-\frac{2}{\hbar} \int_{x_1}^{x_2} [2m(V - E_x)]^{1/2} \, dx\right\}, \tag{1.37}$$

where E_x is the energy associated with the movement in the x-direction. This result is obtained using the Wentzel–Kramers–Brillouin (WKB) approximation, and x_1 and x_2 are the classical turning points where $V(x) = E_x$. The tunneling current through the barrier can be written as

$$J = \frac{2e}{(2\pi)^3} \int\int\int T(E_x)[f(E) - f(E')] \frac{1}{\hbar(\partial E/\partial k)} \, dk_x \, dk_y \, dk_z, \tag{1.38}$$

where $f(E)$ is the Fermi–Dirac distribution function, E is the incident energy, and E' is the transmitted energy.

Evaluation of expressions like (1.38) generally yields an expression which can be interpreted as being a nonlinear resistance, which may have regions in which its value is negative. The standard example of this is the tunnel diode. An interesting question is: What happens if we have two such barriers as shown in Figure 1.9? If the barriers are far apart, where far is defined with respect to the

FIGURE 1.9. The double-barrier problem which may lead to resonant tunneling if the two barriers are sufficiently close.

coherence length of the electrons, then we can model the system by calculating the nonlinear resistance of each barrier separately and adding them as if they were connected in series. When the barriers are brought closer together, however, we no longer can do this and in fact may have tunneling rates that far exceed the simple product. What has happened here is that we have made a quantum well which has a quasi-stationary state as shown in Figure 1.10. If this state aligns in energy with the incident electrons, we can get tunneling through the barriers without attenuation, an effect called resonant tunneling. Resonant tunnel structures exhibit a negative conductance and will be further discussed in Chapter 9.

There is a rather mysterious aspect to this resistance however. We can compute the current-voltage relation for such a system by simply evaluating a transmission coefficient for a purely elastic tunneling process and then integrating the product of this coefficient and an incident electron density (in energy space). Since we have a positive current flow and a positive voltage, such an analysis predicts a positive energy dissipation. However, where is this energy going? The tunneling process plugged into the analysis was elastic. It is believed that the dissipation is introduced by the destruction of the quantum-phase information in the contacts.[20]

There is another debate concerning tunneling. How long does it take for an electron to tunnel from one side of a barrier to another? The answer to this question is still somewhat indefinite, perhaps because the question itself is not well phrased. Kane[21] illustrates this indefiniteness by first arguing that tunneling is instantaneous and then arguing that it is slow. Kane's illustrative argument for instantaneous tunneling examines the phase change of a tunneling wave packet. It can be argued that the wave packet does not experience any phase change associated with the barrier width and that therefore it must traverse the barrier instantaneously. Kane's illustrative argument for slow tunneling is developed by asking how rapidly tunneling can shift energy from one well to another by shifting an electron from one well to another in a double-well structure. Since it can be argued that the rate of energy exchange is low, tunneling must be a slow process. A variety of estimates may be found in the literature.[22-26]

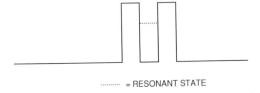

········· = RESONANT STATE

FIGURE 1.10. The double-barrier problem for closely spaced barriers in which resonant tunneling can occur through the intermediate confined state.

The difficulty faced in phrasing this question comes from the fact that a wave packet is needed to describe a spatially localized electron, and we do need spatially localized electrons if we are to ask which side of the barrier the electron is on at a given point in time. Then we must decide how to define times at which a wave packet enters and leaves the barrier region. One definition examines the peak in the incident and transmitted wave packets. Another approach examines the midpoint of the rising edge of the incident and transmitted wave packets. Both definitions have been suggested as measures of the barrier traversal time.[22] Wigner function simulations reveal that these two times are nearly identical for simple rectangular barriers,[27] with the time being proportion to the barrier width and inversely proportional to the wave vector corresponding to the peak in the packet.

1.7. PHONONS AND CARRIER SCATTERING

The various mechanical vibrational modes of the lattices are described by using wave-particle duality in terms of a quasi-particle called the phonon. A phonon is a boson and in equilibrium obeys Bose–Einstein statistics:

$$f(E) = \frac{1}{\exp(-\alpha + E/k_B T_L) - 1},\qquad(1.39)$$

where k_B is Boltzmann's constant and T_L is the lattice temperature. The factor α describes particle conservation. Phonon number, however, is not conserved (although energy must be), and instead of (1.39) we use the simpler expression

$$f(E) = \frac{1}{\exp(E/k_B T_L) - 1}.\qquad(1.40)$$

Phonons fall into several major categories. There are two main branches to the modes, as seen in the dispersion diagram in Figure 1.11. The acoustic branches correspond to mechanical waves in which neighboring atoms move in the identical directions. Optical branches, on the other hand, correspond to mechanical waves in which neighboring atoms move in opposite directions. Similarly, we have transverse and longitudinal modes for both acoustic and optical branches. One of the differences between Si and GaAs is seen in Figure 1.11. For GaAs there is a gap between the two longitudinal modes at the zone edge and between the two optical modes at the zone center, where no such gap exists for silicon. The gap exists in GaAs because of the different masses and charges of the gallium and arsenic atoms, whereas in silicon all the atoms have identical mass and charge.

The second important difference between the two materials with regard to phonons is that of the charge differential. In optical modes the distance between two neighboring atoms changes, and therefore if the atoms are not charge neutral there can be direct electromagnetic interactions between the optical modes and

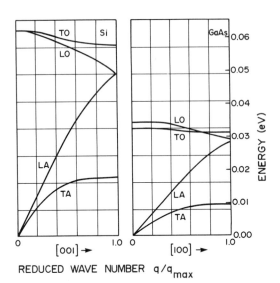

FIGURE 1.11.   Phonon spectra of GaAs and Si. After Refs. 28 and 29.

either electrons or applied fields. GaAs is a polar semiconductor in which these atoms are partially ionized, and therefore such interactions occur. For example, there will be different approaches followed for polar optical phonon scattering at the microscopic level than are used for nonpolar optical scattering. These effects also are associated with piezoelectricity and affect the lattice contribution to the dielectric function, since the atomic charges and polar phonon modes can directly interact with external applied macroscopic fields of appropriate wavelength.[30]

The phonons are a main vehicle for electron and hole scattering in semiconductors. The E-k diagram is obtained by assuming that the ionic cores are at their equilibrium location. The phonons are then treated as a perturbing potential. Generally the analysis proceeds as follows. First, a microscopic potential is chosen to represent the phonon mode of interest. Then, using the Bloch states described by the E-k diagram, we calculate the matrix element $M_{kk'}$ which describes the transition from state k to state k' by the perturbing phonon. It will generally take the form

$$M_{kk'} = \langle \psi_k | V_p | \psi_{k'} \rangle, \tag{1.41}$$

where V_p is the perturbing potential. The actual transition rate per unit time is obtained from the matrix element by using Fermi's golden rule,

$$S(\mathbf{k}, \mathbf{k'}) = \frac{2\pi}{\hbar} g(\mathbf{k'}) M_{kk'}^2 \delta[E(\mathbf{k'}) - E(\mathbf{k})], \tag{1.42}$$

where $g(\mathbf{k'})$ is the density of final states.

There are several points worth mentioning about this procedure. First, consider what happens if the system is subject to two or more perturbations at the same time. The inner-product operation in (1.41) is linear, so the overall matrix element would be the sum of the matrix elements for the individual perturbations. However, the matrix element is squared in the golden rule (1.42), and since the square of a sum does not equal the sum of the squares, the combined transition rate will not be a simple sum of the individual transition rates. Instead, there will be a cross-product term which represents the interference of the two scattering processes. If there are no multiple interactions, even though many types of perturbation are allowed, then we can express the total transition rate as

$$S_{\text{Tot}}(\mathbf{k}, \mathbf{k}) = \sum S(\mathbf{k}, \mathbf{k}'). \tag{1.43}$$

In our problems, we will have many possible perturbations and yet (1.43) is exactly what we normally assume. We therefore have assumed that while there are many possible perturbing influences, only one such influence is seen by the system at a time. This assumption allows us to always add in new scattering rates without readjusting the old. While this is incredibly convenient and virtually universally done, it is also wrong and makes the resulting calculations dependent on the existence of low scattering rates.

The second point of concern is the validity of the golden rule. This rule essentially assumes that the scattering process is allowed to run to completion without interference and thereby allows for the introduction of the energy- and momentum-conserving delta functions. Yet, as both our electronic states or wave packets and our vibrational modes are spread out in space, we know that there is a time required for the interaction to occur. In using the golden rule to obtain our transition rates, we are implicitly assuming that this time, the collision duration time, is sufficiently short that we may assume that scattering processes always run to completion in a time shorter than any other time of interest in our problem. Our collisions then are assumed to occur instantaneously in time and locally in space. Obviously, if we relax the assumption of local instantaneous scattering, we will have multiple perturbations for some periods of time. These two assumptions, however, are not strictly equivalent. There is one perturbation which is always present though, and that is the field. Ordinarily, in calculations of other processes we ignore the field, but when we drop the assumption of an instantaneous event the field can no longer be ignored. There is an intracollisional field effect as the carriers always must interact with the fields even when they are interacting with a phonon. The easiest process by which this particular effect can be handled is to retain our concept of instantaneous scattering but to include in the energy and momentum delta functions contributions which represent the energy and momentum exchanged with the field during a typical interaction. In Figure 1.12 we illustrate the intracollisional field effect.

The main classes of phonon scattering can easily be developed by comparing Figure 1.11 with the energy bands of Figure 1.6. The only phonons which can contribute to an intravalley scattering event are phonons lying near the zone

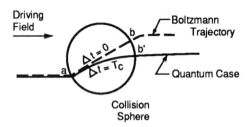

FIGURE 1.12.   The intracollisional field effect.

center. As can be seen, intravalley scattering by acoustic phonons then is a nearly elastic process in which no energy is transferred between the carrier and the lattice. Intravalley scattering by optical modes, on the other hand, is an inelastic process.

Intervalley scattering can occur only with phonons lying well away from the zone center. Here both acoustic and optical modes are inelastic, and we also encounter problems with both theory and experiment in terms of resolving in any detail the electron–phonon or hole–phonon interactions. While in some cases the details are known, in other cases these processes have been treated using adjustable parameters, with the fitting procedure being guided by experimental data on carrier transport coefficients.

In the above discussion, we noticed subdivisions of scattering processes into intravalley and intervalley processes, and into elastic and inelastic processes. There is one more important subdivision: momentum-randomizing and non-momentum-randomizing processes. In many cases $S(\mathbf{k}, \mathbf{k}')$ does not contain any explicit dependence of the actual wave vectors except for their energy. Under these circumstances any $\mathbf{k}'$ which lies on the appropriate constant energy surface is equally probable, and the scattering process is momentum randomizing. In other processes there is an explicit dependence on $\mathbf{k}'$, such as an inverse proportionality of the transition rate to the magnitude squared of $\mathbf{k} - \mathbf{k}'$. Then not all $\mathbf{k}'$ which lie on the final constant-energy surface are equally probable, and the scattering process is said to not be momentum-randomizing.

The most important example of a process which is not momentum-randomizing is the polar optic phonon scattering process. In this process there is a tendency to select small scattering angles, which can lead to a phenomena called polar optic runaway. The physical basis of this is that the field tends to accelerate all carriers in one direction in k-space. If the dominant scattering process is polar optic scattering, there is little change in the general direction of the k-vectors. For a sufficiently large field we then can have a situation in which it is impossible to maintain stable distributions in k-space.[31] This possibility plays an important role in some investigations of the breakdown of silicon dioxide.

In many cases we are interested in knowing the total scattering rate out of a state k and do not really care what the final state is. This rate is determined by integrating $S(\mathbf{k}, \mathbf{k}')$ over $\mathbf{k}'$, remembering that the density of states for $\mathbf{k}'$ has already been built into $S(\mathbf{k}, \mathbf{k}')$ through its appearance in the golden rule. The resulting scattering rate is

$$\lambda(\mathbf{k}) = \int S(\mathbf{k}, \mathbf{k}') \, d\mathbf{k}'. \tag{1.44}$$

Such calculations can be influenced by collisional broadening of the states. Consider an electron lying in a Bloch state $\psi(\mathbf{k})$ which undergoes a total outscattering rate of $\lambda(\mathbf{k})$. The difficulty is that the time dependency introduced by the scattering process cannot be incorporated into the simple Bloch state formalism because we did not develop it using the time-dependent Schrödinger equation. Instead we used the time-independent equation in which there is a suppressed time dependence which takes the form of a phase factor $\exp[-iE(\mathbf{k})t/\hbar]$. To represent the time dependence of the state occupancy, we will need to form a packet of waves with different phase factors, that is, with different energies. Instead of the simple Bloch state, we use a collisionally broadened function of the form

$$\psi(t) = \int A_E \exp\left(\frac{-iEt}{\hbar}\right) dE, \tag{1.45}$$

where the coefficients A_E are defined as

$$A_E = \int \psi(t) \exp\left(\frac{+iEt}{\hbar}\right) dt. \tag{1.46}$$

All we need now is some estimate for the functional form of the time-dependent state $\psi(t)$ with energy E_0. This is determined by considering the role of the scattering process. Since the magnitude squared of ψ is the probability of locating an electron at time t in the appropriate initial state, then the time dependence of this squared magnitude in the presence of scattering out of this state at rate λ is determined by

$$\frac{d}{dt}|\psi|^2 = -\lambda|\psi|^2. \tag{1.47}$$

The solution to this will be an exponential function with argument λt. We therefore use the square root of this function, that is, $\exp(-\lambda t/2)$, as the basic time dependence of our state ψ. Including the energy-dependent phase factor, we write

$$\psi = \exp\left(\frac{-\lambda t}{2}\right) \exp\left(\frac{iE_0 t}{\hbar}\right). \tag{1.48}$$

Substituting this into (1.46) yields the Lorentzian line shape

$$A_E^2 = \frac{\hbar^2}{(E - E_0)^2 + (\hbar\lambda/2)^2}. \qquad (1.49)$$

It is important to note that while such Lorentzian line shapes are commonly used to estimate the collisional broadening of our states, we obviously have ignored the existence of band gaps in the above argument. It is also important to note that while this expression describes some uncertainty in the state energy which arises from the finite lifetime, we at no time made any use of an energy-time uncertainty relation in our argument. Lastly, we obviously would prefer a completely self-consistent analysis in which we would now use the broadening to estimate a new scattering rate λ. It should be noted that scattering rates of the level of 10^{13} events per second, which are not unusually large rates for semiconductors, can produce half-widths on the order of 10 meV, which in turn are substantial fractions of energies such as optical phonon energies.

There is one additional difficulty which can also arise in the application of (1.44). There are systems in which we have singularities in the density-of-states function which if inserted blindly into (1.44) will produce infinite scattering rates. This effect is usually avoided by introducing a self-energy correction which generally has the effect of broadening the infinity into a finite-height function associated with a Lorentzian shape. An example of such a procedure will be found in our discussion of the lateral surface superlattice in Chapter 8.

1.8. SOME SCATTERING RELATED TIMES, PATHS, AND RATES

There are various time and space scales associated with these scattering events. These include the mean free time between collisions, the mean free path, and energy and momentum relaxation times. We can interpret our scattering rate as follows. The number $\lambda(\mathbf{k}) \, dt$ is the probability that our carrier will experience some form of scattering event in the differential time element dt. The mean flight time between collisions for a carrier starting at $\mathbf{k}(0)$ is then

$$\tau_F[\mathbf{k}(0)] = \int_0^\infty t \, \mathrm{Pr}(t) \, dt, \qquad (1.50)$$

where $\mathrm{Pr}(t)$ is the probability that a carrier which scattered at time $t = 0$ will experience its next scattering event at time t. This probability is

$$\mathrm{Pr}(t) = \mathrm{Pr}_{N0}(t)\lambda[\mathbf{k}(t)] \, dt \qquad (1.51)$$

where $\mathrm{Pr}_{N0}(t)$ is the probability that an undisturbed free flight has occurred from time 0 to time t. The only remaining link is calculating this probability, the probability that the carrier did not scatter in the interval $(0, t)$. This is relatively simple to motivate. If we had N carriers to start with, then the fractional number which have not scattered at any point in time is the solution to

$$\frac{dN}{dt} = -\lambda[\mathbf{k}(t)]N(t), \tag{1.52}$$

which is

$$N(t) = \exp\left\{-\int_0^t \lambda[\mathbf{k}(s)]\, ds\right\}. \tag{1.53}$$

This function is the probability of reaching t without scattering. Therefore the mean flight time between scattering events is

$$\tau_F[\mathbf{k}(0)] = \int_0^\infty t\lambda[\mathbf{k}(t)] \exp\left\{-\int_0^t \lambda[\mathbf{k}(s)]\, ds\right\} dt. \tag{1.54}$$

As can be seen, when $\mathbf{k}$ varies with time in the presence of a field and with the functional dependence of λ on $\mathbf{k}$, the computation of even the simplest of our times becomes difficult. We obtain the usually quoted mean free time by incorporating information about the relative probability of various initial states. This would be done by using the distribution function $f(\mathbf{k})$. Our mean free time between collisions then is

$$\tau = \int_{-\infty}^\infty f[\mathbf{k}(0)] \int_0^\infty t\lambda[\mathbf{k}(t)] \exp\left\{-\int_0^t \lambda[\mathbf{k}(s)]\, ds\right\} dt\, dk. \tag{1.55}$$

If we had a simple constant scattering rate λ, then we would find

$$\tau = \frac{1}{\lambda}. \tag{1.56}$$

For this case, however, we have a simple exponential distribution of flight times, and it is important to notice that only about 37% of the carriers will not scatter in this time.

The mean free path is computed by examining the distance traveled during flight time $t_f[\mathbf{k}(0)]$,

$$l_f[\mathbf{k}(0)] = \int_0^{t_f[\mathbf{k}(0)]} v_g[\mathbf{k}(s)]\, ds, \tag{1.57}$$

and integrating over the distribution to yield the mean free path

$$l = \int_{-\infty}^\infty f[\mathbf{k}(0)] \int_0^{t_f[\mathbf{k}(0)]} v_g[\mathbf{k}(s)]\, ds\, dk. \tag{1.58}$$

The inelastic mean free path is often of more interest, however. It is important because of the already noted dependency of the phase of the basic electronic states on the carrier energy. Many of the structures and effects which are discussed here are the results of the wavelike properties of the carriers. They utilize quantum interference effects that can be attained only if phase coherency is maintained. This phase coherency is disrupted by energy changes in inelastic scattering processes but is not disrupted in an elastic scattering. Therefore we expect to see such effects occur on space scales which are less than the inelastic mean free path. This quantity can be estimated by replacing the total transition rate used in (1.58) by the total inelastic transition rate. It should be noted that the inelastic mean free path is often much longer than the de Broglie wavelength associated with the carrier's effective mass and therefore the use of the de Broglie wavelength usually leads to an underestimate of the spatial scale on which quantum phenomena are expected.

Another commonly asked question involves the net rate at which carriers are exchanging energy and momentum with some other physical system. The momentum relaxation rate through scattering is

$$\mathbf{R}_{m,s} = \int f(\mathbf{k}) \int S(\mathbf{k}, \mathbf{k}') \hbar(\mathbf{k}' - \mathbf{k}) \, d\mathbf{k}' \, d\mathbf{k}, \tag{1.59}$$

and the energy relaxation rate through scattering is

$$R_{E,s} = \int f(\mathbf{k}) \int S(\mathbf{k}, \mathbf{k}')[E(\mathbf{k}') - E(\mathbf{k})] \, d\mathbf{k}' \, d\mathbf{k}. \tag{1.60}$$

These rates are not relaxation times. The commonly quoted relaxation times are estimates of the rate at which the system approaches an equilibrium. This topic will be discussed in Chapter 4. The rates in (1.59) and (1.60), however, should be compared with the following rates. The rate at which the carriers are gaining momentum from the field is

$$\mathbf{R}_{m,F} = \frac{e\mathbf{F}}{\hbar} \int f(\mathbf{k}) \, d\mathbf{k}, \tag{1.61}$$

and the rate at which they are gaining energy from the field is

$$R_{E,F} = \int f(\mathbf{k}) \frac{\nabla_k E(\mathbf{k})}{\hbar} \cdot (e\mathbf{F}) \, d\mathbf{k}. \tag{1.62}$$

Note that the momentum exchange rates are vector quantities, while the energy exchange rates are scalar. A steady state can occur when

$$\mathbf{R}_{m,s} = -\mathbf{R}_{m,F} \tag{1.63}$$

and

$$R_{E,s} = -R_{E,F}. \tag{1.64}$$

The consequences of these two equalities are discussed in the next section.

1.9. HOT-CARRIER EFFECTS

A steady-state balance between the energy and momentum exchange with the field and with the lattice is assumed in most textbook discussions of semiconductor devices. When such a balance occurs, it is possible to describe the transport of carriers through the lattice in terms of a set of equations where the parameters are determined by the electric field. The most commonly used model is the drift-diffusion model in which

$$J_n(x, t) = en(x, t)v_n(F(x, t)) + eD_n(F(x, t)) \frac{\partial}{\partial x} n(x, t), \tag{1.65}$$

where v_n is the electron drift velocity and D_n is the electron diffusion coefficient. Both of these quantities are functions of the electric field $F(x, t)$. This field dependence of the drift velocity, shown in Figure 1.13 for Si, InP, and GaAs, is the hot-carrier effect most commonly included in device modeling. For a submicron semiconductor structure the short lengths reduce the voltage required to produce internal fields at which there no longer is a simple linear or ohmic relationship between the carrier drift velocity and the applied field are small. Only 1 V applied across a submicron semiconductor structure is needed to obtain fields in excess of $10 \, \text{kV/cm}$.

The terminology "hot carrier" arises from a consideration of the energy balance described in (1.64). If there were no field present, then $R_{E,F}$ would be zero and $R_{E,s}$ would be zero as well. For this to happen the rate at which the carriers emit optical phonons would have to equal the rate at which they absorb them. When a field is applied, then there is a net flow of energy from the field into the carriers, and to achieve the steady-state there must be a corresponding flow of energy from the carriers to the lattice. This is in turn requires that the probability of emitting an optical phonon must exceed the probability of absorbing one. The probability of absorbing an optical phonon is

$$P_{\text{abs}} \propto N_k \exp\left(\frac{-E}{k_B T_e}\right), \tag{1.66}$$

while the probability of emitting a phonon from the appropriate state is

$$P_{\text{em}} \propto (N_k + 1) \exp\left[\frac{-(E + \hbar\omega_0)}{k_B T_e}\right]. \tag{1.67}$$

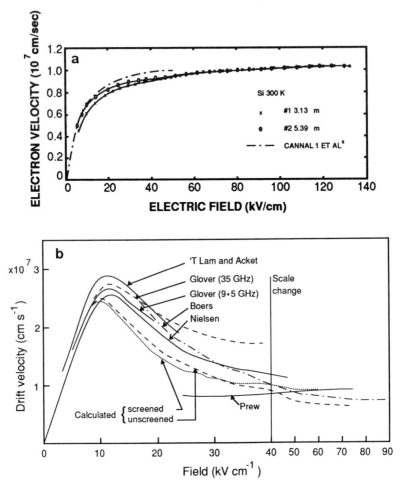

FIGURE 1.13.   Electron drift velocities in (a) Si. After Smith *et al.*[32] (b) InP. After Fawcett and Herbert and references therein.[34] Hole drift velocities in (c) Si. After Smith *et al.*[32] Electron and hole drift velocities in GaAs are shown in (d). After Windhorn[33] and Bauhahn.[52] In (b) Fawcett and Herbert compare several sets of experimental data with several calculations.

Here T_e is the electron temperature, and ω_0 is the phonon frequency, and N_k, the phonon occupation factor for a state with wave vector k, is

$$N_k = \frac{1}{\exp(\hbar\omega_0/k_B T_L) - 1},\tag{1.68}$$

which essentially is (1.40) evaluated for a specific phonon energy. Crucial to our present discussion is the observation that

$$N_k + 1 = N_k \exp\left(\frac{\hbar\omega_0}{k_B T_L}\right),\tag{1.69}$$

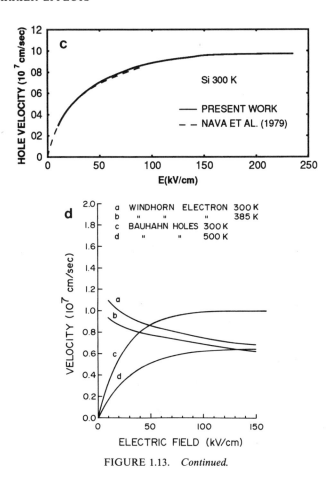

FIGURE 1.13.   *Continued.*

as we then have

$$\frac{P_{\text{em}}}{P_{\text{abs}}} = \exp\left(\frac{\hbar\omega_0}{k_B T_L}\right) \cdot \exp\left(\frac{-\hbar\omega_0}{k_B T_e}\right) = \exp\left(\frac{\hbar\omega_0}{k_B}\frac{T_e - T_L}{T_e T_L}\right). \tag{1.70}$$

Therefore to have the emission rate exceed the absorption rate, we need the electron temperature to be higher than the lattice temperature. Therefore we speak of hot-carrier effects when we enter situations in which, because the electron distribution has taken on a nonequilibrium form, the physical system behaves differently than it does near equilibrium.

While the field dependency of the carrier drift velocities shown in Figure 1.13 is the most commonly seen hot-carrier effect, another important effect occurs at fields lying even higher than those shown in that figure. This is impact ionization or avalanche breakdown. The experimental determination of impact ionization rates is not a straightforward task. Excellent and thoughtful reviews of these difficulties have been provided by Stillman[35] and Chynoweth.[36]

The impact ionization process occurs because sufficiently energetic electrons can excite electrons from valence band states to the conduction band, leaving a hole behind in the valence band in the process. Similarly a sufficiently energetic hole can produce a new electron-hole pair. (The inverse processes are the Auger recombination mechanisms.) While it is obvious that there is a minimum or threshold energy required of the incident carrier if it is to initiate an ionization event, there is an important restriction on the final state as well. A consequence of energy and wave vector conservation is that the group velocities of the resultant particles, including any phonons which may contribute, must be nearly equal.[37] Since phonons move at the speed of sound, they are much slower than electrons and holes, and this condition ensures that any phonon-mediated ionization event yields carriers which lie near conduction-band minima and valence-band maxima. We now outline this argument.

In Figure 1.14 we illustrate a prototypical electron-initiated ionization process. It is important to note that we are using a single effective mass value and that some important modifications must be made in this analysis in its application

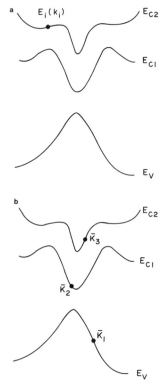

FIGURE 1.14.   Before (a) and after (b) pictures of an electron-initiated ionization even in a system with two conduction bands and one valence band.

to real semiconductors. It does serve our purpose, however, of illustrating the existence of well-defined thresholds for impact ionization. For the process illustrated in Figure 1.14, energy conservation dictates

$$E_i(\mathbf{k}_i) = E_{C1}(\mathbf{k}_2) + E_{C2}(\mathbf{k}_3) - E_V(\mathbf{k}_1) + \sum_j a_j \hbar \omega_{b_j}(\mathbf{k}_j), \qquad (1.71)$$

where j counts phonons and b_j is the branch of the jth phonon. Momentum conservation dictates

$$\mathbf{k}_i = \mathbf{k}_2 + \mathbf{k}_3 - \mathbf{k}_1 + \sum_j a_j \mathbf{k}_j. \qquad (1.72)$$

In (1.71) and (1.72), a_j is positive if the phonon is emitted and negative if it is absorbed. These coefficients will be integer values where the value equals the number of phonons of this sort which were either emitted or absorbed.

While (1.71) and (1.72) are the equations which must be solved if the ionization threshold energies are to be determined, it is possible to restrict the search somewhat. Consider a set of perturbations on the final state particle k-vectors that leaves the total energy unchanged. This requires

$$d\mathbf{k}_1 \cdot \nabla_k E_V(\mathbf{k}_1) = d\mathbf{k}_2 \cdot \nabla_k E_{C1}(\mathbf{k}_2) + d\mathbf{k}_3 \cdot \nabla_k E_{C2}(\mathbf{k}_3)$$
$$+ \sum_j a_j \, d\mathbf{k}_j \cdot \nabla_k [\hbar \omega_{b_j}(\mathbf{k}_j)]. \qquad (1.73)$$

Equation (1.73) can be rewritten in terms of the resultant particle group velocities and becomes

$$d\mathbf{k}_1 \cdot \hbar \mathbf{v}_1 = d\mathbf{k}_2 \cdot \hbar \mathbf{v}_2 + d\mathbf{k}_3 \cdot \hbar \mathbf{v}_3 + \sum_j a_j \, d\mathbf{k}_j \cdot \hbar \mathbf{v}_j. \qquad (1.74)$$

Now, equation (1.74) describes a perturbation of the resultant k-vectors which will preserve energy conservation for a given incident particle. There is a second restriction on this perturbation, however, if we are to still conserve momentum we must additionally require

$$d\mathbf{k}_1 = d\mathbf{k}_2 + d\mathbf{k}_3 + \sum_j a_j \, d\mathbf{k}_j, \qquad (1.75)$$

and this can be used to eliminate $d\mathbf{k}_1$ in (1.74) to yield

$$0 = d\mathbf{k}_2 \cdot (\mathbf{v}_2 - \mathbf{v}_1) + d\mathbf{k}_3 \cdot (\mathbf{v}_3 - \mathbf{v}_1) + \sum_j a_j \, d\mathbf{k}_j \cdot (\mathbf{v}_j - \mathbf{v}_1). \qquad (1.76)$$

Since we can always adjust $d\mathbf{k}_1$ to fit any set of possible k-vector perturbations in (1.76), the only general condition which guarantees the necessary equality is

$$\mathbf{v}_1 = \mathbf{v}_2 = \mathbf{v}_3 = \mathbf{v}_j. \qquad (1.77)$$

Therefore we find that a sufficient condition for impact ionization is that the group velocities of all resultant particles be equal. Since the phonon velocities are very low, any phonon-mediated ionization event satisfying this sufficient condition will produce electrons lying very near conduction-band minima and holes lying very near valence-band maxima. However, since these are minimum energy points for the final state, they also should be related to a minimum initial energy state as well. Therefore we can start our search for an ionization threshold by restricting our immediate attention to a situation with the above resultant electron and hole pair locations. Results obtained from such analyses are provided in Table 1.1.

There are three separate definitions of interest in a discussion of impact ionization. One is the actual scattering rate λ_n, that is, the number of ionization or pair productions initiated by a carrier of fixed energy per unit time. It is this rate that would be predicted by a perturbation model of the actual microscopic process and would be incorporated into a Monte Carlo model. The second definition of interest is encountered when impact ionization appears as a contribution to the electron–hole pair generation term in the continuity equation. We then need to know G_n, the number of electron–hole pairs produced by all of the carriers present in a differential volume of space on a per unit time basis. The third definition is used in experimental studies of the ionization process where the tendency is to describe the process in terms of an ionization coefficient α_n, which is the number of secondary electron–hole pairs produced by a carrier moving a unit distance. The subscript n denotes electrons, and similar quantities α_p, G_p, and λ_p can be defined for holes. These various parameters are interrelated by the equation

$$G_{n,p} = n \int_0^\infty \lambda_{n,p}(E)f(E)\,dE, \tag{1.78}$$

where E is the energy and $f(E)$ is a distribution function normalized to unity, and

$$\lambda_{n,p} = \alpha_{n,p}\langle v \rangle_{n,p}, \tag{1.79}$$

TABLE 1.1. Ionization Thresholds in GaAs

Reference	Carrier	Crystal Orientation	Threshold Energy (eV)
37, 38	Electron	$\langle 100 \rangle$	2.1^a, 2.05^a
	Hole		1.7^a, 1.81^a
	Electron	$\langle 111 \rangle$	3.2^b
	Hole		1.6^b, 1.58^a
	Electron	$\langle 110 \rangle$	1.7, 2.01
	Hole		1.4^a, 1.58^a

[a] The normal conduction or valence band cannot simultaneously satisfy energy and momentum conservation, and initiating particle comes from another band.
[b] The band's energy width is less than the band gap and therefore energy arguments alone prevent a threshold from being found in this band. No threshold attainable in main conduction band.

where $\langle v \rangle$ is the true mean velocity. A common variant on equation (1.79) is

$$\lambda_{n,p} = \alpha_{n,p} \frac{J_{n,p}}{q}. \tag{1.80}$$

Some commonly used expressions and measured values for electron and hole ionization coefficients in Si, GaAs, and InP are provided in Table 1.2.

Just as is the case with the ionization coefficients, the drift velocities, and diffusion coefficients are essentially phenomenological parameters which are determined by a fit to experimental data. A velocity is determined by measuring the time-of-flight of a carrier pulse across a region of known length. The direct measurement of the flight time across regions of length comparable to that of epitaxial layers or those doped at levels appropriate for the devices of interest is difficult. The microwave time-of-flight technique can be applied in these situations. In microwave time-of-flight experiments an electron beam is repetitively swept across one end of a reverse-biased Schottky barrier or *p-n* junction diode in a periodic fashion. Each time the beam sweeps across the diode a pulse of electron–hole pairs is generated near that end of the diode. One species will drift under the influence of the depletion region field to the far end of the device inducing a terminal current, just as discussed in Section 1.3. While the actual transit time is relatively short, the phase delay is large at microwave frequencies. Therefore the phase difference between this current and an external reference current is measured and related to the time-of-flight of the carrier pulses produced by the beam across the depletion region of the diode. Since the depletion region

TABLE 1.2. Commonly Used Expressions and Measured Values for Electron and Hole Ionization Coefficients

$\alpha_{n,p} = A_{n,p} \exp\left[-\left(\dfrac{b_{n,p}}{F} \right)^{m_{n,p}} \right]$					
Reference	A_n	b_n	m_n	A_p	b_p	m_p
		GaInAs				
39		Electron ionization twice the hole rate				
		GaAs				
40	1.84×10^5	6.47×10^5	2	1.84×10^5	6.47×10^5	2
41	2.99×10^5	6.85×10^5	1.6	2.22×10^5	6.57×10^5	1.75
		Si				
42	6.2×10^5	1.08×10^6	1	2×10^6	1.97×10^6	1
		InP				
43	7.36×10^6	3.45×10^6	1	2.04×10^6	2.42×10^6	1

length is determined by the doping profile and bias, the average carrier velocity in the depletion region is then deduced and related to the average electric field found in the depletion region. This technique has been successfully applied to GaAs,[32,33,44] Si,[32] and InGaAs.[33,45] The central disadvantage of the above technique is its reliance on some independently determined velocity-field point as a basis for the interpretation of the relative phase data. Various possibilities of eliminating this have been suggested.[33,46,47]

The two most commonly used diffusion coefficient measurement methods are noise measurements and measurements of the spread of a carrier pulse as it drifts through some known field.[48] Consider an ensemble of electrons, each with a velocity

$$v(t) = v_d(t) + \Delta v(t),$$ (1.81)

where $v_d(t)$ is the ensemble average and $\Delta v(t)$ is the fluctuation term. If we initially have a delta function of carriers in real space, centered at $x = 0$, then it is easily shown that the mean square fluctuation of x away from the average value; that is, the spread in the carrier pulse obeys the Einstein formula

$$\langle \Delta x^2(t) \rangle = 2Dt$$ (1.82)

for large t, where the angle brackets denote an ensemble average and the diffusion coefficient is

$$D = \tfrac{1}{2} \int_{-\infty}^{\infty} \Phi_{\Delta v}(\Theta) \, d\Theta.$$ (1.83)

Here $\Phi_{\Delta v}(\Theta)$ is the velocity fluctuation autocorrelation function. Since the Fourier transform of the autocorrelation function is the noise power spectral density of the current fluctuations created by the individual carrier velocity fluctuations, that is,

$$S(\omega) = \tfrac{1}{2} n^2 q^2 \int_{-\infty}^{\infty} \Phi_{\Delta v}(\Theta) \, e^{j\omega\Theta} \, d\Theta,$$ (1.84)

noise measurements can also be used to determine diffusion coefficients.

In Figure 1.15 we show some measured or commonly used diffusion coefficients for electrons and holes in Si and GaAs as functions of electric field. The diffusion coefficient for electrons GaAs sharply peaks around 3 to 4 kV/cm as a result of the transfer of electrons from the light-mass central-valley Γ-valley to the higher-energy, heavy-mass satellite L- and X-valleys. The negative differential mobility of the electrons in GaAs is prominently seen and widely understood because of its role in the Gunn effect. The peak in the diffusion coefficient curve is not as widely known, but is also a prominent feature of some models of hot-carrier transport in GaAs. As can be seen, the exact value of the diffusion coefficient near this peak is not well known. It has been suggested[49] that this may in fact be an artifact of an experimental error by Ruch and Kino[50] which led to an overestimate of the diffusion coefficient, which in turn has been used by theoreticians motivated by other Monte Carlo diffusion coefficient estimates which do not show this peak.[49,51]

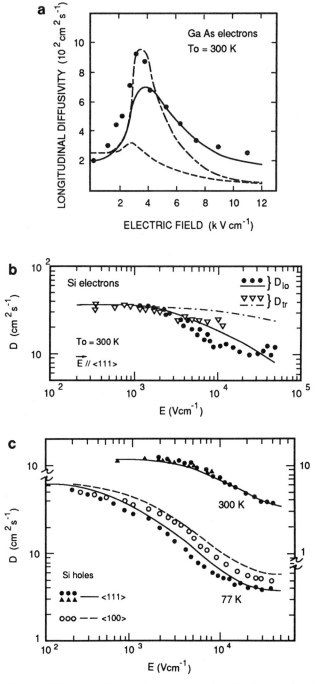

FIGURE 1.15. Electron diffusion coefficients in (a) GaAs and in (b) Si. Hole diffusion coefficients in (c) Si and (d) GaAs. After Canali *et al.*[53] and Joshi and Grondin.[51] [In (a) the solid circles are due to Ruch and Kino,[50] the dashed line is due to Fawcett and Rees,[54] and long-short dashed line is due to Pozela and Reklaitis.[55] In (b) and (c) the data and theory are due to Canali *et al.*[53] In (d) the results are due to Joshi and Grondin.[51]]

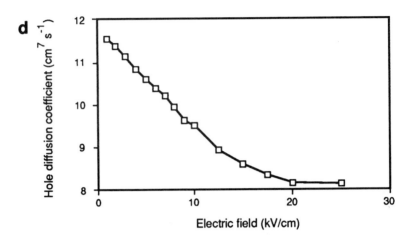

FIGURE 1.15. *Continued.*

Little direct experimental data seem to be available for the field dependency of the hot-hole diffusion coefficient in GaAs. This probably is a result of a common bias that this parameter is uninteresting. However, it does play an important role in IMPATT devices, and we therefore provide the value used by Bauhahn[52] in his study of IMPATTs. This functional form is simply a constant set equal to the low field value determined by the Einstein relation. The situation is somewhat different for silicon, where both electron and hole diffusion coefficients have been measured and computed with good agreement being obtained. Recent Monte Carlo calculations give this low field value, but show that the longitudinal hole diffusion coefficient rapidly falls with field.[51]

1.10. SUMMARY

As we noted at the outset, our goal is to explore the physical processes of importance in the experimental study of small (or fast) semiconductor structures. Throughout the text we will use classical electromagnetic theory and will be concerned about how one completes the description of the systems of interest in terms of Maxwell's equations. We will rarely worry about magnetic effects, but will occasionally concern ourselves with dielectric functions, as our concern will usually be with the redistribution of charge in response to electric fields. We close this chapter with a brief overview of the time scales which control the physical model being used to solve this transport problem.

In Figure 1.16 we summarize the issues in terms of several questions and illustrate an appropriate modeling selection based on the answers to these questions. The first of these is the question of how the rate of change in the system's forcing function, which is the fields seen by the carriers, compares with the rate at which these carriers respond to such changes. When the system is able to track

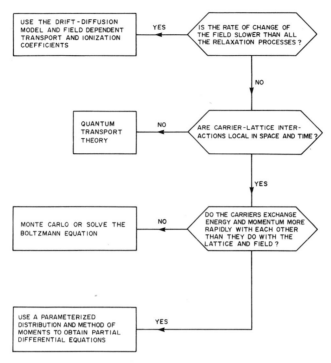

FIGURE 1.16.   Several modeling choices and factors which motivate their selection.

variations in the fields, then the carriers are well represented by the steady-state situation in which the rate of flow of energy and momentum into the carriers from the field is equal to the rate of flow of energy and momentum from the carriers to the lattice. The hot-carrier effects discussed generally are modeled in device systems by such quasi-static models.

   If the fields seen by the carriers are rapidly changing, then other models must be used. In some cases a genuine quantum theory is necessary. This issue is particularly complicated for processes that involve phonon emission or absorption, as the period of a phonon is much longer than a few femtoseconds. Therefore, the typical semiclassical approximation of an instantaneous collision is obviously wrong. Replacing this model, though, is a great challenge. Although we speak of electrons as generating or absorbing phonons, physically an electron has no direct interaction with a mechanical vibration of a crystal lattice. They see the lattice ions through electromagnetic interactions. Even though a phonon may not have been fully generated, the carrier always sees the lattice potential and is interacting with it. It is continuously exchanging energy and momentum with the lattice ions. At this time scale, therefore, we need to be concerned with the separation of our problem into two well-defined systems, one being the electron and the other the lattice. We need to be very careful in our application of such approximations as the effective mass approximation and the adiabatic approximation.

If we can assume that the collisions are local and instantaneous, then we regain the semiclassical model of effective masses, energy-band dynamics, and Fermi-golden-rule-formulated scattering processes. (This latter point, of course, requires that the scattering rates be low enough that we really do not need to concern ourselves with simultaneous multiple scattering processes.) Ensemble Monte Carlo models can be very effective in this regime, but we must be careful with other models that utilize distribution functions. A single time distribution function is almost always used, and as a result memory effects are thrown away. This is discussed in Chapter 5. Models such as Boltzmann transport theory, which utilize single time correlation functions, are best applied on longer time scales in which any correlation between the carrier's initial and final state has been destroyed, a process which takes several or more scattering events.

The above temporal regime is sometimes called the near-ballistic regime. There is an important feature about this regime that is not illustrated in Figure 1.16. In this regime the carrier also does not have its quantum mechanical phase destroyed by scattering processes, and quantum interference effects may be visible. When these phenomena occur, we no longer use the Monte Carlo models just mentioned, because in these models we are using a particle representation of electrons in which wavelike properties related to quantum phase functions are not present. Methods for modeling these quantum effects will be discussed in Chapter 9.

At longer times, yet where several or more scattering events occur in the time period of interest, we can then begin to track macroscopic variables such as average energies and momenta. This gives us a vast reduction in the number of variables present in the model. These models are discussed in Chapter 4. These models contain a set of partial differential equations whose solution describes the time evolution of the macroscopic variables. While these models involve a greatly reduced variable set, they still allow for deviations from the simple drift-diffusion models. In Chapter 5 we discuss these deviations.

APPENDIX

A.1. ELECTRON SCATTERING IN GaAs AND Si

Here we give the expressions for the transition rate $S(k, k')$, the total scattering rate, and the angular dependence of the various scattering mechanisms affecting the electrons in GaAs and Si. For nonumklapp processes the transition rate will usually be expressed as

$$S(\mathbf{k}, \mathbf{k}') = \frac{2\pi}{\hbar} V(\mathbf{q}) G(\mathbf{k}, \mathbf{k}') \delta(E' - E), \qquad (A.1)$$

where $V(\mathbf{q})$ contains the dependency upon $\mathbf{q} = \mathbf{k}' - \mathbf{k}$ of the square Fourier transform of the matrix element and G is the overlap factor. In the rest of this section we present rates for given scattering processes.

A.1.1. Polar Optical Scattering

The matrix element is

$$V(\mathbf{q}) = \frac{e^2}{4\pi\varepsilon q^2} \frac{\hbar\omega_0}{2} \left(\frac{1}{\varepsilon_\infty} - \frac{1}{\varepsilon_0} \right) (N_0 + \tfrac{1}{2} \pm \tfrac{1}{2}), \tag{A.2}$$

where N_0 is given by the Bose–Einstein distribution (1.67), the plus sign refers to phonon emission, and the minus sign is phonon absorption. The total scattering rate is

$$P(E) = \frac{em^*\omega_0}{4\pi\varepsilon_0\sqrt{2}\,\hbar} \left[\frac{1}{\varepsilon_\infty} - \frac{1}{\varepsilon_0} \right] \frac{1 + 2\alpha E'}{\sqrt{\gamma(E)}} F_0(E, E') \times \{N_0 + \tfrac{1}{2} \pm \tfrac{1}{2}\}, \tag{A.3}$$

where

$$E' = E \pm \hbar\omega_0 \tag{A.4}$$

and

$$F_0(E, E') = \log_e \left[\frac{\sqrt{\gamma(E)} + \sqrt{\gamma(E')}}{\sqrt{\gamma(E)} - \sqrt{\gamma(E')}} \right], \tag{A.5}$$

for nonparabolic bands, and we have neglected the admixture of the p-functions in the overlap integral. The angular dependence of the scattering is given by

$$P(\theta)\, d\theta = \frac{\sin\theta\, d\theta}{[\gamma(E) + \gamma(E') - 2\sqrt{\gamma(E)\gamma(E')}\cos\theta]}. \tag{A.6}$$

In a Monte Carlo calculation, as discussed in Chapter 4, we would generate a random number r, uniformly distributed between 0 and A. We set this number equal to

$$r = \int_0^\theta P(\theta)\, d\theta, \tag{A.7}$$

or

$$\cos\theta = \frac{(1 + f) - (1 + 2f)^r}{f}, \tag{A.8}$$

where

$$f = \frac{2\sqrt{\gamma(E)\gamma(E')}}{(\sqrt{\gamma(E)} - \sqrt{\gamma(E')})^2}. \tag{A.9}$$

Once θ is determined from (A.8), the azimuthal angle is determined by using a random number which completely specifies $\mathbf{k}'$.

A.1.2. Intervally Scattering

The matrix element is given by

$$V(\mathbf{q}) = \frac{D_{ij}^2 \hbar^2}{2\rho\hbar\omega_{ij}} (N_{ij} + \tfrac{1}{2} \mp \tfrac{1}{2}), \tag{A.10}$$

where D_{ij} is the deformation potential for scattering for the ith to the jth valley, and the associated phonon frequency is ω_{ij}. The scattering rate from valley i to valley j is given by

$$P_{ij} = \frac{z_j D_{ij}^2 (m_j^*)^{3/2}}{\sqrt{2}\,\pi\rho\hbar^2(\hbar\omega_{ij})} \gamma_j^{1/2}(E')(1 + 2\alpha_j E')(N_{ij} + \tfrac{1}{2} \mp \tfrac{1}{2}), \tag{A.11}$$

where

$$E' = E_i - \Delta_j + \Delta_j \pm \hbar\omega_{ij}. \tag{A.12}$$

The upper sign is for absorption of a phonon. The intervalley scattering processes completely randomize the momentum at each scattering event. Thus the scattering has no preferred direction. The angle θ is determined by using a random number according to

$$\cos\theta = 1 - 2r. \tag{A.13}$$

For intervalley scattering between equivalent valleys $\omega_{ij} = 0$.

A.1.3. Intravalley Nonpolar Optical Scattering

The square of the matrix element is

$$V(q) = \frac{D_0 \hbar^2}{2\rho\hbar\omega_0} (N_q + \tfrac{1}{2} \pm \tfrac{1}{2}), \tag{A.14}$$

where D_0 is the optical deformation potential and $\hbar\omega_0$ is the optical phonon energy. The total scattering rate is

$$P(E) = \frac{D_0^2 (m^*)^{3/2}}{\sqrt{2}\,\pi\hbar^2\rho(\hbar\omega_0)} \gamma^{1/2}(E')(1 + 2\alpha E')(N_0 + \tfrac{1}{2} \mp \tfrac{1}{2}), \tag{A.15}$$

where

$$E' = E \pm \hbar\omega_0 \tag{A.16}$$

and

$$N_0 = \frac{1}{\exp(\hbar\omega_0/kT) - 1}.$$ (A.17)

The angular dependence is determined from (A.13).

A.1.4. Piezoelectric Scattering

The square of the matrix element is

$$V(q) = \frac{\hbar e^2 p^2}{2\rho s(4\pi\varepsilon)^2 q^2}(N_0 + \tfrac{1}{2} \mp \tfrac{1}{2}),$$ (A.18)

where p is the piezoelectric tensor and s is the speed of sound. The total scattering rate is

$$P(E) = \frac{m^*}{4\pi\hbar^2}\left[\frac{eP_z}{\varepsilon\varepsilon_0}\right]^2 \frac{k_B T}{C_e} \frac{1 + 2\alpha E}{\sqrt{2}\, m^*\gamma(E)} \log_e\left[1 + \frac{8mL_D^2}{\hbar^2}\gamma(E)\right],$$ (A.19)

where the Debye length L_D is

$$L_D = \sqrt{\frac{\varepsilon\varepsilon_0 k_B T}{ne^2}}.$$ (A.20)

The angular dependence is given by

$$P(\theta)\,d\theta \sim \frac{\sin\theta\,d\theta}{[\gamma(E) + \hbar^2/4m^*L_D^2 - \gamma(E)\cos\theta]}.$$ (A.21)

The value of θ is determined using a random number r, which gives

$$\cos\theta = 1 + \frac{\hbar^2}{4mL_D^2\gamma(E)}\left\{1 - \left(1 + \frac{8m^*L_D^2}{\hbar^2}\gamma(E)\right)r\right\}.$$ (A.22)

A.1.5. Acoustic Scattering

The scattering is considered elastic at high fields and temperatures. However, at low temperatures they are inelastic. We need to consider both cases. The square of the matrix element is

$$V(q) = \frac{\Xi_1^2 \hbar q}{2\rho s}(N_q + \tfrac{1}{2} \mp \tfrac{1}{2}),$$ (A.23)

where Ξ is the acoustic deformation potential, ρ is the crystal density, s is the speed of sound, and N_q is the Bose–Einstein distribution. Our two cases differ in the total scattering rate. For the case of an elastic process where the energy equipartition approximation holds,

$$N_q = \left\{ \exp\left(\frac{\hbar s q}{k_B T}\right) - 1 \right\}^{-1} \sim \frac{k_B T}{\hbar s q}, \tag{A.24}$$

where $q = |\mathbf{k} - \mathbf{k}'|$. The total scattering rate is then given by

$$P(E) = \frac{(2m^*)^{3/2} k_B T \Xi^2}{2\pi\rho s^2 \hbar^4} \gamma^{1/2}(E)(1 + 2\alpha E). \tag{A.25}$$

The scattering completely randomizes the momentum so that the angular dependence is calculated from (A.13).

For the case of inelastic scattering at low temperature, the equipartition approximation fails and one has to expand N_q carefully. The result is

$$N(q) = \frac{1}{\gamma q} - \frac{1}{2} + \frac{\gamma q}{12}, \tag{A.26}$$

where

$$\gamma = \frac{\hbar s}{k_B T} \tag{A.27}$$

and

$$q = |\mathbf{k} - \mathbf{k}'|. \tag{A.28}$$

A.1.6. Ionized Impurity Scattering

The central difficulty faced in ionized impurity scattering is that the bare Coulomb interaction between the electrons and the impurities yields an infinite scattering cross section. While there are a variety of ways of overcoming this difficulty,[1] the two most commonly used models are the Conwell–Weisskopf and the Brooks–Herring. In the Conwell–Weisskopf model a bare interaction is used, but it is cutoff at the average distance between the impurity ions. Here we will present the rates for the Brooks–Herring model, which uses the screened potential

$$V(r) = \frac{Ze}{4\pi\varepsilon r} e^{-\beta r}, \tag{A.29}$$

where Z is the number of charged impurities, and for a nondegenerate semiconductor β is

$$\beta = \left[\frac{ne^2}{\varepsilon k_B T} \right]^{1/2}. \tag{A.30}$$

The square of the matrix element then is

$$V(q) = \frac{NZ^2 e^4}{(4\pi\varepsilon)^2 (q^2 + \beta^2)^2}. \tag{A.31}$$

The total scattering rate for the Brooks–Herring model is

$$P(E) = \frac{N_I}{32\pi\sqrt{2m}} \left(\frac{Ze^2}{\varepsilon} \right)^2 \frac{1}{[\varepsilon_\beta/4\gamma(E)][1 + \varepsilon_\beta/4\gamma(E)]} \frac{1 + 2\alpha E}{\gamma(E)^{3/2}}, \tag{A.32}$$

where

$$\varepsilon_\beta = \frac{\hbar^2 \beta^2}{2m}. \tag{A.33}$$

The angular dependence is obtained from

$$P(\theta) \, d\theta \sim \frac{\sin\theta \, d\theta}{(4k^2 \sin^2(\theta/2) + \beta^2)^2}. \tag{A.34}$$

Upon using a random number, we get

$$\cos\theta = 1 - \frac{2(1 - r)}{1 + r(4k^2/\beta)}. \tag{A.35}$$

A.2. HOLE SCATTERING IN GaAs AND Si

The total scattering rates are calculated using a two-band model for the valence band. These bands are assumed to be parabolic and spherically symmetric.

A.2.1. Polar Optical Phonon Scattering

For polar optical phonon scattering of holes the total intraband scattering rate is

$$P_{ii}(E) = \left(\frac{B_{po} m_i}{2\pi h^2} \right) \frac{1}{\sqrt{2m_i}} E^{-1/2} \psi_i(E) G_{ii}(E) (N_0 + \tfrac{1}{2} \mp \tfrac{1}{2}), \tag{A.36}$$

where

$$G_{ii}(E) = \frac{1 + 3\Phi_i(\Phi_i - \psi_i^{-1})}{4}, \tag{A.37}$$

$$\psi_i = \log_e \left[\left| \frac{\sqrt{E} + \sqrt{E'}}{\sqrt{E} - \sqrt{E'}} \right| \right], \tag{A.38}$$

$$\Phi_i = \frac{E + E'}{2\sqrt{EE'}}, \tag{A.39}$$

$$N_0 = \left[\exp\left(\frac{\hbar\omega_0}{k_B T}\right) - 1 \right]^{-1}, \tag{A.40}$$

and

$$E' = E \pm \hbar\omega_0. \tag{A.41}$$

The index $i = 1$ for heavy holes and 2 for light holes.

For interband scattering from band i to band j the total scattering rate is

$$P_{ij}(E) = \left(\frac{B_{po} m_j}{2\pi\hbar^2}\right) \frac{1}{\sqrt{2m_i}} E^{-1/2} \psi_{ij}(E) G_{ij}(E)(N_0 + \tfrac{1}{2} \mp \tfrac{1}{2}), \tag{A.42}$$

where

$$G_{ij}(E) = \tfrac{3}{2}[1 - \Phi_{ij}(\Phi_{ij} - \psi_{ij}^{-1})], \tag{A.43}$$

$$\psi_{ij}(E) = \log_e \left[\left| \frac{\sqrt{E} + \sqrt{(m_j/m_i)E'}}{\sqrt{E} - \sqrt{(m_j/m_i)E'}} \right| \right], \tag{A.44}$$

and

$$\Phi_{ij}(E) = \frac{E + (m_j/m_i)E'}{2\sqrt{(m_j/m_i)EE'}}. \tag{A.45}$$

The angular dependence for intraband scattering processes takes the form

$$P_{ii}(\theta)\, d\theta \sim \frac{(1 + 3\cos^2\theta)\sin\theta\, d\theta}{[k^2 + k'^2 - 2kk'\cos\theta]}. \tag{A.46}$$

Similarly for interband scattering processes the angular dependence is ($i \neq j$)

$$P_{ij}(\theta)\, d\theta \sim \frac{(\tfrac{1}{2} - \cos^2\theta)\sin\theta\, d\theta}{[k^2 + k'^2 - 2kk'\cos\theta]}. \tag{A.47}$$

Usually in order to generate the angles after scattering the rejection technique discussed in Chapter 4 is used because of the difficulty in using the direct method.

A.2.2. Nonpolar Optical Phonon Scattering

The scattering rate for intraband processes is

$$P_{ii}(E) = \frac{(2m_i)^{3/2} D_0^2}{4\pi\varepsilon\hbar^3\omega_0} \bar{G}_{ii}(E \pm \hbar\omega_0)^{1/2}(N_0 + \tfrac{1}{2} \mp \tfrac{1}{2}), \tag{A.48}$$

while the interband scattering rates are

$$P_{ij}(E) = \frac{(2m_j)^{3/2} D_0^2}{4\pi\varepsilon\hbar^3\omega_0} \bar{G}_{ij}(E \pm \hbar\omega_0)^{1/2}(N_0 + \tfrac{1}{2} \mp \tfrac{1}{2}). \tag{A.49}$$

Since the matrix element has no angular dependence, it is common practice to assume[2]

$$\bar{G}_{ii} = \bar{G}_{ij} = \tfrac{1}{2}. \tag{A.50}$$

Consequently, this scattering process completely randomizes the momentum. The angles after scattering are selected using random numbers according to (A.13).

A.2.3. Ionized Impurity Scattering

The total intraband scattering rate is given by (assuming heavy-hole band)

$$P_{11}(E) = \frac{\bar{B}_I}{4\pi\varepsilon_\beta^2} \left(\frac{2}{m}\right)^{1/2} \frac{E^{1/2}}{1 + 4\varepsilon/E_\beta} G_{11}(E), \tag{A.51}$$

where

$$G_{11}(E) = \frac{1}{4}\left\{1 + \frac{3}{4}\frac{XY^2}{(E/E_\beta)^2}\left[\frac{1}{x} + \frac{1}{Y^2} + \frac{\log_e[X^{-1}]}{2Y(E/E_\beta)}\right]\right\}. \tag{A.52}$$

$$X = 1 + \frac{4E}{E_\beta}, \tag{A.53}$$

$$Y = 1 + \frac{2E}{E_\beta}, \tag{A.54}$$

$$E_\beta = \frac{\hbar^2\beta^2}{2m_1}, \tag{A.55}$$

and

$$\bar{B}_I = \frac{N_I e^4}{\varepsilon^2}.$$

(A.56)

The angular dependence is obtained from

$$P(\theta)\, d\theta \sim \frac{(1 + 3\cos^2\theta)\sin\theta\, d\theta}{[4k^2(\sin^2\theta)/2 + \beta^2]^2}.$$

(A.57)

REFERENCES

1. T. Ando, F. Stern, and A. B. Fowler, *Rev. Mod. Phys.* **54**, 437 (1982).
2. R. Dingle, *Festkorperprobleme* **15**, 21 (1975).
3. R. Tsu and L. Esaki, *Appl. Phys. Lett.* **22**, 562 (1973).
4. L. Esaki and R. Tsu, *IBM J. Res. Develop.* **14**, 61 (1970).
5. R. T. Bate, *Bull. Am. Phys. Soc.* **22**, 407 (1977).
6. R. K. Reich, R. O. Grondin, D. K. Ferry, and G. J. Iafrate, *IEEE Electron Dev. Lett.* **EDL-3**, 381 (1982).
7. *Picosecond Electronics and Optoelectronics* (G. A. Mourou, D. M. Bloom, and C.-H. Lee, eds.), Springer-Verlag, Berlin (1985).
8. P. A. Blakey, private communication.
9. W. Shockley, *J. Appl. Phys.* **9**, 635 (1938).
10. S. Ramo, *Proc. IRE* **27**, 584 (1939).
11. J. D. Jackson, *Classical Electrodynamics*, Wiley, New York (1962).
12. W. K. H. Panofsky and M. Phillips, *Classical Electricity and Magnetism*, 2nd ed., Addison-Wesley, Reading, MA (1962).
13. O. Madelung, *Introduction to Solid State Theory*, Springer-Verlag, Berlin (1978).
14. J. R. Chelikowsky and M. L. Cohen, *Phys. Rev. B* **14**, 556 (1976); see also M. L. Cohen and J. R. Chelikowsky, *Electronic Structure and Optical Properties of Semiconductors*, 2nd Ed., Springer-Verlag, Berlin (1988).
15. J. B. Gunn, *Proc. IEEE* **62**, 823 (1974).
16. B. K. Ridley, *Quantum Processes in Semiconductors*, Clarendon Press, Oxford (1982).
17. J. M. Ziman, *Electrons and Phonons*, Clarendon Press, Oxford (1960).
18. C. Zener, *Proc. R. Soc. A* **145**, 523 (1934).
19. L. Esaki, *Proc. IEEE* **62**, 825 (1974).
20. M. Büttiker, *Phys. Rev. B* **35**, 4123 (1987).
21. E. O. Kane, in: *Tunneling Phenomena in Solids* (E. Burstein and S. Lundqvist, eds.), Plenum Press, New York (1969).
22. K. Stevens, *J. Phys. C* **16**, 3649 (1983).
23. D. Bohm, *Quantum Theory*, Prentice Hall, Englewood Cliffs, NJ (1951).
24. L. A. MacColl, *Phys. Rev.* **40**, 621 (1932).
25. M. Büttiker and R. Landauer, *Phys. Rev. Lett.* **49**, 1739 (1982).
26. W. E. Hagstrom, *Phys. Stat. Solidi (B)* **116**, K85 (1983).
27. N. Kluksdahl, *A Wigner Function Study of Quantum Electronic Transport in Semiconductor Tunneling Structures*, Ph.D. thesis, Arizona State University (1988).
28. B. N. Brockhouse, *Phys. Rev. Lett.* **2**, 256 (1959).
29. J. L. T. Waugh and G. Dolling, *Phys. Rev.* **132**, 2410 (1963).
30. P. Vogl, in: *Physics of Nonlinear Transport in Semiconductors* (D. K. Ferry, J. R. Barker, and C. Jacoboni, eds.), Plenum Press, New York (1980).
31. W. Fawcett, in: *Electrons in Crystalline Solids*, International Atomic Energy Agency, Vienna (1973).

32. P. M. Smith, M. Inoue, and J. Frey, *Appl. Phys. Lett.* **37**, 797 (1980); P. M. Smith, J. Frey, and P. Chatterjee, *Appl. Phys. Lett.* **39**, 332 (1981).

33. T. H. Windhorn, *Electron Drift Velocities at High Electric Fields in Gallium Arsenide and Indium Gallium Arsenide*, Ph.D. thesis, University of Illinois (1982).

34. W. Fawcett and D. C. Herbert, *J. Phys. C: Solid State Phys.* **7**, 1641 (1974).

35. G. E. Stillman, in: *Gallium Arsenide and Related Compounds (Edinburgh) 1976* (C. Hilsum, ed.), The Institute of Physics, Bristol and London (1977).

36. A. G. Chynoweth in *Semiconductors and Semimetals: Vol. 4, Physics of III–V Compounds* (R. K. Willardson and A. C. Beer, eds.), Academic Press, New York (1968).

37. C. L. Anderson and C. R. Crowell, *Phys. Rev. B* **5**, 2267 (1972).

38. T. P. Pearsall, F. Capasso, R. E. Nahory, M. A. Pollack, and J. R. Chelikowsky, *Solid St. Elec.*, **21**, 297 (1978).

39. T. P. Pearsall, *Appl. Phys. Lett.* **36**, 218 (1980).

40. R. K. Mains, G. I. Haddad, and P. A. Blakey, *IEEE Trans. Electron Devices* **ED-30**, 1327 (1983).

41. G. E. Bulman, V. M. Robbins, K. F. Brennan, K. Hess and G. E. Stillman, in: *Optical Communications Systems, Proc. Fifteenth National Science Foundation Grantee-User Meeting*, MIT (1983).

42. W. N. Grant, *Solid St. Elect.* **16** 1189 (1973).

43. I. Umebu, A. N. M. M. Choudhury, and P. N. Robson, *Appl. Phys. Lett.* **36**, 302 (1980).

44. A. G. R. Evans and P. N. Robson, *Solid St. Elect.* **17**, 805 (1974).

45. T. H. Windhorn, L. W. Cook, and G. E. Stillman, *IEEE Electron Dev. Lett.* **EDL-3**, 18 (1982).

46. P. M. Smith, *Measurement of High Field Transport Properties of Semiconductors Using a Microwave Time-of-Flight Technique*, Ph.D. thesis, Cornell (1982).

47. G. Hill and P. N. Robson, *Solid St. Elec.* **25**, 589 (1982).

48. J. P. Nougier, Noise and diffusion of hot carriers, in: *Physics of Nonlinear Transport in Semiconductors* (D. K. Ferry, J. R. Barker, and C. Jacoboni, eds.), Plenum, New York (1980).

49. T. H. Glisson, R. A. Sadler, J. R. Hauser, and M. A. Littlejohn, *Solid St. Elec.* **23**, 627 (1980).

50. J. G. Ruch and G. S. Kino, *Phys. Rev.* **174**, 921 (1968).

51. R. Joshi and R. O. Grondin, *Appl. Phys. Lett.* **54**, 2438 (1989).

52. P. E. Bauhahn, *Properties of Semiconductor Materials and Microwave Transit-Time Devices*, Ph.D. thesis, University of Michigan (1977).

53. C. Canali, F. Nava, and L. Reggiani, in: *Hot Electron Transport in Semiconductors* (L. Reggiani, ed.) Springer-Verlag, Berlin (1985).

54. W. Fawcett and H. D. Rees, *Phys. Lett. A* **29**, 578 (1969).

55. J. Pozhela and A. Reklaitis, *Sol. State Commun.* **27**, 1073 (1978).

2

Fabrication Techniques for Submicron Devices

It is now possible to fabricate test patterns, which compose the basic parts of elemental semiconductor devices, that are as small as 10 nm in size, and to see individual features on the order of 1 nm in size.[1] This suggests that further device miniaturization and entirely new device concepts are certain to occur. However, coupled with the decrease in device size has been a concomitant increase in the number of devices that are contained on a single integrated circuit. Today's 0.5-μm devices allow as many as 16 million transistors on a single dynamic memory chip or as many as a million devices on a microprocessor chip. Therefore, while it is feasible to use exotic processing techniques to fabricate single devices, the industry requires processing techniques which can fabricate large arrays of chips, with each chip containing this large number of individual devices. Thus, the practicality of a particular process is of utmost importance if it is to be used in the continuing growth of integrated circuit density.

Based upon the progress that has already occurred, as well as on the research that is currently underway, it can be expected that in the mid-1990s, we will be dealing with devices with a critical dimension of 0.35 μm or less. It is important then to make an estimate as to the efficacy of the various options that exist for development of processing steps for progress beyond this level. The choice is not one of, say, lithography that gives the highest resolution, since profitability and yield (which are intimately connected) must also be included as primary factors in technology choice.

Optical lithography and normal diffusion and implantation processes currently hold an almost unassailable position as the primary, indeed almost the only, techniques for processing a semiconductor wafer into integrated circuits. This is even true for circuits with critical dimensions down to 0.25 μm, although the masks for this latter dimensional technology may be processed by electron beams. However, device sizes are rapidly approaching dimensions smaller than this in developmental and research designs, and design rules of 0.5 μm are finding their way into production circuits. Now, the design rules are beginning to be at or below the wavelength of the excitation light used in the exposure process for lithography, and it is important to begin to address the question as to what lithographic techniques will be most appropriate for submicron-dimensioned devices. Indeed, we should be asking just what the entire fabrication process will be for circuits in this regime.

Several factors act together to actually determine the answers to the above questions. In lithography, one crucial question is whether or not we will be able to align sequential mask levels to an accuracy sufficient to guarantee adequate yield in the fabrication process. Second, can we maintain control of the mask with the required thinness of resist layers to achieve the required homogeneity across the wafer, again in order to achieve adequate yield in the process? Third, can the exposure technique produce the required sharpness of line edges at the dimensional scale desired? And fourth, can the pattern be transferred to the wafer on a time scale that is reasonable for mass production of the circuits? Can control of the dopants, either by diffusion or implantation, be obtained to maintain lateral control of device geometries through subsequent processing steps? Can parasitics be controlled to a level to not dominate the overall performance of the circuit?

In this chapter, we do not want to predict what the processing will ultimately entail. Rather, we want to review the techniques—lithography, dry processing, epitaxial growth, metallization, contacts, heterojunctions, etc.—that are currently being used in research environments to fabricate devices below 0.25 μm. Indeed, some of these techniques have been combined to build GaAs metal-semiconductor FETs (MESFETs) and HEMTs with gate lengths as small as 20 nm[2] and Si n-type MOS (NMOS) circuits with effective gate lengths as short as 70 nm.[3] Most of the chapter emphasizes lithography because that is the greatest area of question for the future, but will deal with each of the topics listed above in turn.

2.1. LITHOGRAPHY

It is unlikely that any process will surpass direct-writing electron-beam (e-beam) lithography for straightforward resolution. Even in the cases of optical and/or x-ray lithography, as well as some others, the masks used in the process will probably be fabricated by e-beam techniques. Indeed, the small devices in GaAs and Si mentioned above were developed with e-beam processing. However, direct writing by e-beam techniques has a well-known problem of very low throughput. This latter problem directly affects profitability by holding down the number of wafer levels per day that can be processed, and hence restricts the number of chips actually made. Thus, profitability is a very large driving force. It is not particularly recognized that yield is a driving force on the choice of lithographic techniques until discussions of overlay accuracy are begun. Clearly, techniques that require overlay accuracies beyond the state-of-the-art are not amenable to implementation.

In this section, we want to examine the various lithographic techniques with a goal of estimating their limitations in the fabrication process for submicron integrated circuits. To this end, we will concentrate on their various adaptability to the fabrication process.

2.1.1. Patterning Considerations

There are a number of factors that must be decided as we proceed. Today's processing is primarily done with optical lithography, with both optical and e-beam techniques utilized for the preparation of the masks. However, we note that the actual process may involve whole-wafer projection lithography, contact printing, or direct step-on-wafer (DSW) processing. In particular, the projection technique may also be whole-wafer or DSW, and may also entail reductions such as 1:1 or 10:1. It is not clear at this point whether all of these choices will remain at the smaller scale envisioned for submicron devices, but it is unlikely, for example, that whole wafer exposures will continue to be viable and are little discussed anymore except in the context of x-ray lithography with a high-fluence synchrotron source. Each of the approaches has different constraints when being considered for comparison with other technology approaches. In fact, it will likely turn out that we do not have such a range of choices.

In microcircuit fabrication, the minimum linewidth is set more by the ability to control this linewidth reproducibly across the wafer rather than by just the ability to produce a thin line.[4] In general, a pattern containing a range of linewidths is exposed on many wafers, and hundreds of measurements are made in order to determine the statistical variations before this approach can even be considered for processing. These measurements determine the standard deviation of the line, and a designer must be content to choose a process scale factor for the minimum linewidth that will allow this line to be produced consistently with perfect accuracy. The linewidth that can be used is usually 10–15 times the standard deviation of the lithography. Thus, with a minimum linewidth of 0.1 μm for gates, we require that the process be capable of writing line dimensions with a standard deviation of only 10 nm. This rule of 10–15 times is set so that the devices can be tolerant of processing variations of the order of 3 times the standard deviation.

Errors in the position of one pattern with respect to another are called overlay errors. These are generally also measured during test processes in order to determine the standard deviation. Overlay errors can be produced either from misalignment of the mask or, more often, from differential expansion of the wafer during processing so that mask dimensions no longer match already processed dimensions on the chip. This is a serious problem and will probably eliminate whole-wafer, single-shot processing; e.g., direct DSW techniques that can be locally aligned, and for which the total expansion is much smaller than that of the whole wafer, will be required to control the overall overlay errors. To understand this, we note that a 10-nm change in size across an 8-in wafer is a change of less than 1 part in 10^7. Thermal expansion coefficients of common materials are 10–100 times larger than this (per °C), and the difference in thermal expansion between SiO_2 and Si is 20 times larger than this. Indeed, there is no known technique today to control overlay errors to better than the order of 0.1 μm in a masked technology. On the other hand, if we have to control this overlay error only over the size of a chip (e.g., 0.5 cm), then we gain a factor of 40 and are now in the range where we can begin to consider overlay errors of the proper magnitude. Consequently, it is probable that DSW techniques will have to be utilized for this reason alone, if for no others. However, we note that direct-writing approaches are already classed as DSW, so that their disadvantages compared to other approaches are reduced by this requirement.

2.1.2. Time

The time required to process each wafer level is an important consideration. In comparing the various technologies, it is one point to which we want to play very close attention. This processing time is determined by the intensity of the source, the size of the illumination spot, and the sensitivity of the resist used to reproduce the pattern. We can understand this easily for direct-writing systems. We can illustrate this for a beam technology. To reproduce 0.1-μm lines, we need to consider a spot size that is at least a factor of 2 smaller in a beam technology, or say 50 nm. Then, in order to write the full 0.5 × 0.5 cm pattern, we need of the order of 10^{10} pixels separately written. If each spot required a dwell time of 1 ms, this would require 10^7 s, or more than 100 days. This is a clearly unacceptable writing period. Thus, a much more sensitive resist and/or more intense beam would have to be developed.

2.1.3. Resist Materials

With the exception of direct ion-beam implantation processing, current processing steps require lithographic techniques that depend upon the exposure and patterning of a layer of energy (optical or particle) sensitive material which is placed upon the semiconductor surface. The incident radiation, whether it is electrons, ions, or photons, causes a chemical or structural change in the exposed areas of the resist, thus causing these areas to have different etch properties in a subsequent developer than the unexposed areas. In nearly all cases today the resist is a polymer. In a negative resist, the exposure induces cross-linking between the polymer chains, and the unexposed area is washed away by the developer. On the other hand, a breaking of polymer bonds occurs in a positive resist, and the exposed area is the part that is removed by the developer. The remaining areas of resist, after development, form a stencil mask on the substrate through which subsequent processing steps can proceed. These subsequent steps are usually referred to as *pattern transfer*, and are as important a part of the process as the lithography itself.

Important parameters in the selection of a resist material are the sensitivity of the resist to the exposing radiation, the resolution limits of the resist, the capabilities of the resist to withstand the subsequent pattern transfer step, and its compatibility with the overall process being used. An example of the conflicting requirements which these parameters impose are illustrated by the case of poly(methyl methacrylate), or PMMA. Although many materials have been tested as an electron-sensitive resist, the only practical resist to have a demonstrated resolution better than 5 nm is PMMA.[1] Although other resists exist, which can achieve this resolution on *thinned* substrates with e-beam exposure, or in multi-level resist schemes, they have not been adequately characterized. Moreover, the thinned substrate approach is really impractical for production integrated circuits. Yet, PMMA is generally agreed to have far too low a sensitivity to be really useful in production processing of integrated circuits. On the other hand, there do not appear to be any studies of intrinsic resist resolution that prove that the polymer unit size affects final pattern resolution, particularly in the size scale of 50 nm.

Exposure. Designation of dose and sensitivity has taken a variety of forms over the years. Taylor[5] recently published an excellent review of the various approaches and a more-or-less standardized set of definitions. It is important to note that the exposure, whether by a particle or photon process, is basically a photographic processing step. Thus, just as in photographic film, paramount to the exposure process is a curve of remaining resist thickness as a function of exposure, or the photographic density-exposure curve (D-E curve). Here, the developed thickness of the resist plays the role of density in the photographic film. Examples of sensitivity curves are shown in Figure 2.1 for the positive PMMA resist and a negative high-resolution resist (Shipley SAL 601-ER7). It is important to note two facts about these curves that are crucial to the process. Both of these are related to the subsequent developers used for processing the exposed resist. The developer will also attack and subsequently thin the area that is not to be removed so that the differential etch rate is an important factor in choosing a developer/resist combination. The second aspect relates to the slope of the D-E curve in the transition region. As in high-contrast films, it is quite important that this slope be as large as possible. In general, we define a contrast factor

$$\gamma = \left[\ln\left(\frac{D_1}{D_0}\right) \right]$$

(2.1)

for positive resist, where D_1 is the extrapolated dose required to effect complete removal in a given development time, and D_0 is the extrapolated dose at which development just begins. Contrast factors as high as 9 have been obtained for PMMA in the authors' laboratories,[6] which appears to be the best at this time. On the other hand, the high-resolution negative resist mentioned gives a value only about a third of this (the points at which the D are defined are moved as is evident from the figure). Finally, the crucial exposure quantity required for processing is the dose required for total thinning of the desired area, which is about $2D_1$. This latter value and the contrast factor γ combine to define the

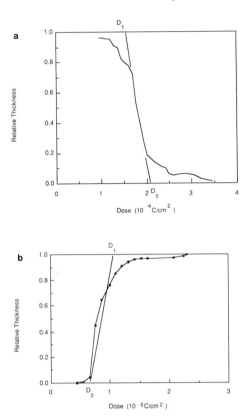

FIGURE 2.1.   The exposure curves for positive (a) and negative (b) resists. The various parameters are discussed in the text.

acceptable fluctuation of the incident exposure beam. If γ is small, then a small variation of the dose will partially expose the resist. On the other hand, if γ is large, and the resist has little tailing around the point D_0, then considerable variation in the dose is allowed without any degradation of exposed area. We will see that the fluctuation in the number of incident particles (or photons) per pixel can be a limiting factor, and proper control over this requires good contrast in the resist.

Let us reiterate these points. In semiconductor processing, it is essential that very high contrast resists be used. The rationale for this lies in those processing techniques that generate scattered radiation or scattered secondary electrons. High contrast is important for achieving sharp, nearly vertical resist edge profiles. These extra particles, or radiation, can partially expose the resist. For ultimate performance, it is desired that the resist have an almost infinite contrast, so that exposure at levels below threshold do not expose the resist, while exposures above the threshold give full desired development. This is especially important when exposing at linewidths approaching the resolution limit of the exposure process. In general, improved resist contrast results from weaker developers, higher exposures, and larger molecular weight of the initial polymer.

Resolution Limits. For electron, ion, and x-ray exposure, it is often the limitations of the resist, in terms of exposure parameters, that set the ultimate limit on pattern resolution usable in processing. On the other hand, for UV-optical exposure, it is clear that the limit is set by diffraction effects of the source-mask combination. PMMA is the most commonly used resist in research studies of very high resolution, although AZ-1350J has also been used fairly often. When PMMA is exposed as a positive resist, the primary exposure arises from secondary electrons. These electrons usually have an energy below 100 eV, whether the exposing radiation is electron, ion, or photon. The effective range of these secondary electrons has been estimated as less than 10 nm.[7-9] This can thus be taken as the lower limit on resolution in this resist. The main drawback of PMMA is its poor sensitivity, and it is not therefore in general use for processing. We can realize this by considering the time required to expose each pixel with this resist. If we use a 0.5-μA beam, with a diameter of 50 nm, then we are depositing 2.5×10^4 (C/s)/cm^2 in the area of the spot. At 50 keV, the sensitivity of PMMA is about 50 μC/cm^2, so that we require an exposure time of 2 ns at each point, and our 10^{10} pixels require 20 s to expose. While this is clearly far better than our overly pessimistic estimate in the previous section, it would still require more than 4 h to expose a 150-mm wafer, and this is an unexceptable time scale. Nevertheless, PMMA remains the standard of comparison in high-resolution, positive resists. In Figure 2.2, we show a set of gold lines produced by lift-off processing in PMMA which was exposed by e-beam techniques. The graininess of the lines is produced by the grain size of the gold particles rather than any fluctuation in the exposure of the resist, and this graininess can be reduced by using a gold-palladium alloy rather than pure gold.

Unfortunately, there is an intrinsic conflict between the requirements of high resolution and high sensitivity due to statistical considerations. If a minimum resolution area $A = \pi d^2/4$ is to be exposed on a thin resist, the average number n of exposing particles, whether electrons, ions, or photons, incident upon this area must be much greater than the statistical fluctuations in this number. Since $n \sim N^{1/2}$, we can set a minimum level of particles as 100 per pixel to avoid these statistical fluctuations (this sets the fluctuation as less than 10% of the dose). Then if we know the fluence of exposing particles, we can set a minimum usable exposure level for each type of particle. We show such a consideration in Table 2.1.

Clearly, from Table 2.1, only ion-beam exposure is approaching a critical level with the relatively slow PMMA resist, although even here there is room for a factor of almost 2 improvement in resist sensitivity (speed). On the other hand, we note that since electrons and x-ray photons interact only weakly with typical resists, a larger dose is tolerable for exposure. On the other hand, we must also be aware that it is the secondary electrons produced which provide much of the exposure. If the secondaries are typically 100 eV, then the electron and ion beams will produce several thousand in the exposure area, but the x-ray photon will

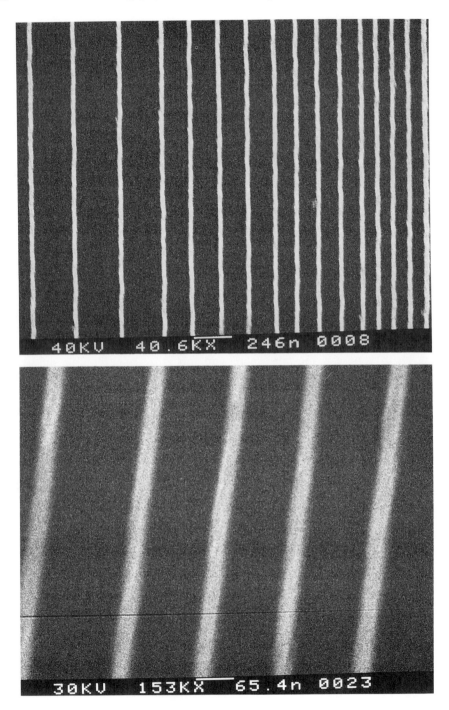

FIGURE 2.2.   Gold lines deposited by lift-off processing with PMMA resist exposed with 40 keV electrons and using a high-resolution, high-contrast developer.

TABLE 2.1.   Minimum Exposure Level for Given Particle
(exposure of 100 particles per pixel of 50 nm)

Source	Dose	PMMA sensitivity level[1,10]
50-keV electrons	0.8 μC/cm^2	50 μC/cm^2
55-keV ions	0.8 μC/cm^2	1.5 μC/cm^2
0.45-nm x-ray photons	23 mJ/cm^2	400 mJ/cm^2

produce only a few hundred. This makes the latter process much more susceptible to fluctuations than at first sight expected.

It is clear from the above data that PMMA can easily be used by almost any exposure tool to achieve adequate resolution without worrying about statistical fluctuations. However, it is just as clear that the margin between the last two columns of Table 2.1, which determines the flexibility for finding a more sensitive resist material than PMMA, is actually very slim for any technique other than e-beam lithography. If we seek a faster resist for, say, x-ray lithography than PMMA, and increase its sensitivity by a factor of 50, then we would have to expect to have fluctuations of some 20% in the exposure dose per pixel. This is below our margin of error discussed above and would endanger the opportunity to produce the circuits with usable yield, an unacceptable result. The situation is worse in the case of ion beams. These results seem to indicate that we cannot use a faster resist in some exposure systems even if we could develop it. Instead, it appears that for these approaches we really need more intense sources and not more sensitive resists if we are to maintain reasonable levels of uniformity.

From the foregoing, we can surmise that we are not really up against any fundamental limits, either of the resist structure or of resolution for lithography in the 0.1-μm dimensional scale regime. However, we will have to worry about the actual exposure dose in terms of the number of particles in each pixel that are used for the exposure. High-resolution exposure requires that a prescribed level of dose is used to ensure full development. Unless specified overexposure is planned for, which is not acceptable in a process that is already limited by requiring too much exposure time, then statistical fluctuations may well limit our choice of resist and, ultimately, of exposure and lithography tools.

Multilevel Resists.   In processing semiconductor devices, we must also face the fact that high-resolution patterns may be very difficult to generate on the nonuniform and nonplanar surface of the already partially processed chip. This is because covering topographical structures may impose a resulting nonuniform resist thickness. Moreover, a spin-cast resist must be thick enough to ensure proper coverage, but such a thickness allows considerable scattering and diffraction in the resist itself, which limits resolution. In contrast, good linewidth control with high resolution requires a uniform resist film that is considerably less than 0.5 μm thick (if we are to pattern on the 0.1-μm level). Consequently, the conflicting requirements imposed on the resist film thickness has led to the development of multilayer resist systems which separate the imaging and step coverage requirements. While two-level systems have been examined, it is apparent that a three-level system is ideal for the high resolution required here. One of the drawbacks of a two-layer resist system is the nonuniform mixing which sometimes occurs at the interface between the two resist layers, resulting in processing difficulties and poor pattern definition.[11]

Three-layer resist systems have been introduced which incorporate a barrier film sandwiched between the two active resist layers. This barrier film prevents mixing between the two resist layers and therefore allows for a wider choice of resist materials for these layers. Often, however, this arises at the expense of increased exposure times and increased number of processing steps. The top layer is normally selected primarily for its sensitivity and resolution, thus providing the basic control over image pattern geometry. The two lower layers of the system do not even need to be normal photoresist material, but can be chosen on the basis of desired chemical and physical properties for adhesion and step coverage. In the classical process, the top layer is the only one that is defined by "photo"-lithography, and the pattern defined in this layer is transferred through the remaining layers by dry processing techniques, such as reactive ion etching. In nearly all cases used for high-resolution work, PMMA remains one of the constituents of the three-level systems, usually the upper-resolution-defining layer. Thus, the three-level systems are already limited to be as slow as normal PMMA single-level processing steps.

As we will see, most processing exposure techniques will require the resolution layer be very thin, say less than 0.1 μm. It is almost impossible to obtain good step coverage of a partially processed wafer with a resist this thin, and therein lies the inherent value of the three-level system. The lowest level is used for step coverage, the second level is a separation level, and the top level can be made quite thin to provide the desired resolution. The inner layer, the separation layer, is quite often composed of an amorphous semiconductor or a metal. While some of the chalcogenide glass semiconductors can be used as resists in their own right, and can provide very high resolution,[12] they are usually quite slow and insensitive as a resist. In the three-level system, this layer is usually processed by dry etching, and so the speed is not a factor.

Dry Processing. Most negative polymeric resists developed in solvents have resolution limited by swelling of the resist during development. While positive polymer resists do not usually show this to any great degree, it is undesirable to continue to use wet processing due to adhesion problems of the resist and the resulting drop in yield. Consequently, there is a desire (and a need) to develop totally dry resist processing in which no liquid solutions ever touch the substrate. Since the swelling problem is more acute with negative resists, most research has been done in this area. However, there is a problem with the dry deposition process for the resists, as this is usually done with chemical vapor deposition or plasma deposition, which results in a highly-cross-linked polymer that is not therefore usable for conventional exposure and development. For these reasons, the resists today are still largely spin-cast onto the substrate and it is not clear when an all dry process can be fully developed for production.

2.1.4. Optical Patterning

Almost without exception, optical lithography is the sole method of exposing resists in use today for production of integrated circuits. This area has seen its own share of progress by which an ever-narrowing series of improvements were made either in the mask-making or the wafer fabrication process. The techniques in use now include near-UV shadow printing (or proximity printing as it is called) and projection lithography, particularly in DSW applications. Currently, 10:1 reduction steppers are the most popular, although 1:1 steppers still find some applications, and the current generation of DSW machines can easily produce circuitry with 0.7-μm design rules.

In discussing the resolution of optical writing systems, the most usual definition of resolution is the Rayleigh criterion, which was developed for telescopes. This criterion is defined around the 80% intensity point of an optical beam or image, and is expressed simply as

$$d = \frac{0.61\lambda}{n_a}, \tag{2.2}$$

where λ is the wavelength of the exposing radiation and n_a is the numerical aperture ($=n\sin(\theta/2)$, where n is the index of refraction of the lens material). In (2.2), d is the separation distance of two diffracted images and θ is the angle of the light cone entering the objective lens. The usual convention for optical systems is that 2.5 times the Rayleigh limit gives an estimation of the resolution. Thus, for a desired resolution of 0.1 μm and a lens with a numerical aperture of 0.61 (which is exceedingly high), we would need a wavelength of 40 nm, which is completely out of the optical region. Conversely, if we use the shortest current UV laser source, the excimer laser at 157 nm, we are really pushing the resolution limit of a lens with a numerical aperture of 0.61 to achieve a resolution of 0.4-μm resolution. In actual fact, this criterion is exceedingly pessimistic, and we can do somewhat better than this (but not by a great deal).

Proximity Printing. Generally, resolution in a shadow image is set by diffraction between the mask and the bottom of the resist layer. Thick resist layers, or gaps between the mask and the resist, both serve to increase the distance to the bottom of the resist and therefore increase the diffraction spreading of the pattern that is trying to be imaged. In practice, the minimum usable linewidth W in proximity printing can be approximated from[4]

$$W = 15\left(\frac{\lambda S}{200}\right)^{1/2}, \tag{2.3}$$

where S is the distance between the mask and the bottom of the resist. Broers[4] has pointed out that (2.3) corresponds to the condition for which the intensity at the center of an isolated line matches the background intensity. This criteria is evolved from the resolution degradation due to Fresnel diffraction through the mask. It is clear from (2.3) that if we are to improve resolution in proximity printing, the mask must be placed closer to the actual substrate itself, although the resolution improves only as the square root of this distance. In fact, this equation suggests that for reasonable resolution with a spacing S greater than almost zero, wavelengths in the x-ray region will be required.

In fact, a 157-nm excimer laser has been used to achieve the writing of lines with a width of 0.15 μm, using an exposure of 0.5 J/cm (for a line) and PMMA.[13] Equation (2.3) would suggest that $S = 0.13$ μm was required, which is in fact hard contact between the mask and a quite thin resist. Problems in this exposure, as with others using the proximity approach, is that it is extremely difficult to maintain tight and controlled contact, or separation, between the mask and the wafer over large areas because of contaminating dust particles, and a few square centimeters of area is probably a practical limit for production. Thus, we are faced with the need to step and repeat the pattern at the chip level, and such direct DSW techniques are generally not amenable to proximity, or contact, printing due to mask and wafer wear. While a 0.25- μm line may be possible in near-proximity situations, with a spacing of 0.35 μm between the mask and the substrate, in an excimer laser exposure, the requirement to hold this spacing across the whole wafer is probably beyond production tolerances.

We can summarize the problems with proximity printing by saying that 0.1-μm linewidths are probably unachievable even with excimer lasers, as they would require resist thicknesses, including the planarization layers, below 0.1 μm. It is unlikely that production tolerances, especially the overlay accuracy required, are achievable. Thus, one can expect that proximity printing may continue to be used for research devices, but probably cannot be used in production.

Projection Patterning. Both reflecting and refracting optics currently find use in projection DSW systems. With the trend to lower wavelength, it is likely that reflecting optics will probably be more generally utilized. The performance that can be expected from various optical systems can be described by the *modulation transfer function* (MTF), given by

$$\text{MTF} = \frac{2}{\pi}[\phi - \tfrac{1}{2}\sin(2\phi)], \tag{2.4}$$

where

$$\phi = \cos^{-1}\left(\frac{\lambda}{4\,Wn_a}\right), \tag{2.5}$$

and W is the linewidth, or periodicity, in the pattern. The MTF defines the degree of contrast, or ratio between minimum illumination and maximum illumination, just as in the exposure D-E curves discussed for the various resists. Thus, as the linewidth becomes much larger than the wavelength of the exposing light, such as in writing 2-μm lines with near-UV illumination, $\phi \rightarrow \pi/2$, and the MTF approaches unity. On the other hand, as the linewidth decreases toward the wavelength of the illuminating light, ϕ decreases and the MTF reduces to a value much less than unity. For example, our previous discussion used 157-nm excimer illumination and a numerical aperture of 0.61. Thus, for these values, a 0.1-μm line would produce only 23% for the MTF, which is a clearly unacceptable contrast.

The above equations assume that a fully incoherent source is used. In practice, higher contrast is obtained by the use of coherent or partially coherent sources, such as the excimer laser source. In practice, 60% modulation has been considered to be necessary in the optical image if satisfactory exposure is to be obtained. Broers,[4] on the other hand, has suggested that with coherent sources, one can "live with" a contrast of as little as 30% at one-half of the desired bandwidth of the source.

The possibility of optical aligners reaching sub-0.25-μm resolution presents a formidable challenge to competitive methods. However, this possibility also presents a formidable challenge to the developers of optical DSW aligners. It is clear from (2.5) that as large a value of the numerical aperture as possible is necessary in the lens system, preferably well above 0.6. On the other hand, most aligners today have numerical apertures only in the range 0.35–0.45. If optical techniques are to be capable of pushing to the desired small device sizes, then advanced lens design is absolutely necessary, and high-repetition-rate excimer lasers (or even continuous-wave (cw) sources) are necessary.

The exposures required for the excimer experiments that have been reported to date are about 0.4 s. This is less than an order of magnitude from the estimate expressed above for that required to write the same pattern with e-beam lithography. This differs from the usual optical techniques, in which the exposure times are well under that required by other methods, as the optical sources write all pixels in parallel and deliver much higher doses to the resist. However, these techniques also require that a precision mask be generated, which is probably accomplished by direct e-beam exposure. The final limitation of the optical exposure systems lies in the resolution obtainable. Even if we can live with a contrast of 0.3 (30%), this limits the linewidths obtainable by excimer laser exposure in an optical system with a numerical aperture of 0.6 at about 0.11 μm. On the other hand, if we require 0.6 for the contrast, then we are limited to linewidths greater than 0.2 μm. We should remark that few are contemplating using optical techniques down to this resolution at the present time, but it seems to be readily possible if the optics can be produced.

2.1.5. X-ray Lithography

X-ray lithography utilizes photons in the wavelength range of 0.4–10 nm to expose more or less conventional resists. Resolutions of the order of 20 nm have been achieved, which approaches that of e-beam techniques. From (2.3), we even note that it has the advantage of being able to use relatively large separations between the mask and the substrate, so that a stepper technology is readily possible (DSW systems actually exist which utilize modest x-ray exposure techniques). Sources of soft x-rays, which are used in this exposure technique, are electron bombardment of a rotating metal anode (rotating for heat dissipation purposes), a hot-plasma discharge, or synchrotron radiation from an electron storage ring.[14,15] Obviously, the first of these is the most common that is in use today, while the latter is relegated to a few research institutions, but several major semiconductor companies are constructing such facilities for production, both here and in Japan. The plasma source is also relatively new and is mostly found in research environments at present. The electron storage ring is probably the ultimate requirement if x-ray lithography is to be put into production for integrated circuits because of the intensity of the source and its relatively broadband nature.

Two factors work to set the range over which x-ray lithography will provide usable resolution, particularly in proximity printing. These are the diffraction effects, discussed above in connection with (2.3), and the range of photoelectrons formed when the energy of the x-ray photon is absorbed in the resist. This latter range is thought to be less than 100 nm, and decreases as the wavelength of the x-ray increases. Indeed, values of less than 10 nm are expected for x-ray wavelengths above 3 nm. Exposure takes place, however, by the generation of these photoelectrons during absorption of the x-ray photons, which leads to chain scission (in positive resists) or cross-linking (in negative resists) in the polymers. Thus, electron-sensitive resists also make good x-ray resists. While longer wavelengths give better resolution, (2.3) reminds us that this also requires placing the mask closer to the surface of the substrate. Thus, the use of x-ray photons for exposure requires that a set of compromises be met, and it is not expected to provide a panacea for the needs of the integrated circuit fabrication at the 0.1-μm range.

The minimum linewidth that can be achieved with x-ray exposure is generally expected to be set by the range of photoelectrons created by the incident photons. This range is thought to be limited to the Gruen range[4]

$$W = \frac{2.57 \times 10^{-11} E^{1.75}}{\rho} \approx 10^{-23} \lambda^{-1.75}, \tag{2.6}$$

where E is the maximum energy of the photoelectrons associated with x-ray wavelength λ, and ρ is the density of the resist (the values given above are for PMMA of relatively large molecular weight, e.g., 950,000). We may now combine (2.6) with (2.3) to find the wavelength at which the minimum linewidths are achieved, and this gives an x-ray wavelength of 3.9 nm for a mask to substrate spacing of 1 μm. This should then be the optimum wavelength for x-ray lithography. However, a shorter wavelength allows a greater mask to substrate spacing; e.g., a 10-μm separation pushes the optimum wavelength to about 2.1 nm. Most storage rings (that are being considered for integrated circuits) produce their peak energy in the wavelength range of 0.4–1.0 nm, and this should allow even greater mask to substrate spacings. However, these wavelengths produce photoelectrons with ranges correspondingly of 0.28–0.05 μm. This would suggest that only the longer wavelength rings of this group will be capable of producing 0.1-μm design rule circuits.

X-ray masks for proximity printing consist of a thin membrane substrate on which an x-ray absorber is patterned. The substrate is typically of the order of 1 μm thick and must have good transmission for the x-ray photons. As we have seen, a resolution of 0.1 μm allows a reasonable spacing to be maintained between the mask and the substrate, so that reasonable resist thicknesses can be utilized. High aspect ratios in the resist, resulting from this reasonable thickness allowed, are a significant advantage for x-ray lithography, when compared to other exposure techniques.

One problem which must be faced, however, is that of statistical fluctuations in the resist if resists significantly more sensitive than PMMA are utilized. Production tools envisaged for laboratory use with a rotating anode have been suggested with a total dose of the order of 250 mJ, giving a dose of 3 mJ/cm^2 over a 100-mm wafer.[16] This requires a resist some 133 times more sensitive than PMMA, but Table 2.1 suggests that only some 15 photons per pixel are deposited, which suggests a 25% fluctuation in the dose. Clearly this will limit resolution due to such statistical fluctuations, and yield will severely suffer. The solution is not to go to a more sensitive resist, but to go to a more intense source or a more concentrated source, such as collimating the source radiation and going to DSW techniques. These require, however, that the slower resist be utilized to eliminate or reduce the statistical fluctuations and that a larger dose be deposited in each pixel or resolution cell. The general view is that current resists are too insensitive for high-throughput DSW applications and more sensitive resists are needed. But we see that this entails significant fluctuations in the dose delivered per pixel, and that the solution is really to find a more intense source, such as the storage rings, which in turn are quite expensive. Without these rings, however, it is likely that the fluctuation problem of more sensitive resists will limit x-ray lithography to laboratory applications.

2.1.6. Electron-Beam Lithography

Electron-beam lithography is the best-developed and most versatile very high resolution pattern generation technique that is currently available. While it is versatile, it is limited in the need to write a very large number of pixels serially, which is the cause of its relatively slow exposure times. On the other hand, since the e-beam is scanned over the area to be written, and alignment marks are also sensed, e-beam technology can also dynamically correct for changes in the exposure area due to thermal cycling of previous pattern levels.[17,18] By sensing these thermally induced distortions in the area to be written, the pattern can be dynamically corrected for these effects. If two fiducial marks (alignment marks) are used for the registration of the pattern to be written, then this pattern can be compensated for wafer translation, rotation, and uniform shrinkage or growth. However, three marks will allow one to also correct for skew distortions, which lead to both linear distortion and shear distortions, and this results in better layer-to-layer alignment of the individual exposure layers and overlay accuracy.

The resolution achieved in most e-beam writing systems is limited by the scattering of the incident electrons in the resist or from the substrate. Forward scattering in the resist layer can be an important limitation, but it is easily made relatively negligible by high beam voltages. The use of trilevel resists with a very thin resolution setting top layer is very amenable to this process. Although forward scattering in the resist can thus be reduced, a further limitation is the backscattering of electrons generated from an average depth of several microns in the substrate or lower layers of the trilevel resist. When writing in the 0.1-μm level, though, these backscattered electrons can be expected to provide only a uniform, and diffuse, background level of "prefog" to the exposure, resulting in an overall reduction of contrast but not of resolution. On larger patterns, these backscattered electrons produce the well-known proximity effect.

Broers[4] suggests that the above-mentioned background exposure by backscattering of primary electrons will in fact limit the contrast to about 50% for dense electron-beam exposure of patterns with lines in the 0.05–2.0-μm level on thick substrates. Combining this fact, with the discussions presented in the section on optical lithography, we arrive at the conclusion that optical lithography provides better contrast and resolution in the dimensional scale above about 0.25 μm, so that e-beam exposure will be fruitful only on a smaller scale.

In general, higher accelerating potentials offer a better resolution and a higher contrast, as the higher-energy electrons are scattered less in the actual resist layers and penetrate deeper into the substrates before undergoing backscattering. On the other hand, resist sensitivity is reduced at higher voltages because less of the energy of the incident electrons is deposited in the actual resist layers. This results in a design trade-off, and is somewhat compensated for since the source brightness is generally greater at the higher beam voltages.

The maximum field size in early systems that could be addressed by the deflection system is limited by the noise bandwidth of the electronics to about $2–5 \times 10^4$ times the beam diameter (or this equivalent number of spots). This is no longer a severe problem, and field sizes containing more than 10^9 pixels are readily obtainable in current high-resolution e-beam systems utilizing 16-bit low-noise analog-to-digital converters. The use of laser interferometry for adjustment of stage translation then allows for the "stitching" of several fields of view to obtain large writing areas as is done in normal DSW patterning.

As mentioned previously, PMMA remains the standard resist for positive exposures in research studies requiring very high resolution. In Figure 2.3, we show test structures obtained in the novalak-based negative resist developed by Shipley (mentioned previously). In this type of resist, exposure causes the generation of an acid internal to the resist, which catalyzes bonding between the novolak and the cross-linker upon heating (the postexposure baking step). The developer used was a proprietary SAL MF-622 developed by Shipley. In these tests, a resist thickness of 0.93 μm was used and resulted in quite high aspect ratios in the developed resist. In Figure 2.4, we show some quite dense patterns

FIGURE 2.3. Example of exposed resist structures obtainable in a high-resolution negative resist. The grid is composed of lines as small as 70 nm on a 100-nm pitch.

written in this negative resist, which still demonstrate quite good resolution, well below the 0.1-μm level.

Several good, plasma-developable resists are also available which have sensitivities an order of magnitude below that of PMMA and are usable as a positive resist. Hiraoka[19] has used polymethacrylic acid (poly-MAA) and a polymerized copolymer of MAA and acrylonitrile as positive resists, and poly-chloroacrylonitrile as a negative resist. With these, he found that the relief image was formed with a heat-treatment fixing, which could be enhanced in an oxygen plasma. These resists required only 5–10 μC/cm^2 at 25 keV, a factor of 5–10 improvement over PMMA. Other results suggest that poly(butone-1-sulfone) (PBS) also gives good sensitivity at this same level,[20] while the common Si resist polymerized copolymer of glycidyl methacrylate and ethyl acrylate is capable of exposure at the level of 0.2–0.4 μC/cm^2. These give sensitivities more than two orders of magnitude greater than PMMA. With good lens design, e-beam manufacturers should be able to approach beam currents of 0.5 μA in a 50-nm beam, for which one should be able to write a single-chip field in the order of 0.2 s. This time makes the DSW process of e-beam lithography really practical, and one could process an entire 150-mm wafer in a time of 1.2 min, which is an attractive alternative to x-ray costs. However, the major time here will remain

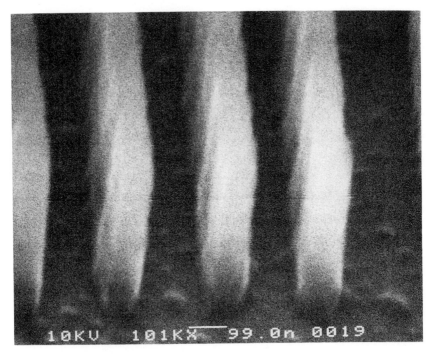

FIGURE 2.4. Dense patterns obtained in a high-resolution negative resist that is 570 nm thick. Slightly greater linewidths are obtained from the close feature proximity, but the aspect ratio obtainable is clearly evident. (See Ref. 12.)

that for alignment of the pattern to the underlying patterns, e.g., the time to achieve a desired level of overlay accuracy.

Several improvements in traditional e-beam columns have also been incorporated which could speed up the processing. First, shaped beams with larger source emitters have been used.[21] These machines have sufficiently high current that the fundamental limitation on resolution is the electron optical column itself and not the electron scattering. In these columns, the current is so high that the electrons are sufficiently close to each other as to interact and blur the edges of the beam.[22]

Finally, in Figure 2.5, we illustrate a GaAs MESFET written by an all-e-beam technology and lift-off processing. The gate length of this device is 37.5 nm, and the device demonstrated an f_T of 167 GHz in microwave testing. This demonstrates the ability to make really small devices with e-beam lithography.

2.1.7. Ion-Beam Lithography

We want to close this section with a discussion of the use of ion beams for lithography. Much use has recently occurred for direct implantation with a focused

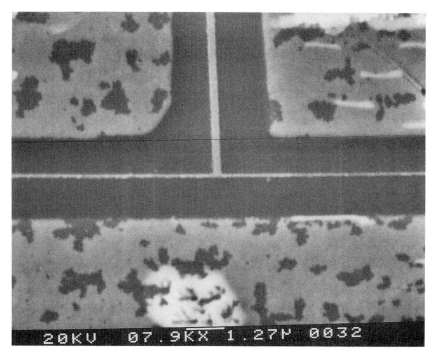

FIGURE 2.5.   A GaAs MESFET, in which all metal layers have been patterned by lift-off. The gate length is about 40 nm.

ion beam, so that the capability to write devices with this technique is also viable. Here, though, we want to concentrate on the use of ion beams to expose a resist structure, although the beams can be used directly to cut patterns in various layers through beam-induced sputtering. The combination of capabilities available to ion-beam processing holds some promise to simplify techniques and reduce overall cost in processing.

As a general rule, resists are more sensitive to ion beams than to electron beams, since the ions tend to deposit more energy per unit volume than the equivalent electrons. Depending upon the mass of the exposing ions, this increase in sensitivity is usually in the range of a factor of 100–300 at 20 keV. By choosing a suitable ion and resist combination, and by carefully adjusting the energy of the ions, the overall process can be optimized. Moreover, ions produce only low-energy secondary electrons, thus there are none of the detrimental backscattered electrons that produce the unwanted prefog reduction of contrast. Thus, the ion-beam written structures are exposed with the primary beam and will in general have a higher contrast in the resist. There are two major ways of writing with ion beams: (a) direct writing as in e-beam lithography, and (b) exposure with a mask.

The main obstacle to developing ion-beam lithography has been the general absence of suitable high-brightness sources. Recently, there have appeared both field-ionization and liquid-metal sources which have alleviated this problem. These latter sources have brightnesses comparable to those available from thermal electron sources, which when coupled with the increased sensitivity of the resist, gives comparable or faster writing speeds. If a column similar to that in an electron-beam machine is used, focused ion beams with diameters well below 100 nm are now available. Improvements should continue to occur, so that we can expect to see ion exposure systems with beam diameters of the order of 10 nm. There remain problems, however, even with these two high-brightness sources. The total current available even from the field ionization source is still only about 100 pA, which is too low for successful lithography. The liquid metal source is limited to those atomic species which yield a liquid metal (typically Ga), and has a relatively large chromatic spread of 5–14 eV, which limits focusing. This is why the beam diameter remains large at present.

As on other forms of lithography, the thickness of the removed resist depends upon the dose provided, the energy of the incident ions, and the development time. Typically, PMMA is still used for high-resolution work and has a sensitivity of 0.5–1.5 $\mu C/cm^2$. As indicated in Table 2.1, this level is such that there is a problem with statistical fluctuation in the dose provided for exposure. Thus, a more sensitive resist cannot really be utilized, as the problem is similar to that in x-ray lithography—a more intense source is required. There are also many less sensitive resists, but these do not get around the basic problem. Direct written exposure by ion beams will remain in the laboratory until better high-brightness sources are obtained.

Masked ion-beam lithography is probably the most interesting and advanced ion-beam approach to date. Indeed, ion projection lithography has appeared and appears to have some definite advantages.[23] In this latter technique, the mask is located some 2 m from the substrate, and the ions are accelerated after passing through the mask, with subsequent focusing onto the wafer. A relatively large dose, 10^{16} (ions/s)/cm^2 allows a 1-cm-square pattern to be exposed in 1 ms, which means that a 150-mm wafer can be written in some 250 ms, to which must be added the overhead time for alignment and stage motion. A commercial masked ion-beam lithography system has produced 0.2-μm lines and 0.1-μm spaces in tests,[24] so that the method certainly seems to offer considerable promise at a fraction of the cost of a storage ring for x-ray lithography.

In principle, the masks that are used in ion-beam lithography are similar to those used for x-ray lithography, in that both are types of stencil masks. However, the relatively short range of the energetic ions complicates the problem, so that true stencil masks must be used, rather than the x-ray masks which use an absorber deposited or grown on a freestanding membrane such as polyimide. While the latter can be used for ion beams, the membrane must be much less than 100 nm thick if the ions are not to be severely attenuated. There is some hope, however, for if a single-crystal material is used as the membrane, the ions can channel through a somewhat thicker membrane, but scattering in such a channeling membrane is still sufficient that linewidths below 0.1 μm do not seem practical. Thus, masked ion-beam lithography at present presents a viable alternative to x-ray lithography in the 0.1–0.25-μm linewidth regime, but considerable further advancements are required in this technology.

2.2. DRY PROCESSING

As the dimensions of integrated circuit devices continue to decrease in response to the demand for increased packing densities, there is a continual motivation to improve the processing and fabrication technology to enhance yield in structures with reduced dimensions. Dry etching technology has become a major factor in current/projected submicron device manufacturing because of its finer resolution capabilities (over wet etching) and high anisotropy factors. Principally, dry etching refers to the use of gas-phase etching in a reduced-pressure environment, and has been used to describe the various combinations of plasma etching, reactive ion etching, reactive ion-beam etching, etc. The processes ordinarily employ inert, but reactive, gases that are activated by an electric discharge to form a plasma consisting of electrons, ions (both atomic and molecular), photons, and neutrals. These new species react with the substrate materials to produce volatile products, which are then removed from the system by the gas flow and vacuum pumps. The combined action of chemically active species and ion bombardment gives rise to anisotropic etching under the proper conditions.[25]

Plasma-assisted etching technology has received concentrated developmental effort because of its applicability to all dry processing technology. Although sputter erosion techniques have existed for more than a century,[26] it was only applied to semiconductor processing in 1954.[27] There are currently at least 13 different approaches to dry etching. Many review articles have tried to delineate a taxonomy of the processes, but the best approach is to classify them according to the mechanism by which the etching occurs. In this way, the various approaches reduce to four basic processes:

1. Physical etching, in which the impact of the incoming ions and neutrals from the plasma is used to sputter away material from the surface. It is obvious that this type of etching can be highly anisotropic, but is in general not very selective. One recent version of this is neutral beam sputtering, which is quite useful in etching nonconducting surfaces because the impacting particles do not charge the surface.

2. Chemical etching, in which a chemical reaction takes place between the incoming ions and the surface ions of the material being etched. This usually results in a volatile by-product which can easily be carried away in the flowing gas stream to the vacuum pump. This approach also tends to be very selective of the surface atoms being removed, but is not particularly directional in nature.

3. Chemicophysical etching, which is generally conceived to be a combination of the above two processes. The detailed mechanisms are not fully worked out in each case, but the process is thought to involve (a)

weakening of the chemical bonds of the substrate due to physical damage, (b) a resultant enhanced chemical reactivity, and (c) chemical sputtering.

4. Photochemical etching, in which photon absorption is used to assist the etching process. There are varieties of this process as well, and the photons can (a) photoexcite the reactive species, giving preferential excitation of the plasma, (b) photoexcite the substrate, causing light-assisted chemical etching, and (c) thermally heat local regions of the surface, causing thermally assisted etching. This process can be quite directional and highly preferential in characteristics.

The processes found most often in processing use today, though, are plasma etching and reactive ion etching. These are both thought to be variations of the third process above. This etching is achieved by using a gas which dissociates in a plasma to give a particular reactive species, which then competes with the ion bombardment from the plasma to cause enhanced etching of a susceptible substrate.[28] The two approaches differ in detail, however. Plasma etching is usually carried out in a relatively weak vacuum, with a pressure above several hundred millitorr, and the substrate is mounted on the undriven electrode. In reactive ion etching, on the other hand, processing usually takes place at lower pressures, of the order of 1–100 mTorr, and the substrate is mounted on the driven cathode. Reactive ion etching has been used to produce very impressive fine-line geometries in many materials, including Si,[29] Al,[30] and GaAs.[31-33] At present, research and development activities are directed at improvements of this technology and toward laser-assisted etching.

Independent control of the ion current density and the ion energy can be achieved in reactive ion-beam etching.[34] In this process, a collimated beam of reactive ions and neutrals is extracted from a plasma source and directed at the sample surface to be etched. Variations in this approach allow separate formation and excitation of the reactive and sputtering components in chemically assisted ion-beam etching.[35-37] There is considerable progress and promise in another variation of reactive ion etching. Most reactive ion etching uses a plasma excited by 13.56-MHz rf sources. However, other frequencies have been used, and there is a growing indication that microwave excitation provides good control of the process. In this latter case, the reactive plasma is extracted from an ion gun, in which the chlorine plasma is excited at 2.45 GHz.[38] This plasma excitation is termed electron cyclotron resonance excitation, since this frequency correlates to this process in the electrons and increases their mean free path in the plasma, which in turn augments their ability to sustain the plasma excitation. A bias voltage applied to a grid over the exit of the plasma gun gives much more sensitive control over the energies of the emerging ions.

Based upon current trends in this area, it is highly likely that reactive ion etching will soon dominate dry etching for pattern transfer. Etching equipment and processing chemistry have been stabilized to a level that can meet the sensitivity and selectivity requirements for submicron device processing. Reactive ion etching does have some problems with regard to etch rates in some materials and with damage problems as well as with residue chemistry. Many of these problems look like they will be overcome with continued reduction in geometry and with the use of electron-cyclotron-assisted etching mentioned above.

2.2.1. Reactive Ion Etching Mechanisms

Anisotropy is a most important feature of dry processing. This anisotropy is usually induced by positive ion bombardment from the plasma. Various mechanisms have been proposed to explain the interaction between ions, neutrals, and the substrate, but these are generally categorized into three major groups.

The first process has been termed a *clearing* process, in which it is proposed that the enhanced etch rate that occurs under ion bombardment is caused by the presence of chemical sputtering. One case of this is thought to be sputter desorption of a fluorinated surface layer in Si processing.[39,40] In this process, enhanced reactivity is found by the use of XeF_2 or atomic fluorine, although this is somewhat debatable on the grounds of kinetic and thermodynamic arguments.[41] In this model, selective chemistry at the surface is thought to follow

$$XeF_2 + Si \rightarrow SiF_2 + Xe, \qquad (2.7)$$

for which all species are gases except Si. It is thought that the SiF_2 molecules are attached to the surface of Si, and that the ion bombardment clears these with the reaction

$$2SiF_2 \rightarrow Si + SiF_4, \qquad (2.8)$$

with the last molecules being carried away in the gas stream.

The second mechanism is a simple damage mechanism, in which impinging ions in the direction normal to the substrate surface produce lattice damage that extends several monolayers below the surface. Reaction chemistry can proceed at these damage sites at a much faster rate than normal, so that etching predominantly occurs in the direction normal to the surface.

Finally, the third mechanism that contributes to the anisotropy is thought to be a surface recombination mechanism. It is thought that many of the radicals and by-products of the plasma excitation may either chemisorb or physisorb on the substrate surface.[42] Consequently, ion bombardment can stimulate reactive, or even nonreactive, desorption of these radicals, yielding a surface for which the steady-state concentration is larger in regions not exposed to the ion irradiation, such as sidewalls. Thus, the sidewalls would have a slower etch rate than the open surface, and anisotropy is achieved.

2.2.2. Etching Rate and Control

The etching rate that arises in a process like reactive ion etching depends on essentially all parameters of the system: e.g., the plasma power (number of reactant species and their densities), background pressure of the plasma (number of species and their densities), substrate bias (ion energy), components of the plasma, etc. It is clear that design and construction of the etching chamber will also affect the etch rates obtained, as it clearly affects the properties of the plasma itself. Thus, it is not generally meaningful to quote detailed numbers for etch rates, as they depend on the reactor geometry used to obtain the rates. Rather, we will discuss the factors that affect the etch rates, and give some typical numbers below that have been obtained in past studies by a variety of groups.

Anisotropy is usually induced by the presence of positive ion bombardment during the etching process, according to the models discussed above. The role of this positive ion bombardment depends on the particular process employed. Usually, the anisotropy is increased by increasing the plasma rf power, which will increase the bombardment energy of the ion species. Adding a negative bias to the substrate has the same effect as it accelerates the positive ions toward the substrate. On the other hand, increasing the pressure in the discharge (with constant flow rate of chemical species), will increase the residence time and particle collision rate (they go hand in hand), while decreasing the average ion energy, so that the anisotropy is decreased while the etching speed is increased.

Etch rate depends on the rate of generation and arrival of the active species, as well as on the rate at which these species are consumed in the surface reactions and the by-products removed from the interaction volume. Since the electron energy determines the generation rate of the active species in the plasma, we can expect that the etch rate will in general increase with a decrease in pressure, an increase in the bias at the cathode, or with an increase in the rf power to the plasma. In addition, an increase in the flow rate of chemical species (at constant power) will also enhance the etch rate. One drawback at the lowest pressures is that depletion of the reactant species can slow the etch rate, but in general the etch rate is enhanced by the same factors that enhance anisotropy discussed in the last paragraph.

Selectivity refers to different etching rates for different materials placed in the same plasma, for example rapid etching of GaAs but slow etching of GaAlAs in preferential etching of layers of a heterojunction. To date, this has been determined entirely empirically, but the selectivity is generally determined by the specific chemistry of the process.

Uniformity control is obtained by maintaining exactly the same processing conditions from wafer to wafer. These factors are generally determined by reactor design, as it is usually concentration gradients of the reactant species over time, and inhomogeneities in plasma properties as wafers are exchanged that affect the uniformity. These factors also affect the homogeneity across a single wafer that is obtained during the etching process. Generally, lower gas pressures tend to enhance the uniformity across a single wafer.

Damage and residual species are again determined by the chemistry of the particular process. In particular, the use of fluorine-containing compounds often results in the deposition of various polymers of fluoridated hydrocarbons on the surface of the substrate material. Subsequent wet etching, required to remove these polymers, defeats many of the reasons for using dry processing. This has caused a significant effort to find non-fluorine-containing processes for use in submicron processing, a process that seems to be more likely with the electron-cyclotron excitation process.

Typical etch rates that have been obtained for GaAs are $0.04\ \mu m/min$, $0.4\ \mu m/min$, and $0.8\ \mu m/min$ at 40 mTorr pressure for gases of pure Ar, CF_4, and CCl_2F_2, respectively,[31] for which the target voltage was -3 kV. In our own laboratories, an etch rate of $0.3\ \mu m/min$ was obtained in a small reactor at 40 mTorr, with the plasma excited at an rf power of $0.8\ W/cm^2$ and a CCl_2F_2 flow rate of 14 sccm. These results indicate the range of etch parameters that can be expected, and the variations from laboratory to laboratory. Similar etch rates, and ranges of etch rates are obtained in Si etching by various gases.

2.3. EPITAXIAL GROWTH

When we refer to epitaxial growth, we are referring to the ability to grow thin layers of semiconductor films in a manner in which atoms impinging upon the surface of the substrate align themselves with the crystalline order of the substrate. Moreover, it is also desired that the growth of an epitaxial layer upon the substrate is accomplished without creating an interfacial layer of defects, traps, or other imperfections which can perturb the transport properties of electrons, either in the epitaxial layer or in the substrate, that are near to the interface. Normally, one would not spend much time with this consideration except that in submicron devices, the active areas can be quite thin—perhaps only of the order of a hundred or so atomic planes. Thus, the ability to controllably grow such thin layers is very important to the ability to process and manufacture integrated systems with this technology.

The oldest epitaxial growth method is, of course, growth from the melt. This cannot be used today to grow thin layers of controlled properties. It is even relatively difficult to grow entire boules of controlled properties by this method. Liquid-phase epitaxy then came into being and has been used to grow layers in a significant number of different heteroepitaxial systems, principally for use in semiconductor laser, and light-emitting, diodes. Layers are thin as a few tens of nanometers have been grown by this method, principally for the central region of quantum well laser diodes. However, control of the process is not such that multiple layers can be grown with thickness, and dopant properties, on the nanometer scale. Rather, the growth methods that have shown the ability to provide the desired growth control are varieties of vapor-phase epitaxial growth, in which the layers to be grown are built up of atoms precipitating from various vapor reactants.

2.3.1. Vapor-Phase Epitaxy

Vapor-phase epitaxial growth is a relatively old technology that is used extensively in Si integrated circuit processing. For this purpose, an Si substrate is heated to the growth temperature and placed in a flowing gas stream of, e.g., $SiCl_4$ and either H_2 or SiH_4. The gas molecules react at the heated substrate, depositing atomic Si species, with the reactant HCl molecules being carried out in the flowing hydrogen gas stream. The process can be carried out either near atmospheric pressure or at a reduced pressure, and the growth rates are controlled by substrate temperature and partial pressures of the reactants.

In growth of III-V compounds, the technique is somewhat more complicated, because, in general, the group III reactants are taken from solid sources. Then, group V reactants have most often occurred either as chlorides or as hydrides. For example, GaAs can be grown either by constructing the flowing gas stream of $H_2 + AsCl_3$ or of $H_2 + AsH_3$, referred to as the chloride or the hydride process, respectively. This gas stream is then passed over a heated melt of Ga, and carries Ga (molecules?) along to the heated substrate at which point the growth reaction occurs. Again, the growth rate depends upon the substrate temperature and the partial pressure of the reactant species.[43,44] The problem is that the doping concentration incorporated into the grown layer also depends on the partial pressure of the reactants, and this is thought to be due to the interaction of the HCl vapor with the quartz walls of the growth vessel, freeing Si which is incorporated into the grown layer.[45] Use of a boron nitride liner in the tube essentially eliminates this problem, but at the loss of control of Si dopant incorporation![46] Dopants can be included by a suitable reactant gas inclusion. On the other hand, control of the grown layer thickness is typically only on the tens of nanometer scale. A recently introduced variation of this process, in which the reactant gases are passed through a diffusant layer to react with the *bottom* of the substrate has shown the ability to grow atomic layers.[47]

2.3.2. Metal-Organic Chemical Vapor Deposition

The metal-organic chemical vapor deposition process (MOCVD) has become an important method over the past few years for the growth of heterostructures of the III-V materials, as well as others.[48] This method is a variation of the vapor-phase growth process in which the reactants involve metal-organic compounds, the reaction at the substrate is mainly a pyrolysis reaction. The most commonly used reaction for this purpose is given by

$$R_n M + XH_n \rightarrow MX + n\,RH, \tag{2.9}$$

where R is an organic radical, M is one component of the resulting semiconductor layer, X is the other component, and n is an integer. Here, for example, $R_n M$ would be trimethylgallium or triethylgallium, while the second reactant would be arsine. An alternative is the substitution of the hydride species with a second metal-organic radical, such as trimethylarsenide, although this usually is restricted to the growth of II-VI compounds. One advantage of the MOCVD process is the relative ease by which ternary and quarternary compounds can be grown, merely by controlling the partial pressures of the various gas streams. Dopants are also readily incorporated in the gas stream.

The control of the grown layers is obtained both by the temperature and gases, but also by the reaction chamber design. The latter also carries over to the elimination of unwanted intermediate by-products.[49] Precise control of the growth rates is obtained by careful control of the vapor pressures of the reactants, by the use of temperature-controlled baths for the metal alkyls and bubbler rates for the hydrides, and the use of high-accuracy mass flow controllers. The growth rate is linear in the metal-alkyl mole fraction as well.[48] Coleman and co-workers,[50] have demonstrated that heterojunctions can be grown between AlAs and GaAs that are abrupt to within one monolayer. This latter group has incorporated these high-quality interfaces into a variety of heterostructure devices, and this abrupt interface is necessary for the reliable operation of most heterostructure devices.

The MOCVD process has found extensive use in nearly all areas in which III-V and II-VI heterostructures are required, such as discussed in later chapters. However, it has been limited mainly to these compound semiconductors, although heteroepitaxy between different material systems (e.g., GaAs on Si) is in principle possible.

2.3.3. Molecular Beam Epitaxy

Molecular-beam epitaxy (MBE) is a departure from the normal trend of the vapor-phase growth techniques. Carried out in an ultrahigh-vacuum environment, MBE is closer to vacuum evaporation in scope. The sources of the reactants for MBE growth come from heated chambers, termed *effusion cells*, whose structure is such as to induce the flux to form an atomic or molecular beam. An array of these cells, all focused onto a heated substrate upon which the layers are to be grown, provides the various species to be incorporated into the grown layer—the atomic species as well as any dopant atoms. This is shown in Figure 2.6. Each of the cells is shuttered, and the flux of growth atoms is governed by the opening and closing of these shutters. The entire process is usually carried out under computer control, and was perhaps the first example of automated crystal growth for multiple layers.[51] To improve thickness and doping uniformity across the wafer, the substrate is usually rotated during growth, and the source-to-substrate distance is increased. Early systems used first 1-in and then 2-in substrates, although modern systems can handle 3-in wafers, and some new systems can take even larger wafers. This growth technique revolutionized the approach to heterostructure devices, as it was the first in which it was demonstrated that abrupt interfaces on the monolayer scale could be achieved in superlattices of alternating monolayers of GaAs and AlAs.[52] Indeed, early studies with heterostructures demonstrated the simple quantum mechanical aspects of quantum wells, both single and coupled.[53]

One problem of MBE, which has been suggested as being critical for its use in production environments (which is controversial), is that the growth rates are quite slow, being of the order of magnitude of microns per hour or less. To overcome this, some have begun to use gas sources for some of the reactants,

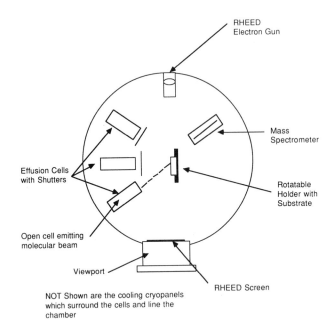

FIGURE 2.6.   A descriptive view of a molecular-beam epitaxial system, in which the effusion cells provide a "molecular" beam of reactant atoms. The growth takes place when these atoms strike and subsequently stick, to the substrate surface. The substrate is heated to ensure sufficient surface diffusion for the growth to be epitaxial and single crystal.

particularly those that can be obtained in the hybrid form, and metal organics are also being used. These lead to the use of the terms *chemical-beam epitaxy* or *MOMBE* to describe the composite system.[54,55] While this process promises to provide somewhat faster growth rates, it is not clear what effect intermediate products will have on the overall grown layer. In pure MOCVD, the gas flow stream carries out these intermediates, while in pure MBE, only the reactants arrive at the substrate. Since there is little flow in the MOMBE process, the action of these intermediates is not at present understood.

2.3.4. Heterostructures

It is clear from the foregoing that several growth techniques are available for preparation of epitaxial films of arbitrarily small thickness (at least down to the monolayer level). It is the availability of these growth techniques, and the ability to grow heteroepitaxially that provides the ability to design unique new submicron structures for special applications. Indeed, this opens the door to *band gap engineering*, a process by which the properties (alloy composition, lattice constant, etc.) are chosen for particular requirements, and layers with these properties are then grown. Whether it is the growth of GaAs on Si, or of *strained-layer superlattices*, in which controlled lattice mismatch (of the isolated compounds) is used to modify the properties by building in strain, either compressive or tensile. We will discuss these band-gap engineering material systems in Chapter 6, and the heterostructure type of devices in Chapter 3. It is only through the presence of these growth techniques that we can conceive of using such ideas to structure unique and different material properties.

2.4. METALLIZATION

Metallization usually refers to the process by which metals are put on the surface of active devices. This can consist of deposition (or growth) of a polysilicon layer, as in Si device processing or direct evaporation of metal. The process is straightforward, and polycrystalline layers of Si (polysilicon) are readily grown on almost any surface by processes as simple as evaporation and as detailed as vapor-phase epitaxy. On the other hand, metals are usually simply evaporated, either by thermal evaporation or by electron-beam evaporation. Our interest here, though, is more than just the simple deposition of the so-called metal but also by its patterning. The metal can be deposited and then patterned by a photolithography process as described in earlier sections, in which the pattern is transferred from the photoresist into the metal by an etching step. On the other hand, the reverse can be done, in which the photoresist is first exposed and developed and then the metal is deposited. Subsequent dissolution of the photoresist removes the metal from all areas except where the photoresist was developed away. This *lift-off* process is depicted in Figure 2.7, and is the preferred method of metallization for most microwave GaAs (or compound semiconductor) devices. In Figure 2.5, we showed an angled view of a microwave GaAs MESFET, in which all metal layers were patterned by lift-off.[56] The gate length in Figure 2.7 is about 40 nm, and the device, with an active epitaxial layer thickness of only 60 nm, exhibited an f_T of more than 160 GHz.

2.5. CONTACTS AND SCHOTTKY BARRIERS

Although the earliest studies of the metal-semiconductor contact were made by Braun in 1874,[57] it was in 1938 that Schottky explained the behavior in terms of stable space-charge regions within the semiconductor.[58] Indeed, many think, and rightfully so, that the discussion of Schottky barriers should be included with a discussion of semiconductor devices. Why then have we included it here, in a chapter on fabrication techniques? In nearly all integrated circuitry, the metal-semiconductor interface is a crucial part of the circuitry. In some cases, it is used to create *ohmic* contacts, while in others it is used to create rectifying junctions. In still other cases, it is desirous to have the metal-semiconductor contact totally passive, as in interconnect metallizations laid upon the bare semiconductor surface, as in GaAs microwave circuits. Thus, the choice for the properties of the various interfaces is part of the design, and fabrication, process

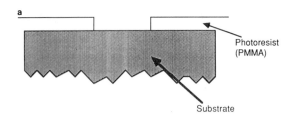

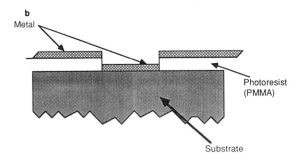

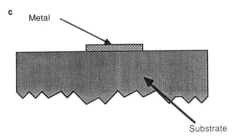

FIGURE 2.7. The various steps of the lift-off process for metallization. In (a), the positive photoresist has been developed, leaving openings for the desired metal patterns. In (b), the metal has been deposited. Upon dissolving the remaining photoresist, the metal on top of it is carried away, leaving only the desired metal patterns (c).

since the resulting properties depend on the processing steps to a larger or smaller degree in all cases, and on the underlying semiconductor properties in all cases. In this section, we treat the built-in potential and discuss the concept of the Schottky barrier itself and its dependence on the properties of the metal and the semiconductor material.

2.5.1. The Potential Interface

The details of the metal-semiconductor interface depend crucially on the properties of the metal, those of the semiconductor, and in many cases those of any intermetallic layers that may form. For example, when Pt is deposited upon the Si surface, any temperature cycling, even if only to a few hundred degrees Celsius, can lead to the formation of the silicide $PtSi_2$ if the reaction goes to equilibrium. This compound is only one of several intermetallic compounds formed from Pt and Si, but is generally believed to be the one that is the stable equilibrium form. If the interface contains a layer of this compound (or any other), then the metal-semiconductor interface is not Pt/Si, but PtSi$_2$/Si. This difference affects the properties of the resulting junction in some cases.

In Figure 2.8, we illustrate the formation of the space-charge layer in the semiconductor. First, in Figure 2.8(a), the separate metal and semiconductor are shown in isolation, but for the case where both are in equilibrium with the environment. Critical quantities are the work functions W_m and W_s, which define the energy difference of the respective Fermi level from that of the free-electron vacuum. While the work function W_m of the metal is a relatively stable value, that of the semiconductor, W_s, depends on the doping of the material, since this moves the Fermi level around in the band gap. On the other hand, the electron

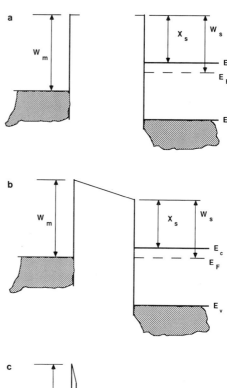

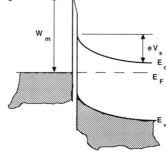

FIGURE 2.8. (a) The metal and semiconductor in isolation, but in equilibrium with the free-electron vacuum. (b) When connection is made between the two materials, the Fermi levels equilibriate, but the situation shown is not in stable equilibrium. (c) The field between the two, even when in contact, causes charge to form at the interface between the two materials. This is shown as a depletion in the semiconductor, and there is a corresponding accumulation in the metal, which causes negligible band bending there because of the high carrier density.

affinity of the semiconductor,

$$\chi_s = E_{\text{vacuum}} - E_c, \tag{2.10}$$

is relatively constant with doping, and is probably more fundamental for semiconductors than the work function. This follows since the ionization energy, $E_{\text{vacuum}} - E_v$, differs from the electron affinity by only the band gap, and both the band gap and the ionization energy are basic properties of the atomic potentials of the core atoms (see Chapter 6).

In Figure 2.8(b), we have "connected a wire to the back side of the metal and the semiconductor." That is, we have brought the two materials into near equilibrium with each other, which gives them a common Fermi energy. This structure is not stable and really requires that the two materials be infinitely far apart, so that the field existing across the indicated gap is infinitely small. The presence of this field leads to the necessity for surface charge to form on each material, which leads to band bending in each material. In the metal, the carrier density is so high that band bending is essentially negligibly small. On the other hand, for most semiconductors, band bending is quite appreciable and leads to the stable space-charge layer that leads to the major diodelike effects.

The final, stable configuration that arises when band bending occurs and the two materials are brought into conjunction with each other is shown in Figure 2.8(c). The end result is that a surface potential ϕ_s exists within the semiconductor. This band bending results in the nonreciprocal current-voltage characteristics, since nearly all current flow mechanisms depend upon the actual height of the potential barrier for electrons that exists at the interface. This barrier is the height of the conduction band in the semiconductor above the Fermi level in the metal. As we will see, the actual barrier height is modified slightly by the image potential seen by electrons moving from the metal to the semiconductor, but this is a correction that can be made later. Here, we would like to address the question of the actual barrier height: What is the actual Schottky barrier height $E_b = \phi_s + W_b$, where

$$W_b = E_c - E_F = E_c - E_{Fi} + e\phi_b = k_B T \ln\left(\frac{n}{N_c}\right), \tag{2.11a}$$

where E_{Fi} is the intrinsic level, and ϕ_b is the extrinsic shift

$$\phi_b = \frac{k_B T}{e} \ln\left(\frac{n}{n_i}\right). \tag{2.11b}$$

In the oldest theories of the Schottky barrier, it was assumed that the barrier energy E_b was simply given by the difference in the work functions of the two materials, or

$$E_b = W_m - W_s. \tag{2.12}$$

However, it was found that this did not seem to hold in many cases. Then Bardeen[59] suggested that one must consider surface states at the interface. If surface states exist in this interface, arising probably from the discontinuity of the semiconductor at the interface, then the charge that forms in response to the electric field discussed above must also include the charging and/or discharging of these interface levels as the Fermi level is moved from its bulk position. In fact, Bardeen suggested that there was probably a dominant interface level lying within the gap of the semiconductor and that this level would *pin* the Fermi energy at its position. Thus, the barrier E_b would be independent of the work function differences but would depend upon the properties of the semiconductor itself, as

$$E_b = -W_b - E_{ss},\qquad(2.13)$$

where E_{ss} is measured from the conduction band (and therefore is <0). Here, there is absolutely no mention of the work functions.

Kurtin, McGill, and Mead set out to test the various theories by measuring Schottky barriers composed of a large variety of metals deposited upon as many different semiconductors as they could. They proposed that the barrier could be defined by a general equation of the form[60]

$$E_b = E_{b0} + S(W_m - W_s),\qquad(2.14)$$

where E_{b0} and S are properties of the individual semiconductors. In general, they deduced that for homopolar (fully covalently bonded) semiconductors, S was very small and the barrier was dominated by E_{b0}, while for ionic semiconductors S was very nearly equal to 1. For compound semiconductors, in which the bonding is a mixture of ionic and covalent bonding, S was found to vary smoothly with the electronegativity difference of the two elements in the compound. This variation went from near zero to almost unity, with a major transition near $\chi_A - \chi_B = 0.4$. The data was later reexamined by Schlüter,[61] who found that it was very difficult to say that S was unity in the ionic compounds, and that values as high as 2 or 3 fit the data equally well. This suggests other than a simple work function argument, unless interfacial compounds exist for these barriers. In fact, the ionic compounds tend to be very active in forming other compounds with metal overlayers, and this could account for the strange values. Indeed, there is evidence that the presence of the interfacial compounds is critical, even to be related to heats of formation of the compounds.[62,63] The acceptance of many of the competing arguments is still quite controversial, and there is no *a priori* prediction from any theory that can be used to suggest the value of the Schottky barrier E_b in a material prior to measurement. In nearly all semiconductors, the barrier height is set by the semiconductor itself, presumably due to surface or interface states. In Si and Ge, the density of these states is low, so that the Fermi level can be moved across the gap. On the other hand, in GaAs, the density of trap levels is very high at a particular point, about 0.9 eV below the conduction band edge, and the Fermi level is quite effectively pinned at this position.[64]

2.5.2. The Space-Charge Region

We wish now to develop an equation which describes the variation of the local potential away from the interface but within the semiconductor region. In the following, we will assume nondegenerate but strongly extrinsic n-type semiconductor material. The extension to degenerate material is straightforward (but tedious), while the extension to intrinsic material or p-type material is also straightforward. To proceed, we introduce the reduced variables

$$u = \frac{e\phi_b}{k_B T}, \qquad v = \frac{e\phi}{k_B T} - u, \tag{2.15}$$

where ϕ is the local potential that gives the carrier density in a manner similar to the bulk potential

$$n(x \to \infty) \equiv n_b = n_i e^u, \qquad n = n_b e^v. \tag{2.16}$$

Obviously, $u > 0$ for the n-type material we will be concerned with here, and $v < 0$ for the depletion region near the metal-semiconductor interface.

To calculate the local potential, and the charge variation in the space-charge region, we begin with Poisson's equation

$$\frac{d^2 V}{dx^2} = \frac{e}{\varepsilon_s}(n - n_b), \tag{2.17}$$

or, in reduced units,

$$\frac{d^2 v}{dx^2} = \frac{1}{L_D^2}(e^v - 1), \tag{2.18}$$

where

$$L_D^2 = \frac{\varepsilon_s k_B T}{n_b e^2} \tag{2.19}$$

is the extrinsic Debye length in the semiconductor. For very small v ($\ll 1$, or a voltage small compared with the thermal voltage), the exponential can be expanded and a simple exponential behavior results for which the potential varies over a distance of the order of the Debye length; e.g.,

$$v \approx -v_s \exp\left(\frac{-x}{L_D}\right), \tag{2.20}$$

where v_s is the surface potential $-\phi_s$ in reduced units. In order to solve (2.18), we first multiply both sides by $2(dv/dx)$ and integrate once, noting that both v and its derivative vanish for sufficiently large x:

$$\left(\frac{dv}{dx}\right)^2 = \frac{1}{L_D^2}[(e^v - 1) - v].\qquad(2.21)$$

For extrinsic semiconductors in general, and for n-type material in particular, $|v| > 2$, and v is <0. For these conditions, we can then write (2.21) as

$$\frac{dv}{dx} = \frac{1}{L_D}\sqrt{|v| - 1},\qquad(2.22)$$

and the potential has a very similar behavior as that found in $p - n$ junctions. Here, we find that the space-charge layer extends into the semiconductor a distance

$$W = \sqrt{|v_s| - 1}\, L_D,\qquad(2.23)$$

which is a distance considerably larger than the Debye length. For all practical purposes, the factor of unity can be ignored in calculating the depletion depth W. The factor of unity corresponds to the thermal voltage that is usually added to the potential (or subtracted in this case) to get the exact depletion depth.

2.5.3. Current Flow

In calculating the current flow from the metal to the semiconductor, or that in the reverse direction, we generally need to know the exact barrier height. It is generally believed that the actual barrier is modified by the image force of the electrons near the metal but in the semiconductor. This effect lowers the actual barrier through the *Schottky effect*, which couples the image potential, the Schottky barrier height, and the field in the semiconductor due to the spatial variation of the space charge. This is shown in Figure 2.9, and results in a lowering of the actual barrier by an amount

$$\Delta\phi = 2Ex_m = \left(\frac{e}{4\pi\varepsilon_s}\right)^{1/2}\left(\frac{n_b e\phi_s}{\varepsilon_s}\right)^{1/4},\qquad(2.24)$$

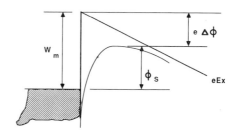

FIGURE 2.9. The modification of the actual barrier height arises from the confluence of the electric field in the semiconductor and the image force for carriers coming out of the metal. The maximum in the potential arises a short distance from the actual interface, and the effective barrier is determined at this point as shown.

where E is the electric field at the surface, and we have used the results of the previous subsection. In general, there is some debate about the use of (2.24) and of the modification of the Schottky barrier height. We note from Figure 2.9 that the field alone would not lower the effective height of the barrier. It is only the field, in conjunction with the Coulomb field caused by the image force, that leads to an effective lowering of the barrier.

There are three major contributors to the net current through the Schottky barrier metal-semiconductor interface junction. These are thought to be thermionic emission, diffusion of minority carriers, and tunneling through the barrier. We note that the first two are concerned with particles traveling *over* the barrier, while the last concerns particles going *through* the barrier. This distinction is critical, for measurements of particles going over the barrier by photoemission generally have to be adjusted for the barrier lowering by $\Delta\phi$,[65] while measurements of photoassisted tunneling on the same samples generally do not show any effective barrier lowering.[66] The difference is thought to lie in the fact that the image force is a dielectric response and has an associated dielectric relaxation time associated with it. Emission of particles over the barrier is thought to be relatively slow, and, therefore, the metal can respond to these particles with an image force correction. On the other hand, tunneling is a very fast process, for which it is thought that the metal does not have time to respond during the tunneling process.[67] Therefore the image force is not present during tunnel emission processes.

The various processes of current flow by thermionic emission and diffusion are treated in many introductory books, so we shall not review the details of that process here. We note that the current may quite generally be written as

$$J = J_s\left[\exp\left(\frac{eV}{k_BT}\right) - 1\right], \tag{2.25}$$

where J_s differs for each of the three mechanisms. For the thermionic emission process, we find that[68]

$$J_{sT} = A^*T^2\exp\left(-\frac{e\phi_{SB}}{k_BT}\right), \tag{2.26}$$

where the barrier is the effective Schottky barrier, including image force lowering, and A^* is the effective Richardson constant in which the mass is the density of states mass for the appropriate semiconductor surface. On the other hand, for the diffusion theory, the saturation current level is given by[68]

$$J_{sD} = \frac{e^2N_cD}{k_BT}\left[\frac{e(\phi_s - V)2n_b}{\varepsilon_s}\right]^{1/2}\exp\left(\frac{e\phi_s}{k_BT}\right), \tag{2.27}$$

where D is the diffusion constant in the semiconductor and the other parameters have their normal meaning. Again, the barrier height is the effective barrier height, adjusted for the image force lowering.

For the tunneling current, such a simple form as (2.25) does not in general exist. Rather, we must calculate the tunneling probability for transition through the barrier and then integrate that over the occupied states. This leads to[68]

$$J_{\text{tunnel}} = \frac{A^*T}{k_B} \int_0^{e\phi_s} [F_m(\eta) - F_s(\eta - eV) T(\eta) \, d\eta, \tag{2.28}$$

where ϕ_s is the effective barrier height, without the image force correction, and η is the reduced energy, measured *downward* from the top of the barrier. In addition, the other currents must be added to this, but each is modified by the tunneling reflection that can occur for energies *above* the top of the barrier. The quantities F_m and F_s are the distribution functions in the metal and the semiconductor, respectively. The tunneling coefficient is approximately given in the form, in the WKB approximation,

$$T(\eta) \sim \exp\left(\frac{-e\phi_s}{E_n}\right), \tag{2.29}$$

where

$$E_n = \frac{3e\hbar}{8} \left(\frac{n_b}{\varepsilon_s m^*}\right)^{1/2} \tag{2.30}$$

accounts for the field in the semiconductor. We note from (2.30) that the tunneling current will increase exponentially with the square root of the doping density in the semiconductor.

2.5.4. Contacts

In the previous subsection, we noted that the tunneling current through the Schottky barrier increased exponentially with the square root of the doping density. This provides a mechanism for the creation of an ohmic contact. For a sufficiently large field in the semiconductor, and hence sufficiently large doping density, the tunneling probability can approach unity and the current density will be quite large even at small bias voltages across the barrier. If we take an effective barrier height of $\phi_s = 0.7$ V, then the tunneling probability will be approximately 0.5 for $E_n = 1.0$ V, but this requires a doping density of 6.5×10^{22} cm^{-3}, which is quite high. However, the standard method of providing ohmic contact to active regions is to dope the surface layer quite high and then use a metal with a relatively low effective barrier height. This works reasonably well for Si, where the doping can be made quite high. For the III-V materials, this does not work as well.

In the III-V materials, such as GaAs, a relatively good ohmic contact has been made with the interaction of the alloy Ni/Au/Ge at the GaAs surface. However, the nature of the contact is not well understood, and it is thought that several intermetallic compounds form at the interface. Some of these are thought to be highly doped compounds of Ge, and the contact process proceeds by a combination of tunneling into these highly doped compounds (which may then make a heterojunction interface with the GaAs) and field emission from points of nickel compounds protruding into the interfacial region.[69] In any case, this metallurgy is something of a black art and not at all well understood.

REFERENCES

1. R. E. Howard and D. E. Prober, in: *VLSI Electronics: Microstructure Science* (N. Einspruch, ed.), Vol. 4, pp. 145–189, Academic Press, New York (1982).
2. G. Bernstein and D. K. Ferry, *Superlatt. and Microstructures* **2**, 373 (1986); J. Han, D. K. Ferry, and P. Newman, *IEEE Electron Dev. Lett.* **11**, 209 (1990); A. Ishibashi, K. Funato, and Y. Mori, *Jpn. J. Appl. Phys.* **27**, L2382 (1988).
3. G. A. Sai-Halasz, M. R. Wordeman, D. P. Kern, S. Rishton, and E. Ganin, *IEEE Electron Dev. Lett.* **EDL-9**, 464 (1988).
4. A. N. Broers, *IEEE Trans. Electron Dev.* **ED-28**, 1268 (1981).
5. G. N. Taylor, *Solid St. Technol.*, June 1984, pp. 105ff.
6. D. K. Ferry, G. Bernstein, and W.-P. Liu, in: *Physics and Technology of Submicron Structures* (H. Heinrich, G. Bauer, and F. Kuchar, eds.), pp. 37–44, Springer-Verlag, Berlin (1988).
7. A. N. Broers, *J. Electrochem. Soc.* **128**, 166 (1977).
8. T. N. Hall, A. Wagner, and L. F. Thompson, *J. Vac. Sci. Technol.* **16**, 1189 (1979).
9. E. Spiller and R. Feder, *Top. Appl. Phys.* **22**, 35 (1977).
10. A. N. Broers, J. M. E. Harper, and W. W. Molzen, *Appl. Phys. Lett.* **33**, 392 (1978).
11. E. Bassous, L. M. Eprath, G. Pepper, and D. J. Mikalsen, *J. Electrochem. Soc.* **130**, 478 (1983).
12. G. B. Bernstein, W.-P. Liu, Y. N. Khawaja, M. N. Kozicki, D. K. Ferry, and L. Blum, *J. Vac. Sci. Technol.* B **6**, 2298 (1988).
13. H. G. Craighead, J. C. White, R. E. Howard, L. D. Jackel, R. E. Behringer, J. E. Sweeney, and R. W. Epworth, *J. Vac. Sci. Technol.* B **1**, 1186 (1983).
14. W. D. Grobman, *J. Vac. Sci. Technol.* B **1**, 1257 (1983).
15. R. P. Haelbich, J. P. Silverman, W. D. Grobman, J. R. Maldonado, and J. M. Warlaumont, *J. Vac. Sci. Technol.* B **1**, 1262 (1983).
16. G. N. Taylor, *Solid St. Technol.*, June 1984, p. 124.
17. E. Tobias and A. Carroll, *J. Vac. Sci. Technol.* **21**, 999 (1982).
18. H. Sewell, *J. Vac. Sci. Technol.* **15**, 927 (1978).
19. H. Hiraoka, *J. Electrochem. Soc.* **128**, 1065 (1981).
20. E. D. Roberts, *Solid St. Technol.*, June 1984, p. 135.
21. H. C. Pfeiffer, *IEEE Trans. Electron Dev.* **ED-26**, 663 (1979).
22. M. Idesawa, T. Soma, E. Goto, and T. Sasaki, *J. Vac. Sci. Technol.* **19**, 953 (1981).
23. G. Stengl, R. Kaitna, H. Löschner, R. Rieder, P. Wolf, and R. Sacher, *J. Vac. Sci. Technol.* **19**, 1164 (1981).
24. G. Stengl, H. Löschner, and E. Hammel, in: *Physics and Technology of Submicron Structures* (H. Heinrich, G. Bauer, and F. Kuchar, eds.), pp. 56–61, Springer-Verlag, Berlin (1988).
25. E. Hu, in: *GaAs Technology II* (D. K. Ferry, ed.), H. W. Sams, Indianapolis, IN (1989).
26. W. R. Grove, *Phil. Trans. Roy. Soc. London* **142**, 87 (1852).
27. R. Castaing and P. Laborie, *Compt. Rend. Acad. Sci.* **238**, 1885 (1952).
28. J. L. Mauer, J. S. Logan, L. B. Zielinski, and G. S. Schwartz, *J. Vac. Sci. Technol.* **15**, 1734 (1978).
29. G. C. Schwartz and P. M. Schaible, *J. Vac. Sci. Technol.* **16**, 410 (1979).
30. P. M. Schaible, W. C. Metzger, and J. P. Anderson, *J. Vac. Sci. Technol.* **15**, 334 (1978).
31. R. E. Klinger and J. E. Greene, *J. Appl. Phys.* **54**, 1595 (1983).
32. E. L. Hu and R. E. Howard, *Appl. Phys. Lett.* **37**, 1022 (1980).

33. J. Vatus, J. Chevrier, P. Deslescluse, and J. F. Rochette, *IEEE Trans. Electron Dev.* **ED-33**, 934 (1986).

34. K. Asakawa and S. Sugata, *Jpn. J. Appl. Phys.* **22**, L653 (1983).

35. M. W. Geis, G. A. Lincoln, N. Efremov, and W. J. Piacenti, *J. Vac. Sci. Technol.* **19**, 1390 (1981).

36. J. D. Chinn, A. Fernandez, I. Adesida, and E. D. Wolf, *J. Vac. Sci. Technol. A* **1**, 701 (1983).

37. Y. Ochiai, K. Gamo, and S. Namba, *Jpn. J. Appl. Phys.* **23**, L400 (1984).

38. V. M. Donnelly, D. L. Flamm, and G. L. Collins, *J. Vac. Sci. Technol.* **21**, 817 (1982).

39. Y. Y. Tu, T. J. Chuang, and H. F. Winters, *Phys. Rev. B* **23**, 823 (1981).

40. U. Gerlach-Meyer, J. W. Coburn, and E. Kay, *Surf. Sci.* **103**, 177 (1981).

41. V. M. Donnelly and D. L. Flamm, *Solid St. Technol.* **24**(4), 161 (1981).

42. H. F. Winters, *J. Appl. Phys.* **49**, 5165 (1978).

43. L. Hollan, in: *GaAs and Related Compounds (1974), Inst. Phys. Conf. Series*, Vol. 24, p. 22 (1975).

44. H. B. Pogge and B. M. Kernlage, *J. Cryst. Growth* **31**, 183 (1975), and references therein.

45. J. V. DiLorenzo and G. E. Moore, Jr., *J. Electrochem. Soc.* **118**, 1823 (1971).

46. D. J. Ashen, P. J. Dean, D. T. J. Hurle, J. B. Mullin, A. Royle, and A. M. White, in *GaAs and Related Compounds (1974), Inst. Phys. Conf. Series*, Vol. 24, p. 229 (1975).

47. H. M. Cox, *J. Cryst. Growth* **69**, 641 (1984).

48. J. J. Coleman, in *GaAs Technology* (D. K. Ferry, ed.), pp. 79–105, H. W. Sams, Indianapolis, IN (1985).

49. J. P. Duchemin, M. Bonnet, G. Beuchet, and F. Koelsch, in *GaAs and Related Compounds, Inst. Phys. Conf. Series*, Vol. 45, p. 10 (1978).

50. S. J. Jeng, C. M. Wayman, G. Costrini, and J. J. Coleman, *Mater. Lett.* **2**, 359 (1984).

51. J. R. Arthur, *J. Appl. Phys.* **39**, 4032 (1968).

52. A. C. Gossard, P. M. Petroff, W. Wiegmann, R. Dingle, and A. Savage, *Appl. Phys. Lett.* **29**, 323 (1976).

53. R. Dingle, A. C. Gossard, and W. Wiegmann, *Phys. Rev. Lett.* **34**, 1327 (1975).

54. M. B. Panish and S. Sumski, *J. Appl. Phys.* **55**, 3571 (1984).

55. W. T. Tsang, *Appl. Phys. Lett.* **45**, 1234 (1984).

56. J. Ryan, Thesis, Arizona State University, unpublished.

57. F. Braun, *Ann. Phys. Chem.* **153**, 556 (1874).

58. W. Schottky, *Naturwiss.* **26**, 843 (1938).

59. J. Bardeen, *Phys. Rev.* **71**, 717 (1947).

60. S. Kurtin, T. C. McGill, and C. A. Mead, *Phys. Rev. Lett.* **22**, 1433 (1969).

61. M. Schluter, *J. Vac. Sci. Technol.* **15**, 1374 (1978).

62. P. Skeath, I. Lindau, P. W. Chye, C. Y. Su, and W. E. Spicer, *J. Vac. Sci. Technol.* **16**, 1143 (1979), and references therein.

63. L. J. Brillson, *Phys. Rev. Lett.* **38**, 245 (1978); *J. Vac. Sci. Technol.* **16**, 1137 (1979).

64. W. Mönch, in *GaAs Technology II* (D. K. Ferry, ed.), pp. 139–178, H. W. Sams, Indianapolis, IN (1990).

65. B. E. Deal, E. H. Snow, and C. A. Mead, *J. Phys. Chem. Solids* **27**, 1873 (1966).

66. Z. A. Weinberg and A. Hartstein, *Solid St. Commun.* **20**, 179 (1976).

67. A. Hartstein and Z. A. Weinberg, *J. Phys. C* **11**, L469 (1978); *Phys. Rev. B* **20**, 1335 (1980); A. Hartstein, Z. A. Weinberg and D. J. DiMaria, *Phys. Rev. B* **25**, 7194 (1982).

68. S. M. Sze, *Physics of Semiconductor Devices*, 2nd ed., Wiley, New York (1981).

69. T. S. Kuan, P. E. Batson, T. N. Jackson, H. Rupprecht, and E. L. Wilkie, *J. Appl. Phys.* **54**, 6952 (1983).

3

Heterojunctions and Interfaces

The presence of an inversion or accumulation layer of charge, either electrons or holes, at the surface of certain heterojunctions was predicted many decades ago. This is true whether the heterojunction is between two semiconductors, such as in the high-electron-mobility transistor, or between a semiconductor and an oxide (or free space), such as in an MOS device. Of course, this latter method of controlling surface (or interface) charge is the preferred approach to current very large scale integration (VLSI) in silicon, where the control over the charge is exercised by the gate MOS structure in the MOSFET. In contrast to this, early transistors in GaAs were prepared using the metal-gate technology for MESFETs, primarily because of the lack of a good oxide technology in GaAs. The approach to GaAs changed in 1978, when Dingle *et al.*[1] demonstrated that very high mobilities could be achieved in modulation-doped structures through growth by molecular-beam epitaxy (MBE).

Synthetic semiconductor superlattice structures are of both fundamental and technological interest. However, most of the early work focused upon multilayer heterojunction, or periodic doping of homojunctions, in the growth by MBE. However, in the GaAlAs/GaAs heterojunctions, the first material has a somewhat larger band gap than GaAs (see Chapter 6 for details). As a consequence, dopant atoms placed in the former material will become ionized with the free electrons falling into the narrower-band-gap material. As a result, the electrons are spatially separated from the compensating ionized impurities so that there is a drop in the strength of the Coulombic scattering of the carriers. Resultant Hall mobilities are thus larger than that of either the bulk GaAs or of uniformly doped heterostructures. The higher achievable mobility was immediately recognized as beneficial for transistors, whether these were to be used for logic or for high-frequency applications. Both the Japanese[2] and the French[3] were quick to recognize this potential and to demonstrate actual FETs incorporating the modulation doping approach. Since this time, the highest-speed room-temperature logic devices are produced with this technology (less than 8 ps at room temperature).

One of the very important points about the HEMT technology is that the basic operation of the device is essentially the same as that of the Si MOSFET.

In both cases, the conduction electrons (or holes in *p*-channel devices) are constrained to lay in a quasi-two-dimensional layer at the interface between the active layer (GaAs or Si) and the wide-band-gap "insulator" (GaAlAs or SiO_2, respectively). The devices can be made in either enhancement or depletion mode and in either *n*- or *p*-channel configurations. As a result, the similarities of these devices are greater than their differences, and we shall treat the general theoretical developments together in this chapter. First, the approach of modulation doping to get the inversion electrons at the heterostructure interface is developed in the next section. Then the equivalent MOS structure is discussed for the Si devices. We then turn to a discussion of two-dimensional effects in the channel and the quantization that leads to them. This will include a discussion of interface roughness scattering that is particularly important for the MOSFET device. We then turn to a treatment of the two devices themselves.

3.1. MODULATION DOPING

As discussed above, the two-dimensional electron gas (2DEG) that exists at the interface between GaAs and the wider-band-gap GaAlAs exhibits a very high mobility at low temperatures. Even at room temperatures, the mobility is larger than that of bulk GaAs. Two factors contribute to this higher mobility, both arising from the selective doping of the GaAlAs layer rather than the GaAs layers in which the carriers reside. The first is the natural separation between the donor atoms in the GaAlAs and the electrons in the GaAs. The second is the inclusion of a "spacer" layer in the structure. This latter layer arises from the intentional insertion of an undoped GaAlAs layer adjacent to the GaAs itself. Consequently, the structure is actually quite complicated, but relatively easy to fabricate with MBE techniques. (In modern HEMT devices, the spacer layer is now usually omitted, as the higher mobilities are not that relevant to actual high-frequency devices. However, our treatment assumes the spacer layer is included, as its omission is a straightforward modification of the resulting equations.) In Figure 3.1, a typical high-electron-mobility structure is illustrated by a cut through the gate region and the channel. The typical structure begins with a bulk GaAs wafer, upon which is grown a GaAs buffer layer, which is also the active layer. This layer is grown primarily to provide a layer that has higher material quality than the normal GaAs substrate provides. Variations on this growth will sometimes place a GaAs/GaAlAs superlattice between the substrate and the buffer layer, primarily to act as a barrier to the out-diffusion of impurities and defects from the substrate. In any case, these techniques are used to provide an active GaAs layer with the highest possible quality. On top of this active layer is grown a thin, undoped layer of GaAlAs. This latter layer is the so-called spacer layer, and its thickness is typically between 2 and 20 nm. The purity of the spacer layer is required to prevent scattering of the channel carriers by the ionized impurities in this layer, so that doping levels below 10^{14} cm^{-3} are typically required. Finally, over the spacer layer is grown the doped GaAlAs layer. This layer is usually

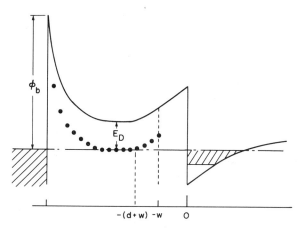

FIGURE 3.1. A schematic diagram of the energy bands for a GaAlAs-GaAs heterojunction HEMT in thermal equilibrium. It is assumed here that the GaAlAs layer is sufficiently thick that the depletion regions from the gate and the modulation doping region do not overlap.

between 20 and 50 nm thick, depending on the mode of operation of the device desired.

There are several factors that must be considered in actually choosing the parameters for the active device. One of these is the composition x of the $Ga_{1-x}Al_xAs$ that is used in the various layers. If x is too small, then the band discontinuity at the GaAlAs/GaAs interface is too small and the carriers are not well confined. On the other hand, for $x > 0.2$, a dominant defect begins to appear in the GaAlAs. This defect is termed the D–X center and is responsible for persistent photoconductivity in the material at low temperatures. In any case, it presents a significant trap that must be accounted for in the material. Several suggestions have been made to avoid this problem. One is just to keep $x < 0.2$, but this does not provide sufficient band discontinuity. Another is to replace the GaAlAs with a superlattice. In this latter case, many alternating layers of AlAs and GaAs are grown with the layers adjusted to yield an average x of the desired level. Dopant atoms are placed only in the GaAs, so that the defect center has no chance to form. This will also apparently work if the AlAs is replaced with an alloy of GaAlAs, but again with the dopants in the GaAs to limit the defect formation. In these cases, we remark that it is apparently the dopant Si atoms (for n-channel devices) that are involved in the formation of the D–X complex, which is thought to be centered around the donor itself. Placing the dopants in the GaAs quantum wells minimizes the number of donor atoms that are in the GaAlAs or the AlAs. In any case, we will continue as if the overlayers, the spacer layer and the doped layer, were actually a single alloy, but the reader should recognize that the actual case may be more complicated. The analysis will not be changed, since the structure of the superlattice, if used, is to provide an average material corresponding to the alloy. Moreover, the individual layers are sufficiently thin that the quantum wells are strongly coupled, and carrier transport between the various layers of the superlattice is easily facilitated. Thus, on the whole, the design of the superlattice doping layer provides an "average" material that behaves as an alloy, but without some of its disadvantages.

3.1.1. Carrier Density

Normally, the electrons in a semiconductor remain close to their donor atoms. In the HEMT, however, the band discontinuity between the GaAlAs and the GaAs creates a different possibility. This was shown in Figure 3.1. The discontinuity in the conduction band (for an n-channel device, which we will utilize throughout), and the required constancy of the Fermi level in equilibrium, lead to the band bending shown in the figure (surface bending arises from the Schottky barrier). The conduction band of the GaAlAs is forced upward (toward higher energies), ionizing the donor atoms near the interface. On the other hand, the conduction band edge in the GaAs is forced downward, leading to an accumulation of electrons in the potential well formed at the interface. In many respects, this potential well is similar to that we will see later in the Si MOSFET, except that the GaAlAs is not a wide-band-gap insulator like SiO_2, and the band discontinuity is an order of magnitude smaller. However, the potential wells are quite similar, and the extent of the electron wave function in both cases is constrained by the potential well, leading to quantization and quasi-two-dimensional behavior.

In Figure 3.1, the surface potential has been made sufficiently small that the surface Schottky barrier depletion region does not punch through to the interface region. In this way, we can use the known doping levels to calibrate the channel charge density itself. To achieve this, we can write an energy balance equation for the various levels in the band diagram of Figure 3.1. We note that this figure is illustrated for low temperatures, in that the Fermi level is taken to lie right at the donor ionization level. We return to this point below, but note that for this to occur, even at low temperatures, only a fraction of the donors must be ionized. If there were no donors ionized in the bulk region of the GaAlAs, the Fermi level would lie midway between the donor level and the conduction band edge. On the other hand, if all of the donors were ionized, the Fermi level would lie below the donor level. For any combination of temperature and fraction of ionization, the Fermi level lies between these two extremes. Only at very low temperatures, and fractional ionization, is the Fermi level actually pinned at the donor level itself, as shown in the figure. While we show the Fermi level, which is our reference level, as being at the donor level, we will correct this later.

Beginning at the Fermi level in the bulk region of the GaAlAs-doped layer, we are just the donor ionization energy away from the conduction band edge. The band edge rises through the charge depletion layer of the ionized donors by the amount

$$eV_d = \frac{e^2(N_D - N_A)d^2}{2\varepsilon_s} = \frac{e^2 N_s^2}{2\varepsilon_s(N_D - N_A)}, \tag{3.1}$$

where $N_D - N_A$ is the net donor concentration in the intentionally doped layer of GaAlAs, ε_s is the permittivity of the semiconductors (assumed to be the same for GaAs and AlGaAs, although the latter value is to be used here), d is the width of the depletion region, and N_s is the total charge in the GaAs and is the sum of the inversion charge in the channel and any depletion charge due to the band bending in this material. Usually, the bulk GaAs is either undoped or slightly p-type, so that the depletion charge is extremely small and can be ignored.

In the spacer layer, the potential varies linearly since there is no free charge in this region (or at least not very much by design). The field in this layer is then constant at a value determined by the free-carrier concentration in the inversion channel, which in turn is determined by the doping level in the modulation-doped layer. The electrons from the donors in the GaAlAs move into the inversion layer at the GaAs side of the interface, as well as neutralizing any depletion charge in the GaAs layer. Thus, the charge on each side of the spacer layer is given by N_s, and the field is

$$F_{SL} = \frac{eN_s}{\varepsilon_s},$$

(3.2)

and this leads to a potential drop of

$$V_{SL} = \frac{eN_s w}{\varepsilon_s},$$

(3.3)

which depends linearly on the spacer layer thickness w. This is not quite a linear dependence as indicated, because as the spacer layer thickness is increased the amount of charge that can be transferred actually decreases somewhat.

At the interface, the conduction band is offset by the amount ΔE_c. In early work on the multilayer superlattices, it was estimated that this quantity was some 85% of the total energy-gap discontinuity (we discuss this in some detail in Chapter 6). There is currently some uncertainty in this number, but the best estimates are now that only 68% of the gap discontinuity lies in the conduction band. The conduction band thus drops by this amount at the interface, as indicated in Figure 3.1. In the channel itself, the quasi-two-dimensional electron gas is quantized and the Fermi energy lies in the conduction band. We must still calculate the quantization energy E_0 and then the Fermi energy.

The lowest subband level has usually been calculated by utilizing arguments equivalent to those reviewed by Ando et al.[4] In this case, the substrate is typically weakly p-type, and a triangular potential well may be assumed. The wave functions in this simple approximation are then Airy functions, and the lowest subband is given by

$$E_0 = \left(\frac{\hbar^2}{2m}\right)^{1/3} \left(\frac{9\pi e F_s}{8}\right)^{2/3},$$

(3.4)

where

$$F_s = e(N_{depl.} + 0.5 N_{inv})/\varepsilon_s \tag{3.5}$$

is the average field in the inversion layer. In many HEMTs, however, the substrate is not slightly p-type but is n-type, and the Airy functions are not good solutions since the charge penetrates deeply into the substrate. In this latter situation, the wave functions are fractional-order Bessel functions, with the eigenenergies being given by transcendental relations. We shall not pursue this case, but return to a more extensive treatment of the wave functions and quantization.

In the typical case of n-channel HEMTs, the structure is quite like that used in Figure 3.1, so that the simple analysis gives quite good results. Thus, the total charge in the GaAs is predominantly from the free carriers in the channel. For a depletion charge of 5×10^{10} cm^{-2}, the lowest subband energy varies from 40 meV above the conduction band edge, for a free-carrier density of 2×10^{11} cm^{-2}, up to 120 meV above the conduction band minimum, for an inversion density of 1.25×10^{12} cm^{-2}.

Finally, we must estimate the position of the Fermi level above the subband minimum in these structures. The statistics of the carriers is governed by the two-dimensional nature of the transport. In this case, the number of carriers in the subband is given by

$$N_{inv} = N'_c \log \left\{ 1 + \exp \left[\frac{E_F - E_0}{k_B T} \right] \right\}, \tag{3.6}$$

where

$$N'_c = \frac{m k_B T}{\pi \hbar^2} \tag{3.7}$$

is the effective density of states. For the density range mentioned in the preceding paragraph, the Fermi level varies at room temperature from -2.5 meV (with respect to the bottom of the lowest subband) to 11 meV. It is clear that plotting the actual inversion density achieved in the heterostructure as a function of the spacer layer thickness for a series of samples allows an alternative approach to determine the conduction band discontinuity ΔE_c, since all other parameters are known in principle. This is important, for the reverse process as well, in that the ability of the gate voltage in the HEMT to vary the free-carrier density in the channel is crucially dependent on the actual band offset that occurs.

We can now put together the various contributions to the energy balance circuit that we have traced. Thus, beginning at the donor energy, we have

$$E_D + e V_d + e V_{SL} - \Delta E_c + E_0 + E_F = 0, \tag{3.8}$$

where E_F is now measured from the subband edge E_0. As the spacer layer is increased, V_{SL} and E_F both change. But V_{SL} cannot increase linearly with w, since E_F in general does not follow as rapidly. This then requires a reduction in the N_s to allow E_F to decline and balance the increase in V_{SL}. Consequently, the carrier density in the channel is usually reduced for larger values of the spacer layer. For arbitrary doping levels and layer thicknesses, (3.8) constitutes a non-linear equation for the channel density N_s, which can be solved for this latter quantity. In Figure 3.2, we illustrate this with a number of plots for various dopings and spacer layer thicknesses. In these plots, we have also corrected for the fact that the Fermi level does not sit on the donor level in the bulk GaAlAs, as discussed next.

The argument that is usually used is that there are always some donors ionized to compensate residual acceptors in the GaAlAs, but this only pins the Fermi level at the donor level at $T = 0$. At real temperatures, there are also electrons excited into the conduction band, and these are the electrons transferred into the inversion layer at the heterostructure interface. In bulk material, we could simply write a balance equation between the electrons excited to the conduction band and those localized in the impurity levels and thus determine the position of the Fermi level from the doping level and the temperature. The situation is not much more complicated in the HEMT, since the bulk region where the electrons are excited to the conduction band remains space-charge neutral even in this case. Only in the depletion region does the charge actually transfer to the GaAs side of the heterostructure. Therefore, there is no reason we cannot actually use the bulk arguments to determine the position of the Fermi energy relative to the donor level in the bulk GaAlAs section. Thus, the number of electrons excited to the conduction band in a partially compensated material is[5]

$$n = \frac{2(N_D - N_A)}{W + [W^2 + (8/N_c)(N_D - N_A)\exp(\varepsilon_d)]^{1/2}}, \tag{3.9}$$

where

$$W = 1 + \left(\frac{2N_A}{N_c}\right)\exp(\varepsilon_d), \qquad \varepsilon_d = \frac{E_d}{k_B T},$$

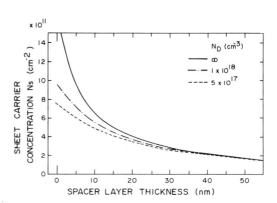

FIGURE 3.2. The sheet carrier density induced in the inversion layer by modulation doping as a function of the undoped spacer layer thickness.

N_c is the effective conduction band density of states in a three-dimensional solid, and E_d is the donor ionization energy. Thus, for high temperatures and small ionization energies, $n \simeq N_D - N_A$. However, for GaAlAs, where E_d can be 60–90 meV, n is a good bit smaller than $N_D - N_A$. In most applications to HEMTs, we can ignore the contributions of the acceptor atoms as being small (except at the lowest temperatures), and (3.9) simplifies to

$$n = \frac{2N_D}{1 + [1 + (8N_D/N_c)\exp(\varepsilon_d)]^{1/2}}, \qquad (3.10)$$

and the Fermi energy can be found from

$$E_F = k_B T \ln\left\{\frac{n}{N_c}\right\}. \qquad (3.11)$$

The Fermi level will pin at the donor ionization energy only if the number of free conduction electrons is small compared with the number of compensating acceptors (where we retain the form of (3.9) in (3.11)). For a doping level of 10^{18} cm^{-3}, typical for HEMTs, the number of free electrons is about 10^{17} cm^{-3}, and the Fermi level lies some 15 meV above the donor energy. Indeed, even with modest compensation, the Fermi level will not pass below the donor energy until the donors are completely ionized. For this reason, proper design of the HEMT will have the gate depletion almost overlap the interface depletion, even in a depletion-mode device. Still, the shift of the Fermi energy from the donor level is significant and requires an accurate determination of the donor ionization energy itself. Numbers from 60 to 90 meV, for GaAlAs in the $x = 0.2$–0.4 range, have been cited in the literature, but with considerable uncertainty. We have used 75 meV for Figure 3.2 and subsequent figures

In Figure 3.3, the inversion density is plotted as a function of the doping (assumed to be uniform in the GaAlAs, other than for the spacer layer) concentration. It is clear that the dependence is less than linear, due to the strong degeneracy

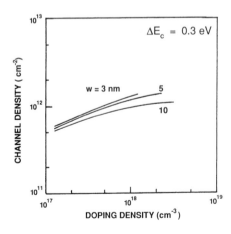

FIGURE 3.3. The sheet carrier density induced in the inversion layer by modulation doping as a function of the doping level in the GaAlAs for various spacer layer thicknesses.

of the inversion layer. If the doping were to be placed in a thin layer, the above problems of free carriers in the GaAlAs would be eliminated, and this has been proposed for these structures by a variety of investigators, although the basic idea is usually attributed to Les Eastman of Cornell. What this means is that if the thickness of the doping layer is thinner than the depletion layer that would form at the channel side of the doping layer, then the thin doping layer is completely depleted. In this case, the entire quantity of electrons from the doping layer falls into the active channel, without Fermi-level constraints. Since the doping layer is fully depleted, the position of the Fermi level moves to one dominated by other defect levels or to the intrinsic level. In this case it is difficult to ascertain the energy balance used above, but the number of inversion carriers is easily determined from the total doping introduced. In Figure 3.4, we show the thickness of the depletion layer itself, as a function of the doping level. If the actual doping region is thinner than this, then it will be fully depleted, and the channel charge can be calculated by determining the total sheet concentration of dopants.

3.1.2. Band Discontinuity

It is clear from the above discussion that the dependence of the carrier density in the channel on the gate potential is critically dependent on the value of the conduction-band discontinuity at the interface. This is also true for holes in the valence band for a p-channel device. Although the GaAs/GaAlAs structure is our best-known interface, and was thought to be well characterized, recent results have reopened the early accepted values. In addition, there is some question as to the value of the heavy-hole mass. Consequently, there will always be some question as to the actual doping that can be obtained in the quantum well at the interface of the heterostructure. In Chapter 6, we shall return to this discussion with the values that can be expected from simple band structure arguments.

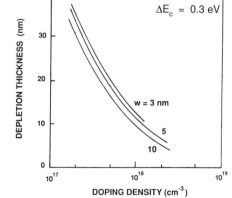

FIGURE 3.4. The thickness of the depleted layer of the doping in the GaAlAs for various doping densities and spacer layer thicknesses. If the doping layer is actually thinner than this depletion thickness, all of the charge is transferred to the inversion layer.

3.2. SEMICONDUCTOR–INSULATOR INTERFACES

In the case of a wide-band-gap insulator, such as SiO_2, grown or deposited upon the surface of a semiconductor, we must modify our heterojunction approach somewhat. To be sure, the band offsets at the conduction and valence bands are much the same as those in the heterostructure, but the interface is seldom as good. In the growth of GaAlAs on GaAs, atomic layer abruptness can be easily achieved by the use of MBE or metal-organic chemical vapor deposition (MOCVD). Thus there is a very abrupt transition between one semiconductor and the other. In the oxide-semiconductor interface, the transition region is a few atomic layers thick. The best current estimates of this, on both theoretical and experimental grounds, conclude that there is a transition layer which may be as thin as 0.5 nm. The composition of the layer provides the lattice transition between the relatively uniform SiO_2 structure, which has excellent short-range order but no long-range order, and the single-crystalline Si.

There is a great deal of empirical information at hand, but our knowledge of the atomic arrangements of localized electronic structure at or near the interface is still largely speculative. Most theoretical models consider that the interface remains largely ordered and is not significantly disordered in comparison with the bulk SiO_2. One of the most widely accepted models is due to Herman and Batra.[6] In this model, the interface is formed by connecting the (100) interface of Si with a $\pi/4$ rotated (100) face of an idealized SiO_2 crystal. This model is able to match the lattices closely, but the structure may not be realistic. If one considers the initial oxidation of Si where the oxygen atoms try to fit in the center of the Si—Si dangling bonds, it is unlikely that the interface bonds rotate such that the interface SiO_2 tetrahedra are bonded with the distant-neighbor Si surface atoms instead of the second-nearest-neighbor surface atoms. A more realistic model is obtained by considering alternate layers of oxygen and silicon atoms on the Si(100) surface arriving in such a manner that the bond angles and bond lengths are adjusted to obtain an ordered interface retaining two-dimensional periodicity. In this case, there is a modification of the actual bond lengths and angles from those expected in bulk SiO_2, but the resultant calculated density of states agrees well with interfacial studies.[7] Even here though, the average bond lengths and angles still fit the expected averages found experimentally. Moreover, new density-of-states peaks are found in the interfacial structure, which agree quite well with careful Auger measurements of the interface.[8,9] New lines are found about -4.5 eV from the main Si peak for the bulk. These lines are not characteristic of the Si bonded in SiO_2, and are thought to be from the Si lying in the interfacial region and connected with a bridging O between two surface orbitals. These results tend to indicate that the initial oxidation of the Si(100) surface, as well as the characteristic Si–SiO_2 interface states in well-developed oxides, is primarily due to atomic oxygen bonded to the Si dangling orbitals. This model, and others, tend to produce a narrow interface region between the well-developed SiO_2 and the bulk Si. The average thickness of this interfacial region is currently felt to be less than 0.5 nm, or about two atomic layers. In

Figure 3.5, we show a very high resolution lattice-plane image of the Si–SiO$_2$ interface. Even from this image, which actually averages through approximately 5–7 nm of the thinned sample, the interface looks quite abrupt.

There are two major features of the Si–SiO$_2$ interface that we can ascertain from Figure 3.5. The first is the relatively abrupt interface, as discussed above. The second is the random variation of the interface itself. By this, we call attention to the fact that the interface is not planar. Although the actual interface, at any lateral position along the interface, is quite abrupt, the interface appears wavy. This latter quality, the variation in interface along the surface, is referred to as the interface roughness quality. To develop a complete picture of the statistical properties of the rough interface, one must average over quite a large number of pictures taken from different parts of the interface. Such an approach has been carried out by Goodnick *et al.*[10] This average was carried out by taking sequences of pictures in which interfacial regions up to 50 nm could be examined. These were then digitized, and the statistical parameters of rms interface roughness amplitude Δ and correlation length L were then obtained. Several different statistical models were used for the interface, and the results were also deconvolved for the finite thickness of the sample.

The one-dimensional roughness that appears in the high-resolution transmission electron microscope (TEM) pictures is thought to be characterized as a first-order autoregressive (Markovian) process, albeit in the presence of a nonstationary component representative of the very long wavelength fluctuations in the interface. This first-order autoregressive model is characteristic of an exponential decay in the autocovariance model for the two-dimensional interface, rather

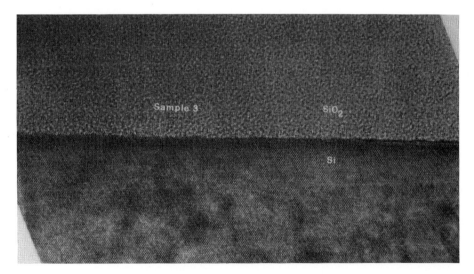

FIGURE 3.5. The lattice-plane image of a Si–SiO$_2$ interface. The image is in the (111) plane, while the interface is a (100) plane. The well-ordered part of the image is the Si material. From S. M. Goodnick, C. W. Wilmsen, D. K. Ferry, Z. Lilienthal, D. Fathy, and O. L. Kiwanek, *Phys. Rev. B* **32**, 8171 (1985).

than a Gaussian model. Such a model of the roughness is more compatible with the discrete nature of the roughness which occurs at atomic steps along the interface than is the assumption of a Gaussian covariance. Further, the results of the analysis suggests that normally grown SiO_2 results in an interface with an rms roughness of $\Delta = 0.14$–0.18 nm and a correlation length (in the exponential model) $L = 0.7$–1.0 nm. These parameters are necessary later for the calculation of the scattering in the inversion layer from the interface roughness.

The picture we now obtain of the insulator–semiconductor interface is also one in which the transition region is small compared with the wavelength of the individual electrons. For purposes of the quantum mechanics of the electron waves, the transition is exactly abrupt. The resulting roughness of the interface then acts as a scattering center, which introduces transitions between the individual electron states. Other defects in the oxide, which are near to the interface, can act like trapping centers if they introduce localized states in the Si band gap and are close enough to interact with the electron wave functions.

3.3. TWO-DIMENSIONAL QUANTIZATION

The most obvious question is just why do we need to treat the electrons in the inversion (or accumulation) layer at the heterostructure interface by a quantum mechanical process? The answer to this lies in the relative size of the electronic wave function with respect to the classical extent of the inversion layer. Classically, we relate the density of charge at the interface to the surface potential via

$$n = n_s \exp\left(-\frac{e\psi_s}{k_B T}\right),$$ (3.12)

where n_s is the peak density right at the interface and ψ_s is the band bending of the conduction band. The density falls by $1/e$ for a voltage variation of $k_B T/e$, and the surface field is related to the peak density by $F_s = en_s/\varepsilon_s$, so that the classical distance is

$$w_c = \frac{\Delta\psi_s}{F_s} = \frac{\varepsilon_s k_B T}{n_s e^2}.$$ (3.13)

We note here that this is not the Debye length, since n_s is a sheet concentration rather than a bulk density. Typical numbers are $\varepsilon_s = 12\varepsilon_0$ and $n_s = 1.0 \times 10^{12}$ cm^{-2}, so that at room temperature we obtain $w_c = 1.7$ nm.

On the other hand, the wavelength of the electron is related to its momentum, and hence to the temperature, in the direction normal to the interface. The wavelength is

$$\lambda = \frac{h}{p} = \frac{h}{(mk_b T)^{1/2}},$$ (3.14)

where we have used the thermal momentum in the de Broglie relation. For the case of the lowest subband (discussed below), $m = 0.91 m_0$ in the direction normal to the interface, and we obtain $\lambda = 11.3$ nm. In Figure 3.6, we plot the values of the wavelengths of electrons in various materials at room temperature. Here, the variable is the effective mass in the direction normal to the interface.

Clearly, one is unlikely to be able to squeeze an electron, whose wavelength is 11.3 nm or larger, into the classical inversion-layer thickness of 1.7 nm. Consequently, we must solve the problem quantum mechanically. The quantization arises from the presence of the potential well, as it constrains the motion of the carriers in the direction normal to the interface. Thus, the quantized energy levels for motion in this direction arise from solving the proper equations to find the constrained values of the electron wavelength and, hence, its energy.

3.3.1. Surface Subbands

A description of surface quantization from first principles is formally quite difficult, in that it requires a detailed model of the chemical bonding and local atomic configuration at the interface. The presence of any reconstruction of the lattice at the interface affects the local configuration and, hence, the details of the quantization of the electron (or hole) wave functions. While significant progress has been made in the understanding of the details of local atomic structure at interfaces, the situation is usually different for each interface structure. This then highlights the differences between materials rather than the similarities. For most applications, the actual local atomic structure differs from the bulk materials only over one or two atomic layers, such as the Si–SiO$_2$ interface. Thus, in nearly all cases, the use of the effective mass approximation has proved sufficient to describe the effect of the interface potential in forming quantized electronic states.

In the effective mass approximation for a single band, the wave function of an inversion (or accumulation) layer electron (or hole) is written as a product of the normal Bloch function corresponding to that band, $e^{i\mathbf{k}\cdot\mathbf{r}}u_k(\mathbf{r})$, and an

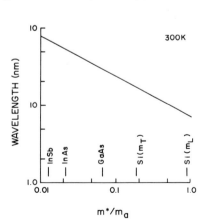

FIGURE 3.6. The wavelength of a quasi-particle electron in a semiconductor as a function of its effective mass at room temperature.

envelope function $\Psi(\mathbf{r})$, where the latter satisfies the effective mass Schrödinger equation

$$\left[-\left(\frac{\hbar^2}{2m}\right)\nabla^2 + V(z) - E\right]\Psi(\mathbf{r}) = 0. \tag{3.15}$$

Here, m is the appropriate isotropic effective mass for the band in question, and $V(z)$ is the surface potential, such as that shown in Figure 3.1 for the HEMT structure. There will be subtle details in the form of $V(z)$ that differ for inversion layers (n-channel in p-type material) and accumulation layers (n-channel in n-type material). These differences arise primarily due to the depth of the potential well and the presence of depletion charge. For the main III-V semiconductors of interest, the lowest conduction-band minimum, the main one of interest, and the valence band maximum both lie at the Γ-point in the Brillouin zone and are characterized by an isotropic (albeit nonparabolic) effective mass. Thus, the complications of an effective mass tensor, as occurs in Si (discussed below) and Ge, are not encountered. We will not consider the upper valleys at present, but we return to them later.

The conduction band of Si is characterized by six equivalent minima located along Δ approximately 83% of the way to X from the Γ-point (see Chapter 6). The interface potential $V(z)$, and the resulting quantization, splits these six ellipsoids because of the anisotropic effective mass. For the more common (100) interface, two of the ellipsoids show the heavy longitudinal mass m_L ($= 0.91m_0$) in the direction normal to the interface. The other four ellipsoids are characterized by the light transverse mass ($= 0.19m_0$) in the direction normal to the interface. Since the quantized energy level is inversely proportional to the effective mass, as shown, e.g., in (3.4), the conduction band splits into a lower doubly degenerate set of subbands and a higher fourfold degenerate set of valleys.[4] This then gives a double set of quantized levels. The lowest set, however, is spherically symmetric with an isotropic effective mass described by the transverse mass. The upper set of subbands is anisotropic, but is usually described by a single average density-of-states mass that is an average of the longitudinal and transverse masses.

In keeping with the idea that we are treating an interface normal to one of the principal axes, we assume $V(z)$ depends only on this direction normal to the interface. Hence, the solution to (3.15) is separable as

$$\Psi(\mathbf{r}) = \zeta_i(z)\exp(i\mathbf{k}\cdot\mathbf{r}'), \tag{3.16}$$

where $\mathbf{r}'$ is the two-dimensional vector parallel to the interface. The quantity $\zeta_i(z)$ is an envelope function which satisfies the new equation

$$\left[-\left(\frac{\hbar^2}{2m}\right)\nabla^2 + V(z) - E_i\right]\zeta_i(z) = 0, \tag{3.17}$$

where the values E_i are the eigenvalues that represent the allowed values for the energy levels of the quantized motion normal to the interface. The total energies form a ladder of parabolic (or nonparabolic in some cases) two-dimensional subbands which satisfy the dispersion relation

$$E(k) = E_i + \frac{\hbar^2 k^2}{2m_i}, \tag{3.18}$$

where we now recognize k as the two-dimensional momentum wave vector in the direction parallel to the interface and m_i is the density-of-states mass in this direction. The density of states for a 2DEG (for parabolic bands) is constant in energy as opposed to the three-dimensional case. Each subband contributes a constant density of states, such that the total density may be written as

$$\rho(E) = \sum_i \left(\frac{s_i m_i}{2\pi\hbar^2}\right) u_0(E - E_i), \tag{3.19}$$

where s_i is the product of the spin degeneracy and the valley degeneracy, and $u_0(x)$ is the Heaviside unit step function ($u_0(x) = 1$ for $x \geq 0$; $u_0(x) = 0$ for $x < 0$). A constant additional number of states is added to the total density as the energy rises past each subband eigenenergy.

The energy levels are usually denoted in ascending order by the index with the numbering $E_0, E_1, E_2, \ldots$. In the III-V compounds, there is only a single valley involved with the quantization for electrons, so $s_i = 2$ for spin degeneracy only. In the case of Si, however, the lowest set of subbands arises from the two valleys which have their longitudinal axis perpendicular to the interface, so $s_i = 4$. In addition, there is a second set of subbands which arises from the other four valleys of the conduction band. In this case, $s_i = 8$. The subband labels mentioned above are usually assigned to the twofold set of levels, while those of the fourfold set are given the labels $E_0', E_1', E_2', \ldots$.

The surface potential $V(z)$ represents contributions from the space-charge layer, the inversion electrons, and the image potential which arises from the different dielectric constants on the two sides of the heterostructure interface. The electronic contributions to $V(z)$ depend on the spatial distribution of the inversion electrons through Poisson's equation. In the Hartree approximation, the average potential due to the electron distribution depends on the magnitude of the envelope function $\zeta_i(z)$, summed over all occupied subbands, and thus (3.17) must be solved self-consistently. Numerical self-consistent solutions of (3.17) have been performed in connection with Si, GaAs, InAs, and InP inversion layers.[4,11] In addition to this, one must also consider the electronic contributions of exchange and correlation when computing the subband energies. Although these latter two energies tend to be on the level of 10–20 MeV, this is also the magnitude of the various subband spacings so that the effects are relatively large in this case, especially for Si. The lighter effective masses in the III-V materials make this effect somewhat smaller, but the degree to which this is true is not fully known at this time.

3.3.2. Approximate Solutions

By making a suitable approximation for the self-consistent electronic potential $V(z)$, (3.17) may be uncoupled from Poisson's equation and analytic solutions are then possible. This neglect of the detailed form of the electronic potential is usually only valid at low inversion or accumulation densities such that the variation of the surface electric field through the inversion layer is relatively small. There are two model potentials that have appeared in the literature, which give exact solutions for (3.17). These are the triangular potential well and the exponential potential. The latter is far more accurate for accumulation layers, while the former is a better approximation for inversion layers. We will treat only the former, but will return to an improved solution available through a variational approach.

The triangular well potential arises from the neglect of quadratic terms which appear in the usual formulations for the space-charge potential. In moderately doped material, where the depletion charge is comparable to the inversion charge, this is not a bad approximation, especially since the inversion charge is confined to a region less than 10 nm from the interface. The approximation is not as good in, say, HEMTs where the depletion charge is not large compared with the inversion charge. However, this approximation has become quite widely used and can be improved if the *average*, rather than surface, electric field, is used in computing the triangular potential. The average field arises since the actual field in the surface potential well differs on the two sides of the inversion charge by the amount en_{inv}/ε_s. Thus, the average field in the quantum well, in the region occupied by the inversion charge, is $e(N_{depl.} + 0.5n_{inv})/\varepsilon_s$. This differs from the actual surface value of the field by the factor of $\frac{1}{2}$ multiplying the inversion charge. This is clearly the value of field that has been used in (3.4), which is the result for the triangular well potential. Thus, the approximation is that the surface potential is just $V(z) = eF_s z$, with F_s being interpreted as the average surface field. Then, the solutions to (3.17) are the Airy functions

$$\zeta_i(z) = \text{Ai}\left[\left(\frac{2m_z eF_s}{\hbar^2}\right)^{1/3}\left(z - \frac{E_i}{eF_s}\right)\right],\tag{3.20}$$

where m_z is the mass in the direction perpendicular to the interface, and the energy eigenvalues are determined by the boundary conditions. In truth, the wave function actually penetrates somewhat into the wide-band-gap material (the GaAlAs or the SiO_2). However, this effect is small, and one usually assumes that the wave function vanishes at the interface. Then, the energy eigenvalues are determined by the requirement that a zero of the function $\text{Ai}(z)$ appear at $z = 0$. The wave function (3.20) is offset, so that the region $z > 0$ contains $i + 1$ oscillations of the Airy function before it becomes exponentially damped. Thus, the values for the energy are given by

$$E_i = \left(\frac{\hbar^2}{2m_z}\right)^{1/3}[1.5\pi eF_s(i + 0.75)]^{2/3}.\tag{3.21}$$

In Figure 3.7, the fraction of carriers in the lowest subband and the average width of the inversion layer are shown as a function of inversion density for several different temperatures for the HEMT structure in GaAs. Similar curves are shown in Figure 3.8 for an inversion layer in the Si–SiO$_2$ system. For the Airy function solutions, the average width of the inversion layer is $2E_i/3eF_s$, so that the layers become narrower as the inversion density increases. For $n_{inv} \gg N_{depl}$, the curves are amenable to scaling with a parameter a as

$$T = aT_1, \qquad n_{inv} = a^2 n_1, \qquad \langle z \rangle = \frac{z_1}{a}, \qquad (3.22)$$

where T_1, z_1, and n_1 are, respectively, the values of temperature, average width, and inversion density shown in Figure 3.7, and $a = m_z/m_1$ (with $m_1 = 0.067m_0$ for GaAs). From Figures 3.7 and 3.8, it is apparent that, for a range of density, increases of the density cause an increase in the effective field, which leads to a deeper potential well. This in turn causes the subbands to be moved further apart, causing the higher-lying subbands to be emptied of carriers.

The above results considered the case of an infinite barrier. In real systems, however, this barrier is not infinite and the wave functions penetrate into the wide-band-gap material. This can be severe in the GaAs/GaAlAs case, as the barrier may only be about 0.3 eV. This leads to a problem in matching the wave functions at the interface and will shift the energy levels to slightly lower values. This correction has been estimated to be only a few percent, which is less than the general error expected from the triangular well approximation.

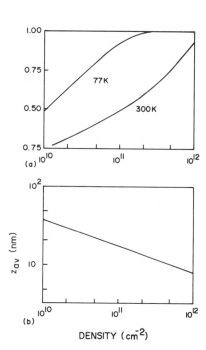

FIGURE 3.7. (a) The fraction of carriers residing in the lowest subband as a function of channel density, for the triangular well approximation, for GaAs. (b) The average width of the inversion layer (lowest subband) as a function of density using the parameters from the previous part.

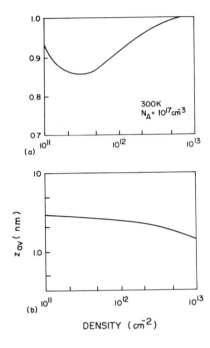

FIGURE 3.8. (a) The fraction of carriers residing in the lowest subband as a function of channel density, for the triangular well approximation, for Si. (b) The average width of the inversion layer (lowest subband) as a function of density using the parameters from the previous part.

An alternate approach to obtain solutions to (3.17), which includes the self-consistent corrections generated by Poisson's equation, is through the use of variational wave functions. Such solutions are useful due to their functional simplicity. The usual form adopted for the wave function in this approach is due to Fang and Howard,[12] who considered an exponential function for the lowest subband

$$\zeta_0(z) = \left(\frac{b^3}{2}\right)^{1/2} z \exp\left(-\frac{bz}{2}\right), \tag{3.23}$$

in which b is the variational parameter. In this case, (3.23) is used to calculate the charge density in the lowest subband, and this is in turn used to calculate $V(z)$ with Poisson's equation. Finally, $V(z)$ is then used in (3.17) to obtain an equation for E_0 as a function of b. This is then minimized by varying b. Such a procedure gives the minimization of the energy level with a value of

$$b = \left(\frac{12 m_z e^2}{\varepsilon_s \hbar^2}\right)(N_{\text{depl}} + \tfrac{11}{32} n_{\text{inv}})^{1/3}. \tag{3.24}$$

We can clearly see a somewhat different effective field in this solution, in that the factor of $\frac{1}{2}$ has been replaced by $\frac{11}{32}$. Moreover, the functional dependence of the average thickness $\langle z \rangle$ ($= 3/b$) has changed as well. For higher-lying subbands, one can add additional terms in the prefactor to introduce oscillatory behavior and in the exponent to modify the decay. One ensures that ζ_1 is orthogonal to ζ_0 as one step, and then uses both to enhance the minimization of the two energy levels. Thus, the first level is characterized by

$$\zeta_1(z) = A^{1/2} z \left[1 - \frac{(b + b_1)z}{6} \right] \exp\left(-\frac{b_1 z}{2} \right),$$

$$A = \frac{3b_1^5}{2(b^2 - bb_1 + b_1^2)},$$

(3.25)

and $b_1 \simeq 0.754b$.

3.3.3. Effects of Nonparabolicity

The discussion so far has been relevant only to very wide band-gap materials where nonparabolicity is not a factor. However, for most materials of interest, strong nonparabolicity of the bulk dispersion relation tends to mix the motion of electrons parallel to the surface with the normal motion. If the surface potential is on the order of the semiconductor band gap, additional mixing of valence and conduction wave functions is expected. This may give rise to splitting of the spin degeneracy due to the reduction of the crystal symmetry at the interface.

There are several techniques to investigate this[13] effect, but the general effect on the actual energy levels themselves is quite small, on the order of 10% or so. In keeping with the other approximations, many of which will cause competing shifts in the energy levels, this degree of error is not large. So for all practical interests, the nonparabolicity can be confined to just the motion in the transverse direction (parallel to the interface).

3.4. INTERFACE SCATTERING MECHANISMS

If the heterostructure interface is not perfect, say local charge and no atomically smooth and abrupt transition layer, then additional scattering centers can exist at the interface. These scattering centers will strongly affect the transport of the carriers along the interface. The presence of these topological imperfections creates charged scattering centers, while the presence of the interface roughness leads to perturbations of the potential well which give a random potential scattering. In addition, there may also be additional phonon modes, which are two-dimensional interface modes that arise from the dissimilarities in material properties at the interface. Here, we will concentrate on the Coulomb scattering that arises from charge centers and on the interface roughness scattering because of the effect these have in the Si–SiO$_2$ system.

In many cases, the scattering cross section may be reduced to a two-dimensional problem through suitable averaging of the perturbing potential over the envelope functions. However, nonvanishing matrix elements between different subbands result in intersubband transitions which couple the motion of electrons in different subbands. This may be problematical in most materials where the number of interface defects is small, but can occur in heavy scattering regimes.

3.4.1. Coulomb Scattering

Disorder of the SiO_2 matrix in the vicinity of the interface gives rise to localized defects in the neighborhood of the interface. In many cases, these defects are associated with dangling bonds and can lead to charged trapping centers which scatter the free carriers through the Coulomb interaction. In addition to these "surface states," ionized impurities incorporated on either side of the interface may be present, and these add to the number of Coulomb scattering centers. This Coulomb scattering of the inversion-layer electrons differs from the case of bulk impurity scattering due to the reduced dimensionality of the carriers.

Coulomb scattering in the context of surface-quantized carriers was first described by Stern and Howard[14] for electrons in the $Si-SiO_2$ system. Since then, many treatments have appeared in the literature, which differ only little from this early approach. In general, the interface is treated as being abrupt and as having an infinite potential discontinuity in the conduction (or valence) band, so that problems with interfacial nonstoichiometry and roughness are neglected. These can be treated in turn by surface roughness scattering.

To approach a treatment of this scattering, it is most convenient to use the electrostatic Green's function for charges in the presence of a dielectric interface. In this way, the image potential is correctly included within the calculation. The scattering matrix element involves integration over plane-wave states for the motion parallel to the interface, and thus we are led to consider the two-dimensional Fourier transform of the Green's function[14]

$$G(\mathbf{q}, z - z') = \frac{1}{2q\varepsilon_s} [\exp(-q|z - z'|) + \alpha \exp(-q|z + z'|)], \qquad z' > 0,$$

$$= \frac{1}{2q\varepsilon_a} \exp[-q(z - z')], \qquad z' < 0, \tag{3.26}$$

where $\mathbf{q}$ is the two-dimensional wave vector, $\alpha = (\varepsilon_s - \varepsilon')/(\varepsilon_s + \varepsilon')$, $\varepsilon_a = (\varepsilon_s + \varepsilon')/2$, and ε' is the permittivity in the wide-band-gap region of the heterostructure interface. Equation (3.26) assumes that the scattering center is located a distance z' from the interface ($z = z' = 0$) and has an image at $-z'$. If the interface were to be nonabrupt, then α would in general be a complex function of $\mathbf{q}$.

For two-dimensional scattering in the simple Born approximation, the scattering cross section is determined by the matrix element over the bare Coulomb potential of a charge located at z', and

$$\langle \mathbf{k}| V(z')|\mathbf{k'}\rangle \simeq e^2 \int_0^\infty |\zeta_0(z)|^2 G(\mathbf{q}, z - z') \, dz', \tag{3.27}$$

where $q = |\mathbf{k} - \mathbf{k}'|$ is the difference in the incident and the scattered wave vectors.

In the linear screening approximation, the scattering potential is simply the Fourier transform of the bare potential (3.27) divided by the static dielectric function for the inversion electrons, which in the random phase approximation for a single subband may be written as

$$\varepsilon(q) = \varepsilon_0 \left[1 + \left(\frac{q_0}{q} \right) F(q) \Pi(q) \right]. \tag{3.28}$$

(The source of this function is treated in detail for the general three-dimensional case in Chapter 7.) Here, q_0 is the inverse screening length, and $F(q)$ is a slowly varying function which accounts for the deviation of the inversion layer from a true two-dimensional system. $\Pi(q)$ is a wave-vector-dependent function which has the value 1 at $q = 0$ and decays to zero as $q \to \infty$. In strongly degenerate systems, $\Pi(q)$ has a discontinuity at $q = k_f$, where k_f is the Fermi wave vector. The limiting case of $q \to 0$, and q_0 evaluated with the Debye formula, gives the normal Debye screened interaction. For weakly degenerate systems, we can approximate this as

$$\Pi(q) \sim \left[1 + \left(\frac{q}{q_0} \right)^2 \right]^{-1}. \tag{3.29}$$

The inverse screening length q_0 depends on the density of states and is given, for only a single subband being occupied, as

$$q_0 = \frac{n_s m e^2}{4 \pi \hbar^2 \varepsilon_a}. \tag{3.30}$$

The relative magnitude of q_0, when compared with the scattering wave vector q, is a measure of the effectiveness of screening in the inversion layer.

It is not a bad assumption to consider only scattering from charges located at the interface itself, since this is the region at which the density of scattering centers is large. In this idealization, charges are assumed to be uniformly distributed in the plane $z' = 0$. The scattering rate in the Born approximation then becomes, for the ground subband,

$$\Gamma_c(k) = \Lambda \int_0^{2\pi} \left\{ \frac{(1 - \cos \theta) A^2(q)}{[q + q_0 F \Pi]^2} \right\} d\theta, \tag{3.31}$$

where

$$\Lambda = \frac{e^4 m N_{\mathrm{sc}}}{8 \pi \hbar^3 \varepsilon_a^2},$$

N_{sc} is the number of Coulomb scattering centers per unit interface area, and $q = 2k \sin(\theta/2)$. The function $A(q)$ arises from (3.27), and is just

$$A(q) = \int_0^\infty dz \, |\zeta_0(z)|^2 \exp(-qz). \tag{3.32}$$

Two limiting cases may be found in the result (3.31). For $q_0 \ll q$, we find an unscreened behavior, and the scattering rate varies as k^{-2}, and the mobility is proportional to the inversion charge density. This reflects the decrease in the scattering cross section with the increase in the average energy of the carriers at the Fermi energy, since this latter quantity increases with the carrier density. In the other extreme, $q_0 \gg q$, the scattering is very heavily screened out by the charge carriers themselves. Here, the wave-vector dependence is determined by the ratio of $A(q)/F(q)$, which is not a strong function of the inversion charge density. Even in the totally screened case, however, the scattering rate for Coulomb charge centers tends to decrease as the inversion density increases, and thus the mechanism tends to dominate the mobility at low carrier densities.

In the GaAlAs/GaAs heterostructure, the charge centers are removed from the carriers in the inversion layer due to the spacer layer. Thus the Coulomb scattering is reduced by this effect. But, the small mass of the electrons also contributes to ensure that the scattering is heavily screened by the carriers themselves. Thus, both of these effects serve to reduce the role of Coulomb scattering, and it is effectively not observed to be a factor at any temperature. In the Si–SiO$_2$ system, on the other hand, the scattering centers are effectively right at the interface, and are more effective scattering centers due to the heavier mass of the electrons in this inversion layer. Thus, Coulomb scattering dominates the mobility of the carriers at low inversion densities, especially at low temperatures.

3.4.2. Surface-Roughness Scattering

In addition to Coulomb scattering, short-range scattering associated with the interfacial disorder limits the mobility of electrons at the interface as well. On a microscopic level, the interface is never truly abrupt, and this disorder can extend over one or two atomic layers, as discussed in the preceding section. This local atomic environment has a random variation which, when coupled to the surface potential, gives rise to fluctuations of the subband energy and, hence, to a random potential scattering. This scattering has strong similarities to alloy scattering, discussed in Chapter 6, and can limit the mobility. At present, a calculation of the scattering rate based upon the microscopic details of the interface does not exist. The usual models instead rely on a semiclassical approach which uses a phenomenological surface roughness. While early treatments relied upon descriptions in terms of specular and diffuse reflections of electron waves, these approaches are inappropriate for the quantized electrons in the interface inversion layers.

In current surface roughness models, the displacement of the interface from a perfect plane layer is assumed to be describable by a random function $\Delta(\mathbf{r}')$, where $\mathbf{r}'$ is the two-dimensional position vector parallel to the interface. This model assumes that $\Delta(\mathbf{r}')$ changes slowly over atomic dimensions so that the boundary conditions on the wave functions can be treated as abrupt and continuous. This assumption is obviously in error when surface fluctuations occur on the atomic level. However, the model has proved to provide quite good agreement with measured mobility variations in a variety of materials and interfaces.

The scattering potential may be obtained by expanding the surface potential in terms of $\Delta(\mathbf{r}')$ as

$$\delta V(\mathbf{r}', z) = V[z + \Delta(\mathbf{r}')] - V(z) \simeq eF(z)\Delta(\mathbf{r}) + \cdots, \tag{3.33}$$

where $F(z)$ is the electric field in the channel itself. The scattering rate for the perturbing potential (3.33) is usually treated in the Born approximation, but including the role of correlation between the scattering centers along the interface. This correlation is discussed in more detail in relation to the short-range alloy scattering in Chapter 6. The scattering matrix element may be found to be

$$\langle \mathbf{k}|\delta V|\mathbf{k}'\rangle = e\Delta(q)\int_0^\infty dz\, F(z)|\zeta_0(z)|^2 = \frac{e^2\Delta(q)(N_{\text{depl}} + 0.5n_{\text{inv}})}{\varepsilon_s}, \tag{3.34}$$

where $\Delta(q)$ is the Fourier transform of $\Delta(\mathbf{r}')$ and we have introduced the average electric field discussed earlier.

In the Born approximation, only the statistical properties of $\Delta(q)$ need be considered. Thus, we may introduce the descriptors discussed in Section 3.2, and although the experimental correlation is an exponential, the deviation from a Gaussian is quite small. The major variation lies in the tail of the correlation, where the Gaussian provides too rapid a falloff of the correlation in comparison with the actually observed exponential behavior. However, the majority of the scattering strength lies in the central part of the correlation function, and here there is little variation between the exponential and Gaussian behaviors. Moreover, the actual calculation of the scattering cross section is somewhat easier with the Gaussian, and it is this form that appears throughout the literature. Thus, we treat the correlation as

$$\langle \Delta(\mathbf{r}')\Delta(\mathbf{r} - \mathbf{r}')\rangle = \Delta^2 \exp\left(-\frac{r^2}{L^2}\right), \tag{3.35}$$

and this leads to the Fourier-transformed result

$$|\Delta(q)|^2 = \pi\Delta^2 L^2 \exp\left(-\frac{q^2 L^2}{4}\right). \tag{3.36}$$

We recall that the parameter Δ is the rms height of the fluctuation in the interface and that L is the correlation length for the fluctuations. In a sense, this latter quantity is the average distance between "bumps" in the interface. We may then combine the above to yield the scattering rate

$$\Gamma_{sr} = \left[\frac{\Delta^2 L^2 e^4 (N_{depl} + 0.5 n_{inv})^2 m}{2 \varepsilon_s^2 \hbar^3}\right] \left\{ I_0\left(\frac{k^2 L^2}{2}\right) - I_1\left(\frac{k^2 L^2}{2}\right) \right\}$$

$$\times \exp\left(-\frac{k^2 L^2}{2}\right). \tag{3.37}$$

The explicit dependence of the scattering rate on the square of the average field results in decreasing mobility with increasing surface field (or inversion density), which agrees with the trends observed in the experimental mobility data of most materials. This decrease in mobility with surface density qualitatively arises from the increased electric-field dispersion around interface discontinuities at higher surface fields, which in turn gives rise to a larger scattering potential.

The roughness parameters Δ and L are usually found by fitting experimental mobility data. Typical values obtained in this manner are 0.2–0.5 nm for Δ and 1.5–4.0 nm for L for Si, and larger values for the various III-V materials. Only in the case of Si have high-resolution TEM studies been carried out, as discussed above, and these give somewhat smaller values than those inferred from the mobility data. In Figure 3.9, we show the mobility obtained for an oxide-InAs interface. The rollover at very low inversion density is that expected for Coulomb scattering and the general fit for higher density follows that expected for surface roughness scattering. The parameters used for this latter scattering are $\Delta = 1.5$ nm and $L = 2.9$ nm.

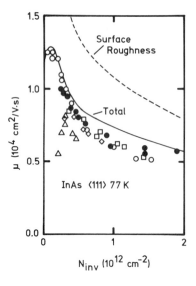

FIGURE 3.9. A plot of the theoretical and experimental mobility in an inversion layer at the surface of InAs at 77 K.

In the case of the Si–SiO$_2$ system found in MOSFETs at room temperature, however, the parameters found from high-resolution TEM (HRTEM) studies point out that the limiting mobility due to surface roughness scattering is much too high to contribute much over most of the operating range of these devices (room temperature and above). Moreover, the variation of the mobility with effective interface field is much too rapid with the F^2 dependence of Γ, in that the data on mobility seems to show a decrease as only the cube root of the effective field. Consequently, surface roughness scattering is found to be largely ineffective in these devices except in the very high surface field cases, where $F > 10^6$ V/cm, or at very low temperatures. This is shown in Figure 3.10. The data in the figure include the studies by Sun and Plummer[15] and Sabnis and Clemens,[16] as well as a few other submicron devices (discussed in Ferry[17]). Coulomb scattering sets in at even lower inversion densities and, consequently, at lower values of the effective field.

The present formalism for roughness scattering is weak in several respects. The expansion of the potential in terms of the quasi-continuous roughness function assumes that one may treat the fields classically at the interface. Yet the typical fit parameters that one obtains from the mobility, and from the HRTEM studies, are less than a few or a few tens of atomic spacings. This suggests that atomic fluctuations are contributing to the scattering and perhaps it is more

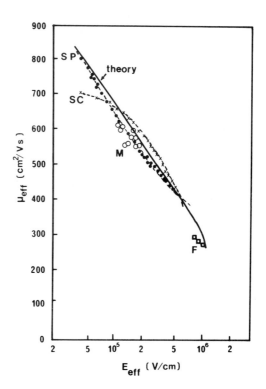

FIGURE 3.10. Comparison of the theoretical mobility, including bulk and interface phonon mobilities for a Si inversion layer at room temperature. Surface roughness scattering accounts only for the strong downturn in mobility for $F_s > 10^6$ V/cm.

appropriate to treat the problem in a manner analogous to that of the disordered or amorphous alloy; i.e., multiple scattering is very important. In compound semiconductors, the situation may be complicated even more by the interface nonstoichiometry effects. Yet, it appears that in both the Si inversion layer (at room temperature) found in submicron devices, and in the HEMT structure for GaAs, the role of scattering by interfacial defects is minimal, at least within the models described by our current understanding.

3.5. THE MOS DEVICE

In the past several years, the integration density of silicon circuits has increased steadily and dramatically. A significant part of this steady increase lies in the reduction of channel lengths of the individual devices, and this reduction has been supported by several technological developments, such as more accurate process control and fine-line lithography. However, as the channel length is reduced, many effects which heretofore were of second-order importance, become of primary importance and dominate device and circuit performance. The reduction of device size in order to achieve greater performance has followed a scaling principle, but this approach is limited by physical and practical problems.

In order to gain a better insight and understanding of device behavior for the submicron device, more general and accurate two-dimensional (at a minimum) solutions are required which are generally only obtainable by numerical techniques. The solution to the two-dimensional Poisson equation represents no conceptual difficulties, and the major physical effects are dominantly related to the manner in which the charge fluctuations and current response are coupled to the local electric field. These are formally related through the continuity equation. However, these computations often give only detailed results that verify our more direct intuitive ideas on the principal physics of operation of the device. Between our intuition and these detailed calculations, a rather good qualitative (and often quite good quantitative) understanding of the physics germane to submicron devices has been obtained, and it is this qualitative understanding that we wish to utilize here. One reason for this lies in the fact that the details of the transient response of the carriers to sudden changes in the electric field occur on such a small time scale in Si, that these effects are really not important in MOSFETs until the total channel length is reduced far below 0.1 micron at room temperature. Other factors may then dominate the device performance as well. Thus, it is quite adequate to utilize a modified one-dimensional gradual channel approximation to investigate the basic physics of the device, and this approach has been demonstrated to yield quite good agreement with experimental devices with effective gate lengths as short as 0.15 micron.[18]

A typical n-channel MOSFET structure is shown in Figure 3.11. Two n-type regions are introduced (by diffusion or ion implantation) into the p-type substrate. These regions form the drain and source contacts, and their diffusion under the actual gate metallization during processing leads to an effective channel length

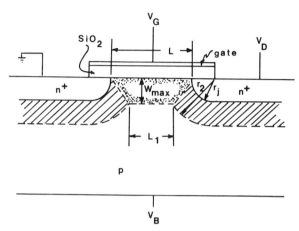

FIGURE 3.11.   The MOSFET. The conduction channel is formed by a surface inversion layer under the MOS capacitor.

much shorter than the actual gate length. In fact, effective channel lengths as short as 60 nm have been achieved by careful control of the diffusion under the gate.[19] The gate structure is essentially combined with the p-type substrate to form an MOS diode. If the gate is biased positively, a negative surface charge appears at the semiconductor surface next to the oxide. For a sufficiently large forward bias, an inversion layer of electrons is induced at the interface between the Si and the oxide. This inversion layer forms a narrow channel between the source and the drain contacts, and this inversion channel conducts current from the source to the drain contacts. The MOS structure modulates this current by varying the surface charge and, hence, the conductance of the channel itself. The depth of the quantized channel is determined by the gate voltage and the drain voltage, since it is the difference in potential between the applied gate voltage and the induced channel voltage that determines the interface charge density at any given point. We may then see that the voltage across the oxide layer and the channel is a decreasing function of x, as x increases from the source to the drain. Pinchoff occurs first at the drain end of the channel, when the drain potential is equal to the gate potential, so that no net potential appears across the interface. At this value of drain voltage, the drain current saturates in the simple model. Actually, drain current saturation is induced earlier by velocity saturation, and the model must be modified to account for this effect.

The basic mode of operation, the enhancement mode, utilizes the approach discussed above. It is possible, however, that the work-function difference and the oxide charge, or even an implanted n-type layer in the channel region, can lead to a surface channel existing when no gate voltage is applied. In this case, the device is termed a depletion-mode device since gate voltage must be used to turn the device off rather than on. The depletion-mode device is regularly used as an active load device in integrated logic circuits and is traditionally fabricated by ion implantation techniques.

3.5.1. The Gradual Channel Approximation

In order to understand the MOSFET in a relatively exact manner, it is necessary to account for the charge variation along the channel by writing differential equations. The incremental voltage drop along the channel is represented as a function of the current through the differential resistance at each point of the channel. Integration of this equation leads to a relationship for the drain current in terms of the applied voltages. We assume that the gradual channel approximation is valid; that is, the fields in the direction of current flow are much smaller than those normal to the interface. In this case, we can break the device into linear sheets normal to the current flow and decouple Poisson's equation for parallel and normal components. This assumption leads to a one-dimensional analysis for the carrier concentrations and the dimensions of the depletion region under the gate. In practice, fields normal to the interface are generally at least an order of magnitude larger than fields along the channel, except in the pinched-off portion of the channel adjacent to the drain, where the current is saturated. This is not a large concern, for we shall see that the onset of velocity saturation essentially decouples the longitudinal and transverse contributions in any case, which actually improves the approximation leading to the gradual channel approximation.

Strong inversion at the interface will occur when the minority carrier density becomes equal to the majority carrier density in the bulk, or for p-type material when n_s ($= n_{inv}$) $= p_0$. This occurs when the surface potential reaches

$$\Psi_s = -2\phi_b = 2\ln\left(\frac{N_A}{n_i}\right), \tag{3.38}$$

where N_A is the acceptor doping concentration in the p-type material, n_i is the intrinsic concentration, the reduced bulk potential is

$$\phi_b = \frac{E_F - E_{Fi}}{k_B T} \tag{3.39}$$

($\phi_b < 0$ for p-type substrates), and

$$\Psi_s = \phi_s - \phi_b = \frac{eV_s}{k_B T} \tag{3.40}$$

is the reduced surface potential. Thus, the critical turn-on voltage for the channel is

$$V_T = \frac{k_B T \Psi_s}{e} = \left(\frac{2k_B T}{e}\right)\ln\left(\frac{N_A}{n_i}\right). \tag{3.41}$$

The effects of work-function differences and oxide charges can be incorporated through the introduction of an ad hoc flatband voltage shift V_{FB}. Pinchoff occurs when the drain voltage increases to a value such that $V_G - V_D \simeq V_T$ at the drain end of the channel. If we have a silicon substrate with $N_A = 10^{16}$ cm^{-3}, we find that $V_T - V_{FB} = 0.72$ V, so that this would be the turn-on voltage in the absence of any interface states or work-function differences.

We can now calculate the current by a procedure in which the current flow through the channel is related to the drain voltage with the differential resistance. For MOSFET operation, it may be assumed that the gate voltage is such that the channel is turned on. Then, an increment of resistance at a point x along the channel is given by

$$dR = \frac{dx}{Z\mu\rho(x)}, \qquad (3.42)$$

where $\rho(x)$ is the inversion charge density at the interface, Z is the width in the lateral dimension (along the interface, but perpendicular to the direction of current flow), and μ is the effective electron mobility (see Figure 3.10). The total charge at the interface is composed of contributions from the depletion region and the inversion layer, so that

$$\rho = Q_T - Q_{depl} = \rho_T - eN_Aw, \qquad (3.43)$$

where w is the thickness (width) of the depletion layer at point x. At the onset of inversion, w becomes nearly constant, but still varies somewhat and affects the current equations. In fact, this variation is an important factor in short-channel corrections to the gradual channel approximation. The applied gate voltage V_G divides between the capacitance of the oxide layer and the surface potential V_s as

$$V_G - V_{FB} = \frac{\rho_T}{C_0 + V_s}, \qquad (3.44)$$

where we have added V_{FB} to account for extraneous charge-induced shifts of the flatband voltage. We can combine (3.43) and (3.44) to give the net inversion charge as

$$\rho = C_0(V_G - V_{FB} - V_s) - eN_Aw. \qquad (3.45)$$

The depletion width is just

$$w = \left[\frac{2\varepsilon_s}{N_Ae}\left(V_s - V_B - \frac{2\phi_bk_BT}{e}\right)\right]^{1/2}, \qquad (3.46)$$

where the surface potential V_s is evaluated at point x in the channel, and V_B is the substrate bias which may be applied. For simplicity, we write

$$V_{TB} = -\frac{2\phi_b k_B T}{e} - V_B. \tag{3.47}$$

In the following, we will replace the surface potential V_s by the explicit local potential $V(x)$ as well. We may then combine the above equations for the charge density, and

$$\rho = C_0[V_G - V'_T - V(x)] - \{2\varepsilon_s N_A e[V(x) + V_{TB}]\}^{1/2}, \tag{3.48}$$

where $V'_T = V_T + V_{FB}$. The increment of channel resistance is then just

$$dR = dx\, Z\mu\{C_0[V_G - V'_T - V(x)] - \{2\varepsilon_s N_A e[V(x) + V_{TB}]\}^{1/2}\}^{-1}. \tag{3.49}$$

For convenience, we will use the source end of the channel as the reference electrode, so that $V = 0$ at $x = 0$.

The voltage drop along the increment of channel length arises from the resistance drop in the incremental length dx, and

$$dV = I_D\, dR. \tag{3.50a}$$

Of course, the current must be constant throughout the device, so that we can just integrate the voltage and resistance over the length. Thus,

$$Z\mu C_0 V_D\left[V_G - V'_T - \frac{V_D}{2}\right] - 0.67 Z\mu(2\varepsilon_s N_A e)^{1/2}[(V_D + V_{TB})^{3/2} - V_{TB}^{3/2}] = I_D L,$$

or

$$I_D = \frac{Z\mu C_0}{L}\left[V_G - V'_T - \frac{V_D}{2} - \frac{Q'_B}{C_0}\right]V_D, \tag{3.50b}$$

where

$$Q'_B = \frac{1}{V_D}\int Q_b(x)\, dV(x)$$

$$= 2(2\varepsilon_s N_A e)^{1/2}\frac{(V_D + V_{TB})^{3/2} - V_{TB}^{3/2}}{3 V_D} \tag{3.51}$$

is an average depletion charge along the channel, and our result (3.50b) holds only for $V_D < V_{D\,\text{sat}}$, the pinchoff voltage.

When the drain voltage is increased to a level such that the charge at the drain end of the channel is reduced to $\rho(L) = 0$, pinchoff occurs. At this point, the current saturates at $I_{D\,\text{sat}}$, and this occurs for $V_D = V_{D\,\text{sat}}$. The value for the saturation drain potential may be found from this condition on the charge by using (3.48), or

$$C_0(V_G - V'_T - V_{D\,\text{sat}}) = [2\varepsilon_s N_A e(V_{D\,\text{sat}} + V_{TB})]^{1/2},$$

or

$$V_{D\,\text{sat}} = V_G - V'_T + K^2 \left\{ 1 - \left[\frac{1 + 2(V_G + V_{TB} - V'_T)}{K^2} \right]^{1/2} \right\}, \qquad (3.52)$$

where $K = (eN_A e)^{1/2}/C_0$ is related to the average depletion charge along the channel. The saturation current is found by using (3.52) in (3.50b). Thus,

$$I_{D\,\text{sat}} = \left(\frac{Z\mu C_0}{L} \right) V_{D\,\text{sat}} \left[V_G - V'_T - \frac{Q'_B}{C_0} - \frac{V_{D\,\text{sat}}}{2} \right], \qquad (3.53)$$

where Q'_B is evaluated at $V_D = V_{D\,\text{sat}}$. For a doping level of 10^{17} cm^{-3} in Si and an oxide thickness of 20 nm, $K \approx 0.25$.

3.5.2. Field-Dependent Mobility

In the above discussion of the gradual channel approximation, it was observed that the drain current, and hence the transconductance, was a function of the effective mobility of the electrons in the channel. In general, the effective mobility is reduced in these devices when compared with the mobility of the bulk material. The results of studies of the inversion-layer mobility suggest that additional scattering processes, or the constraints on the carriers due to the quantization of the carriers in the channel, dominate the mobility and lead to a reduction in its value at high-oxide fields. This was shown already in Figure 3.10. The falloff of the mobility, with oxide fields, has been suggested to follow a variety of voltage-dependent functional forms. There is no clear reason to accept one form over the other at the present time, although a power-law behavior has some credence.[17] One of the more general forms, which is readily amenable to incorporation into the gradual channel approximation, takes the form

$$\mu = \mu_0[1 - \theta(V_G - V'_T)]^{-1}, \qquad (3.54)$$

although this is inexact inside the MOSFET due to the variation in surface potential along the channel, and θ is a constant that is adjusted to give a reasonable fit to the mobility variation shown in Figure 3.10. Since the mobility varies with the *field* rather than the potential, the parameter θ depends on the oxide thickness of the actual device and is not a universal constant as may be suspected.

There is still another field dependence of the mobility, and this is due to the channel (source-drain) field which causes heating of the carriers in the channel. The relevance of hot carriers to the characteristics of MOSFETs became clear when the typical size of the devices began to become small. For example, a device of 1.0-micron channel length operating at 5 V has an average field of 50 kV/cm in the channel, whereas a device of 0.1-micron channel length operating at 1 V has an average field of 100 kV/cm, and these fields can heat the electrons far above thermal equilibrium. There is ample evidence that these effects are significant in today's devices. In comparison with bulk semiconductors, hot-electron effects in the channel seem less in magnitude and occur at high electric fields, although the resultant saturated velocity v_s is not considerably reduced. This is shown in Figure 3.12 for electrons in a Si inversion layer.[20] It is currently felt that the velocity saturation is similar to that for bulk Si, that it is dominated by bulk intervalley phonons, and that the weaker hot-electron effects arise primarily due to the reduced interface mobility. There is, however, some indication in the optical excitation data of Cooper and Nelson[21] that the saturated velocity in the inversion layer may be slightly lower than in the bulk, due probably to the presence of interfacial phonon modes that lead to additional scattering mechanisms, but this is still speculative.

The variation of the carrier mobility with the high channel field can be modeled by a formula similar to that used for oxide fields, as

$$\mu = \mu_0 \left[1 + \frac{\mu_0}{v_s} \frac{dV}{dx} \right]^{-1}. \tag{3.55}$$

Equations (3.54) and (3.55) may be combined to achieve a single expression for the mobility, but the result is not particularly amenable to inclusion within the general development of the gradual channel approximation. Further approximations must be made on the integrations, particularly for the oxide-field contribution. One simple result, however, gives the channel current as

$$I_D = \frac{I_{D_0}}{1 + \theta(V_G - V_T') + \mu_0 V_D / v_s L}, \tag{3.56}$$

where I_{D_0} is given by the gradual channel result (3.50) or (3.53). In the latter case, V_D is replaced by $V_{D\,\text{sat}}$ in (3.56). It is clear from this expression that for sufficiently large drain bias, the velocity becomes saturated in the channel and no further increase in current occurs. In this case, the gate voltage simply modulates the channel conductance in a linear fashion.

The role of velocity saturation can be explored somewhat further. In the general case of a constant mobility, the increasing field along the channel creates an increasing velocity which counteracts the decreasing carrier density. Thus, the total current through the channel can be maintained at a constant level. On the other hand, when the velocity saturates, the variation in density along the channel must be changed from that of the gradual channel approximation. The gradual channel approximation is based on quasi-equilibrium carrier statistics. With velocity saturation present, the carrier density must increase over that expected in quasi-equilibrium. Since the product of vn remains constant, in order to have a uniform current along the channel, saturation of the velocity induces a nonequilibrium density in the channel. Solutions for the details of this nonequilibrium density are quite complex, but we can rely upon the fact that the low-field region dominates the actual current. In this regard, the region near the source, for which the velocity is not saturated, serves as the metering region that sets the current level in the device. It is for this reason that the mobility in the region near the source, as indicated in (3.56), appears as the dominant area for reduced mobility.

3.5.3. Short-Channel Effects

In the above treatment, we note that the drain current depends on the actual channel length L. However, this length changes during variation of the drain bias due to depletion region spreading into the channel. For increases of V_D above the saturation value, the end of the ohmic channel (the pinchoff point) moves away from the drain toward the source. Although this effect is reduced in devices in which current saturation actually arises from velocity saturation, it remains a cause of the actual increase of output conductance in the saturation regime. One model that has been utilized is to replace L by L', which is reduced from L by the increase in the depletion region assuming a simple Schottky-type quadratic dependence. This model, however, overestimates the channel shortening, primarily because of the dominance of velocity saturation in the channel in limiting the drain current.

A second, and much more important, short-channel effect arises from the sharing of the depletion charge between the gate and the drain regions. This is shown in Figure 3.11, where the depletion charge introduced by the gate potential is taken to lie in the trapezoidal area. This is less than that in long-channel results because of the trapezoidal rather than rectangular area. The size of the depletion charge is then given by

$$\rho_{depl} = \frac{ewN_A(L + L_1)}{2}, \tag{3.57}$$

where L_1 is the length on the substrate side of the depletion region. For long channels, L_1 approaches L in length and the short-channel effect disappears. For short channels however, L_1 can become quite small and the effect is large. In fact, punch-through actually occurs first in the substrate region as $L_1 \rightarrow 0$, and subthreshold currents can also be found at this point. The drain depletion width, in the corner region, is given by

$$r_2 - r_j = \left(\frac{2\varepsilon_s V_{DB}}{eN_A}\right)^{1/2}, \tag{3.58}$$

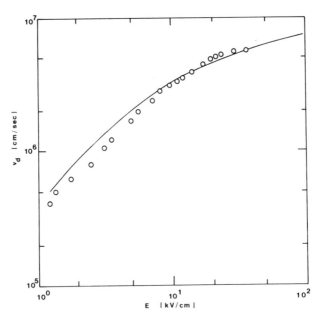

FIGURE 3.12. The typical drift velocity for electrons in a Si(100) surface inversion layer. The solid curve is a theoretical fit, while the dots are typical of data obtained by a variety of investigators.

where $V_{DB} = V_D - V_B + V_{bi}$, V_{bi} is the built-in junction potential, r_2 is the drain depletion region boundary radius, and r_j is the drain metallurgical junction radius. A simple geometrical argument then gives

$$\frac{\rho_{\text{depl}}}{\rho_L} = 1 - \frac{r_j}{L}(g-1) = f, \tag{3.59}$$

where ρ_L is the long-channel value, and

$$g^2 = 1 + \frac{2}{r_j}\left(\frac{2\varepsilon_s V_{DB}}{eN_A}\right)^{1/2} + \frac{1}{r_j^2}\left\{\frac{2\varepsilon_s V_{DB}}{eN_A} - \frac{2\varepsilon(V_s - V_B)}{eN_A}\right\}. \tag{3.60}$$

The parameter f is a unique function of the device geometry and material characteristics. As such, it actually affects the effective threshold voltage for the channel (it reduces the amount of depletion charge that must be switched), and

$$V_T = -\frac{2\phi_b k_B T}{e} + V_{FB} + \frac{f}{C_0}\{2\varepsilon_s N_A e[V - V_{TB}]\}^{1/2}, \tag{3.61}$$

where $V = V_D$ at the drain end of the channel. The presence of the charge sharing leads to a reduction in the threshold voltage in short channels, and this is shown in Figure 3.13.

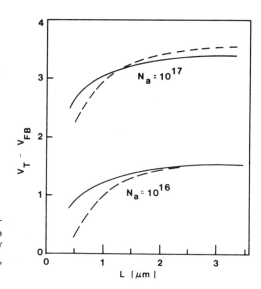

FIGURE 3.13.   Reduction in threshold voltage in a short-channel device according to the theory in the text. A substrate bias of -3 V is assumed. The solid curves are for $V_D = 0$, and the dashed curves are for $V_D = 3$ V.

In modern submicron MOSFETs, a number of fabrication techniques are used to minimize much of the short-channel effects. First, channel implantations are often used to limit the range of the depletion region and to adjust the threshold voltage of the device to a controlled, and uniform, level across the wafer. These implantations place highly doped layers just at the line marked L_1 in Figure 3.12, and they actually serve to significantly reduce the effects of the above short-channel effects. Indeed, by careful attention to details, it has been possible to make MOSFETs with effective gate lengths as short as 0.15 micron that still obey the gradual channel approximation, albeit with significant velocity saturation effects.[18]

3.6. THE HIGH-ELECTRON-MOBILITY TRANSISTOR

In an earlier section, we described the manner in which modulation doping creates carriers in an inversion layer at the GaAs/GaAlAs interface. The comparison of this structure to the Si/SiO$_2$ MOSFET is very striking, in that the principles of operation are quite the same. To make the GaAs/GaAlAs heterostructure into a FET-type structure, we need only add the gate metallization. This metal gate creates the Schottky barrier at the surface, and this barrier introduces additional band bending into the GaAlAs. The depletion region under the gate pushes carriers out of the ternary, modulation-doped region. Indeed, for sufficiently large negative bias on the gate, the conduction bands are pulled far enough above the Fermi level that even the charge channel at the interface is depleted. In this manner, the amount of charge in the inversion channel can be modulated by the potential applied to the metal gate. In this section, we want to look at the manner in which this effect is put to work in the HEMT. We will

discover that it is strikingly similar to the MOSFET, and that the basic gradual
channel equations are the same. Nevertheless, we will follow a slightly different
formulation primarily to give a different point of view and different insights into
the method of operation of both of these devices.

We shall first look at the manner of charge control under the gate itself for
an arbitrary channel potential. Then, the variation of this channel potential along
the active channel under the gate will be investigated to determine the actual
current-voltage curves of the device. To achieve the latter, we shall have to take
into account the presence of velocity saturation in the channel, which is much
more pronounced in this device, compared to the MOSFET, due to the higher
mobility of the carriers in the channel and to the lower field at which saturation
occurs in GaAs. We shall actually divide the channel into three regions to describe
different effects. Again, this approach should be compared and contrasted with
that for the MOSFET.

3.6.1. Charge Control by the Gate

In Figure 3.14, the band diagram of the heterojunction structure and the
Schottky barrier is shown. This figure is very similar to that of Figure 3.1, but in
this case we take care to illustrate the various dimensions and potentials that we
use in the equations below. It is clear that for this device to work as a transistor,
it is necessary that the two depletion regions, one from the gate and one from
the modulation doping interaction, overlap and interact with each other. For this
to occur, we must have either a sufficiently high reverse voltage on the metal gate
or the thickness of the GaAlAs layer must be sufficiently thin. Of course, a

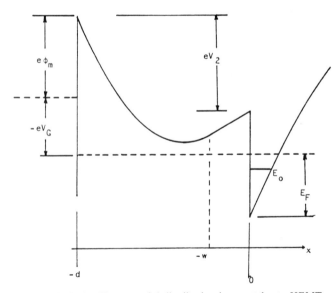

FIGURE 3.14. The potential distribution in a metal-gate HEMT.

properly designed device will trade off the thickness with the bias levels in order to achieve operation at appropriate circuit voltage levels.

In the regime where the inversion charge is controlled by the gate, the GaAlAs layer is totally depleted of charge. In this region, the potential is determined by Poisson's equation, with the doping given by

$$N_2 = N_D - N_A = 0, \qquad -w < z < 0,$$
$$= N_{2D}, \qquad z < -w. \tag{3.62}$$

We have taken the origin at the interface, where we shall also take $V_2(0) = 0$ for convenience. This latter potential is defined by a double integration of the one-dimensional Poisson equation, as

$$V_2(z) = -F_s z - \frac{e}{\varepsilon_s} \int_0^z dz' \int_0^{z'} N_2(z'') \, dz'', \tag{3.63}$$

and the voltage at the gate is

$$v_2 = -V_2(-d) = \frac{e N_{2D}}{2\varepsilon_s}(d - w)^2 - F_s d. \tag{3.64}$$

The first term on the right may be recognized as the voltage necessary to completely deplete the doping region. If we denote this voltage by V_D, we can rewrite (3.64) as

$$\varepsilon_s F_s = \frac{\varepsilon_s(V_D - v_2)}{d}. \tag{3.65}$$

In writing the voltage as zero at $z = 0$, we are, of course, referencing it to the conduction-band edge on the wide-band-gap side of the heterojunction. By examination of Fig. 3.14, we can relate this voltage to the built-in potentials as

$$v_2 = \phi_m - V_G + E_F - \Delta E_c, \tag{3.66}$$

and (3.65) then becomes

$$\varepsilon_s F_s = \frac{\varepsilon_s(V_D - \phi_m - E_F + \Delta E_c + V_G)}{d}. \tag{3.67}$$

It is fairly well established now that there is very little interface charge at the heterojunction interface, so that the field discontinuity across the interface is solely determined by the dielectric discontinuity and the inversion charge. We can therefore write the inversion and depletion charges in terms of this field, as in the MOSFET, and

$$en_s = e(N_{depl} + n_{inv}) = \frac{\varepsilon_s(V_D - \phi_m - E_F + \Delta E_c + V_G)}{d}. \tag{3.68}$$

In general, the Fermi energy is relatively small when compared with the other potentials in the system. This is not always true, but will certainly be true near channel pinchoff where there is little charge in the channel. For all practical purposes, we can ignore its variations with gate voltage (the variations arise because of the self-consistent terms in the degenerate Q2D gas). In addition, in contrast to the MOSFET, there is very little depletion charge so that almost all of the surface charge is in the inversion layer. In the MOSFET, we had to account for the variation of this depletion charge along the channel and it represented a sizable portion of the i-v characteristics. Here, however, the substrate is usually so lightly doped that this charge is quite small and can reasonably be treated as a constant (and ignored). By not concerning ourselves with these two sources of weak nonlinearities, we can then write (3.68) in the more familiar form

$$en_s = \frac{\varepsilon_s(V_G - V_T)}{d} = C_s(V_G - V_T), \tag{3.69}$$

where we have introduced the semiconductor "capacitance" and the turn-on voltage

$$V_T = \phi_m - \Delta E_c - V_D, \tag{3.70}$$

below which there is no charge in the channel. Ignoring the exact position of the Fermi energy in (3.69) and (3.70) means that we do not need to know the exact levels of the subbands in the potential well. However, near threshold this is a good approximation, but fails to be true when the channel is heavily occupied by carriers. On the other hand, in this latter situation, the Fermi level varies very little. Thus, the approximation is reasonably good at both extremes of operation. The turn-on, or threshold, voltage here is more analogous to the pinchoff voltage in a MESFET, rather than a MOSFET, but the terminology is used here because of the more general similarities in the two devices.

There is a second critical voltage in the HEMT. For a sufficiently large value of (hopefully forward) bias, the depletion region no longer punches completely through the GaAlAs layer. The gate can then no longer control the inversion charge, since the heterojunction region is in equilibrium with the modulation doping considerations given earlier in this chapter. For proper operation of the device, this critical voltage must be designed to be out of the operating region defined by the voltage swing. This second critical voltage can be found by equating the equilibrium value of the Fermi energy found in Section 3.1 with that in the above equations. This yields the "gate control" voltage, at which value the gate loses control,

$$V_{GC} = \phi_m - E_D$$
$$- \left[\left(\frac{eN_{2D}d^2}{2\varepsilon_s} \right)^{1/2} - \left(\Delta E_c - E_D - E_{F0} + \frac{eN_{2D}w^2}{2\varepsilon_s} \right)^{1/2} \right]^2. \tag{3.71}$$

Equation (3.71) represents a major limitation on the control of the inversion charge by the gate. Clearly, the equilibrium voltage is maximized when the term in the brackets is zero. The combination of dimensions, for which the bracketed term in (3.71) is zero, is plotted in Figure 3.15, where we can see that only a narrow doped region will satisfy this condition. However, such a region decreases the total capacitance of the gate and so reduces the transconductance of the device itself. For voltages above V_{GC} (a negative quantity as is V_G in general), the gate no longer influences the amount of charge in the conduction channel, and there is significant charge residing in the GaAlAs, which can also contribute to conduction and lowers the transconductance of the device. For the case in which the bracketed term is made to be zero, V_{GC} is about -0.8 V, which is primarily below the built-in Schottky barrier height. Thus, for proper designs, the normal Schottky barrier will prevent this voltage from being reached except in forward-gate-bias situations. However, in many cases this proper design is not followed, or even appreciated, so that V_{GC} is as large as -1.5–2.0 V, and a swing to low values of the gate voltage brings the operating point into this region. This is primarily in the "on" state, so that the reduction of transconductance is in the most disadvantageous region of operation.

3.6.2. The Current–Voltage Relation

In treating the *i-v* characteristics of the HEMT, we must incorporate the actual potential in the channel due to the source-drain bias. We call this potential $V(x)$, as in the last section. Then, the gradual channel approximation for the charge density along the channel becomes

$$en_s = \varepsilon_s[V_G - V_T - V(x)]. \qquad (3.72)$$

In analogy with the MOSFET, we can now write the channel current as

$$I_D = n_s ev(x)Z, \qquad (3.73)$$

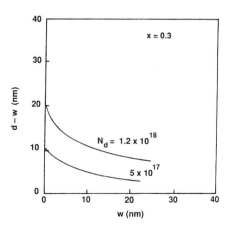

FIGURE 3.15. Parameter agreements required to maximize the equilibrium voltage at which the gate loses control of the inversion charge.

where Z is the gate width and $v(x)$ is the velocity in the channel. As discussed previously, the HEMT is much more susceptible to velocity saturation limitations than the MOSFET, and we will build these in right from the start. We divide the channel into three regions: (1) a linear region where the electric field $F < F_s$, where the latter field is the saturation field defined by the saturation velocity through $v_s = \mu F_s$; (2) a saturated region where the velocity is saturated at v_s; and (3) a pinched-off region of the channel. It is possible that some of these regions will overlap, but it is only in the first two regions that we can solve for the detailed current–voltage relationship of the device.

In region 1, the velocity rises in the channel as the electric field is increased in response to the decrease of the charge along the channel. The latter decreases due to the rise in potential along the channel and consequent decrease in the right-hand side of (3.72). The resulting increase in field and potential go together to ensure a constant current exists along the channel. In this region, the linearity of mobility accounts for the increasing velocity with decreasing density. We can now write the current by combining the above two equations to give

$$I_D = \mu Z C_s [V_G - V_T - V(x)] \frac{dV}{dx} + Z D C_s \frac{dV}{dx}, \tag{3.74}$$

where we have explicitly replaced the field by the derivative of the potential along the channel. The second term in (3.74) represents the diffusion contribution to the current with the density given by (3.72). We now integrate this, just as in the previous section, over a distance from the source to a point $x = s$, at which position $F = F_s$, the field at which the velocity saturates. This gives, with the Einstein relation $D = k_B T \mu / e = V_t / \mu$,

$$I_D = \frac{\mu Z C_s}{s} \left[V_G - V_T + V_t - V_s - \frac{V_1}{2} \right] V_1, \tag{3.75}$$

where V_s is the potential at the source end of the channel and V_1 is the potential at $x = s$, with respect to the source end of the channel. If we take $V_s = I_D R_s$, we can explicitly include the effect of source resistance, and

$$I_D = \frac{\mu Z C_s V_1}{s(1 + V_1 R_s \mu C_s Z / s)} \left(V_G - V_T + V_t - \frac{V_1}{2} \right), \tag{3.76}$$

where we have rearranged the result to give a single value for the drain current. In treating the above equations, we have included explicitly the diffusion current, although this is not normally done. However, it may be an important part of the total current, and is significant in regions 2 and 3. In addition, we have explicitly included the source resistance, which was not done in the MOSFET treatment but may easily be incorporated in a similar fashion.

In region 2, the velocity is saturated, so it cannot continue to increase as the density decreases along the channel with continuing increases in $V(x)$. Consequently, the current must be kept constant by two effects. First, a larger proportion of the total current is carried by diffusion current; second, an increase of the carrier density (over the equilibrium value determined by the gate voltage) occurs. This latter effect clearly shows up in full two-dimensional models of all devices in which velocity saturation occurs. However, the effect is numerically small when considering the effect on the channel potential. We can therefore actually use only the former effect to determine the current. The current is now given by

$$I_D = Zn_s e v_s - ZeD' \frac{dn_s}{dx}$$

$$= ZC_s v_s [V_G - V_T - V(x)] + ZD'C_s \frac{dV}{dx}, \tag{3.77}$$

where D' is the diffusion parameter in the saturated region. We cannot just assume that D' is a constant saturated value, as is the velocity, since this does not relate properly to the mobility (as we see in later chapters, there is no true Einstein relationship in hot-electron conditions). A constant D' would actually imply a carrier temperature increasing linearly with the electric field. In fact, most hot-electron studies show that the temperature increases quadratically with the electric field. This would give a linearly increasing D' (with field), except that most hot-electron studies also show that D' actually decreases with increasing field in the saturation regime. Thus, great care must be taken in trying to relate velocity (or mobility) and the diffusion parameter in the hot-carrier, far-from-equilibrium regime. We shall avoid this confusion, and take a geometric mean between a constant value and a linear increasing value expected for velocity saturation, as

$$D' \sim v_s \left(\frac{\mu}{v_s}\right)^{1/2} \left(\frac{k_B T}{e}\right) \left(\frac{dV}{dx}\right)^{-1/2} = v_s \beta \left(\frac{dV}{dx}\right)^{-1/2}. \tag{3.78}$$

We may then write the current as

$$I_D = ZC_s v_s [V_G - V_T - V(x)] + ZC_s v_s \beta \left(\frac{dV}{dx}\right)^{1/2},$$

and for the pinchoff condition $V(x) = V_G - V_T$ at $x = p$,

$$I_D = ZC_s v_s \left[V_G - V_T - V_1 + \frac{\beta^2}{p - s} \right] \tag{3.79}$$

is found for the drain current. However, this value must be the same as obtained at the point $x = s$, since the current is conserved through the channel. Thus, we may equate (3.79) with (3.76) in order to find a relationship between the values for s and p, as

$$\frac{p}{s} = 1 + 2\left(\frac{V_t}{V_1}\right)G, \tag{3.80}$$

where

$$G = \frac{3(1 + \alpha)}{3 + 2\alpha - (\alpha/3V_t)(V_G - V_T)}, \qquad \alpha = \frac{v_s R_s C_s Z}{2} \tag{3.81}$$

is a correction factor for the source resistance ($G = 1$ for $R_s = 0$). In actual usage, the pinchoff point is found from the voltage along the channel to achieve this effect. This determines p, and (3.80) can be used to find s, which in turn determines the current. In most cases, however, $V_1 = E_s s/2$ (the velocity increases linearly), and we can simplify the results. Although the linear portion (region 1) determines the total current in the device, the resultant equation for the current depends only on the gate potential as

$$I_D = \frac{ZC_s v_s(V_G - V_T + V_t)}{2 + v_s R_s Z C_s}. \tag{3.82}$$

More complicated treatments have been introduced in the literature, but the end result is the same. Equation (3.82), derived for the linear region, is the actual current that is valid throughout the device, regardless of the particular region in which the current is flowing. We reach the important conclusion from this that the HEMT device characteristics are dominated by the low-field, linear regime of the device. Even in the saturated velocity and pinched-off regions of the device, the current level is set by that metered into the channel by the linear region. It is clear that the performance of the low-field region is helped by improvements in the mobility and this is achieved by modulation doping.

In region 3, the current is carried entirely by diffusion of carriers injected into the space-charge region from the pinched off channel. The current is still determined by the previous two regions, and the space charge serves merely as a resistive layer to drop the remaining drain potential. For a 1.0-μm-gate-length device, with a channel mobility of 8000 cm^2/V-s and an effective saturation velocity of 1.5×10^7 cm/s, we find that $V_1 = 0.094$ V. This is the maximum value that this parameter can take, as shorter values of s will arise as the drain current is increased in the device, in response to the increasing drain potential. Thus, this potential is comparable to the thermal voltage which appears in the above equations and may even be smaller over much of the operating regime. As a result, the current level is primarily determined by the gate voltage and the saturation velocity, even though it is region 1 that is metering the current into the channel. The high mobility serves only to make the metering more efficient and to reduce the effective source resistance.

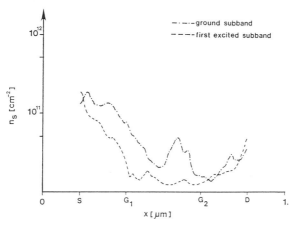

FIGURE 3.16. Plot of the charge density in the channel of a HEMT from a full two-dimensional solution of Poisson's equation.

3.6.3. A Two-Dimensional Result

Full two-dimensional numerical solutions to Poisson's equation for the HEMT are not very plentiful at this time. One has been carried out, however, in a case that treats the transport by a full-ensemble Monte Carlo method.[22] The primary results are in keeping with the conclusions of the above section. In Figure 3.16, we show a plot of the charge density in the channel and adjacent regions. It is apparent from this picture that there is an increase in charge in region 2 (the saturated velocity region) over the equilibrium value expected there. Moreover, it is also apparent that there is significant charge in the GaAlAs itself.

Another detailed calculation of the switching of the device has been carried out by Hess.[23] This result yields quite interesting results for the turn-on of the device. Prior to switching the gate voltage, channel inversion charge exists in the region between the source and gate and the region between the gate and the drain. When the gate voltage is switched, turning on the channel, the initial charges flowing into the channel under the gate come from the charge between the source and the gate, depleting this region. This depletion remains through much of the switching cycle. It is clear from this result that the initial conjecture of Kroemer[24] on current metering by the source–gate space-charge potential is correct. This depletion between the source and gate clearly shows the importance of the source resistance in the operation of the HEMT. It is strongly suggested from these results that self-aligned structures, which minimize the spacing between the gate and the source as in a MOSFET, are important to overcome these effects.

3.7. REAL-SPACE TRANSFER

In all inversion-layer devices, whether these are the GaAs HEMT or the Si MOSFET, there is a problem with scattering of the carriers out of the channel at high drain bias. In these cases, the carriers are heated by the drain field, which also leads to the saturated velocity of the carriers, and can gain sufficient energy to surmount the potential barrier between the channel and the "insulator." In the GaAs HEMT, this leads to carriers residing in the GaAlAs. These carriers contribute to the total source–drain current, but they have a much smaller mobility. Hence, the transconductance, and the total conductance, is reduced by these carriers. In the case of Si, the carriers are predominantly trapped at localized centers in the SiO_2 near the interface. This trapping provides for localized charge near the interface, which in turn causes a shift of the threshold voltage for turning on the device. This shift in threshold voltage is a cumulative effect over time and can thus lead to a significant degradation in operating characteristics of the MOSFET. Very little trapping occurs in the HEMT, since the bias levels in this device keep the defect (impurity) levels empty except at significant levels of forward gate bias.

At first thought, one would expect that the threshold voltage on the drain, for which injection of charge over the barrier from the channel occurs, would be the discontinuity in the conduction-band edge, namely ΔE_c in the heterojunction case and $\simeq 3.2$ V in the Si–SiO_2 case. In fact, the level is somewhat lower than this. Conflicting processes come into play in this, but we can argue through a consistent idea for the value of the threshold level.

Although the injection will primarily occur near the drain end of the channel, where the carriers achieve their maximum energy of eV_D, the actual barrier energy over which they must be injected is strongly affected by the dynamics near the source end of the channel. At the source end of the channel, where the inversion density is the greatest in the gradual channel approximation, the carriers actually reside *above the bottom of the conduction band*. This is because of the two-dimensional nature of the quantized electrons. In the 2DEG, the bottom of the lowest subband is raised above the bottom of the conduction band by the quantization energy E_0. In addition, for strong inversion, the Fermi level lies above this energy. The most dynamic carriers, those which will reach the highest energies and velocities, will come from those carriers which reside near the Fermi energy. As we move along the channel from the source toward the drain, the inversion density is reduced by the channel voltage until we reach the velocity saturation point. However, V_1 is a small quantity in comparison with the gate voltage, so that the density is reduced only a few percent. This is reflected in the reduction of E_0 and E_F, but again this is only a few percent. Beyond the velocity saturation point, the actual density in the channel is not further reduced because of the requirement that the total current density $J = nev$ be a constant with position down the remainder of the channel. Since $v = v_s$ is constant, we must also have the density n be a constant. On the other hand, the increasing channel potential would try to lead to a reduced density within the gradual channel approximation. We are therefore led to the conclusion that there is a very sizable, nonequilibrium density that appears as a significant accumulation layer in the velocity-saturated regime. We must therefore consider that there is a quasi-Fermi level that describes this nonequilibrium density, and that this level differs from

the actual Fermi level by the quantity $V(x) - V_1$. The quasi-Fermi level maintains
the relative position of the dynamic carrier energy levels at much the same point
as at the velocity saturation point in the channel. Thus, the actual barrier that
must be overcome for injection into the oxide, or the GaAlAs insulator, is not
the band discontinuity but the reduced quantity $V_b \simeq \Delta E_c - E_F$, where E_F is
measured from the conduction-band edge, at the velocity saturation point. For
a Si MOSFET, with a 15-nm-gate oxide, 5 V on the gate, and a substrate acceptor
concentration of 3×10^{17} cm^{-3}, we find that E_F is about 0.14 V. On the scale of
the 3.2-V difference in the conduction-band edges, this is not a large amount. On
the other hand, there are experimental data that injection actually begins for
drain potentials as low as 2.5 V,[25] which suggests that we must even begin to
consider the thermal spread in the distribution. Here, the carriers that are actually
injected into the oxide are a small percentage of the total and predominantly
come from the thermal tail of the distribution. The amount is large, even at 2.5 V,
because the thermal tail is not at 300 K but is at an electron temperature characteristic of the very hot carriers in the saturated velocity regime. This nonequilibrium temperature may be 10–20 times the lattice temperature, as we see in later
chapters.

In the case of the HEMT, the role played by the quantization may be much
larger. In Figure 3.17, we plot the various energy levels in the GaAs/GaAlAs
heterojunction for a particular case of band offset, assuming a 60:40 split of the
band-gap difference among the conduction and valence bands. For an inversion
density of 10^{12} cm^{-2}, which is typical in these devices, E_F lies about 42 meV

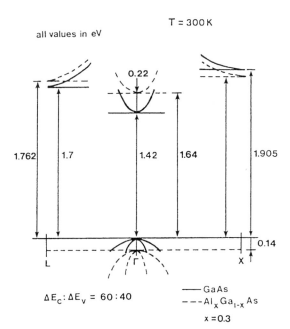

FIGURE 3.17. The positions of the
various valleys in the GaAs-GaAlAs
heterojunction.

above the bottom of the conduction band. Although this is considerably smaller than in the Si case, it is still about one sixth of the conduction-band discontinuity. In the HEMT, the situation is further complicated by the position of the L-valleys. Momentum-space transfer to these valleys by hot electrons accentuates the real-space transfer since the tunneling probability of these electrons into the equivalent GaAlAs valleys is 0.7–0.9. It is almost certain that this is a transconductance degradation effect which occurs in almost all of these devices.

REFERENCES

1. R. Dingle, H. L. Stormer, A. C. Gossard, and W. Wiegmann, *Appl. Phys. Lett.* **33**, 665 (1978).
2. T. Mimura, S. Hiyamizu, T. Fujii, and K. Nanbu, *J. Appl. Phys.* **19**, 1225 (1980).
3. D. Delagebeaudeuf, P. Delescluse, P. Etienne, M. Laviron, J. Chaplart, and N. T. Linh, *Electron. Lett.* **16**, 667 (1980).
4. T. Ando, F. Stern, and A. B. Fowler, *Rev. Mod. Phys.* **54**, 437 (1982).
5. J. Blakemore, *Semiconductor Statistics*, Pergamon Press, New York (1962).
6. F. Herman and I. P. Batra, in: *Physics of SiO₂ and Its Interfaces* (S. T. Pantelides, ed.), p. 333, Pergamon Press, New York (1978).
7. T. Kunjunny and D. K. Ferry, *Phys. Rev. B* **24**, 4593 (1981).
8. C. R. Helms, Y. E. Strausser, and W. E. Spicer, *Appl. Phys. Lett.* **33**, 767 (1978).
9. J. Wager and C. W. Wilmsen, *J. Appl. Phys.* **50**, 874 (1979).
10. S. M. Goodnick, D. K. Ferry, C. W. Wilmsen, Z. Liliental, D. Fathy, and O. L. Krivanek, *Phys. Rev. B* **32**, 8171 (1985).
11. S. M. Goodnick and D. K. Ferry, in: *Physics and Chemistry of III-V Compound Semiconductor Interfaces* (C. W. Wilmsen, ed.), p. 283, Plenum Press, New York (1985).
12. F. F. Fang and W. E. Howard, *Phys. Rev. Lett.* **16**, 797 (1966).
13. G. E. Marques and L. J. Sham, *Surf. Sci.* **113**, 131 (1982).
14. F. Stern and W. E. Howard, *Phys. Rev.* **163**, 816 (1967).
15. S. C. Sun and J. D. Plummer, *IEEE Trans. Electron Dev.* **ED-27**, 1497 (1980).
16. A. G. Sabnis and J. T. Clemens, in: *Proc. 1979 Int. Electron Devices Mtg.*, p. 18, IEEE Press, New York (1979).
17. D. K. Ferry, in: *Proc. 1984 Int. Electron Devices Mtg.*, p. 605, IEEE Press, New York (1984).
18. W. Fichtner, R. K. Watts, D. B. Fraser, R. L. Johnston, and S. M. Sze, *IEEE Electron Dev. Lett.* **EDL-3**, 412 (1982).
19. S. Y. Chou, D. A. Antoniadis, and H. I. Smith, in: *Proc. 1985 Int. Electron Devices Mtg.*, p. 562, IEEE Press, New York (1985).
20. D. K. Ferry, *Phys. Rev. B* **14**, 1605 (1976).
21. J. A. Cooper, Jr. and D. F. Nelson, *IEEE Trans. Electron Dev.* **ED-27**, 2179 (1980).
22. U. Ravaioli and D. K. Ferry, *IEEE Trans. Electron Dev.* **ED-33**, 677 (1986).
23. K. Hess, private communication.
24. H. Kroemer, unpublished notes.
25. L. A. Akers, M. A. Holly, and C. Lund, in: *Proc. 1985 Int. Electron Devices Mtg.*, p. 80, IEEE Press, New York (1984) and subsequent private communication.

4

Semiclassical Carrier Transport Models

4.1. CARRIER TRANSPORT

In this chapter we begin our study of charge and carrier transport through small semiconductor devices. We are confronted with a rich variety of possible transport models. Each of these models makes a different set of simplifying approximations concerning the physics of carrier transport and therefore has its own domain of validity. The overall problem which these various models solve in various ways is the response of a cloud or ensemble of mobile charges, embedded in a media which subjects them to a statistically varied set of scattering events and to a set of externally applied fields. The detail needed in the statistical description of the carrier distribution in phase space generally guides the choice of a transport model. At one extreme, we try to preserve all of the statistical information and solve for a distribution function or use Monte Carlo methods. At the other extreme, we decide that only the mean values are needed and generally solve for the carrier density. (Of particular interest here are a set of models in which hot-carrier phenomena are ignored while nonequilibrium densities are retained.) Lying between these two extremes are a variety of models in which one retains hot-carrier physics in the formulation of carrier current densities as well as nonequilibrium concentrations. If we are interested in noise calculations, we must retain some information concerning statistical fluctuations about the mean value for one or more parameters, and questions then arise as to how one separates the mean motion from the overall motion and how one preserves or represents the fluctuations.

Additional choices exist. Do these carriers interact? Obviously, as they are charged, they must, but is this interaction important? How do the time and space scales of the problem compare with the scattering rates? Will our carriers experience few or many scattering events on the time and space scales of interest? In this chapter we assume that the carrier-carrier interaction is not important and, usually, that the time scales are long with respect to the scattering rates. We progress from the simplest (and most commonly used) models to more complex models.

4.2. DRIFT-DIFFUSION LAWS AND BROWNIAN MOTION

Great use is made inside semiconductor electronics and transport theory of ideas and techniques originally developed for the study of Brownian motion and other diffusion processes. This is the connection by which the Einstein relation enters the field of solid-state electronics, a fact which certainly would have been no surprise to Einstein. In 1906, he suggested that one vehicle for experimentally testing his theoretical conclusions concerning Brownian motion would be the observation of charge transport through a conductor.[1] Classical Brownian motion is the motion of particles in a viscous fluid, and the semiconductor analogy is an ensemble of electrons or holes moving through a phonon bath. These particles (charge carriers) scatter or interact in a random fashion with the constituent molecules (phonons) of the viscous fluid, and as a result they have a random component to their velocity. The main thread of analysis we follow here subdivides the underlying particle motion into average and fluctuating components. We are primarily interested in determining the average motion, and we initially study a drift-diffusion equation for this average motion. This type of analysis works well in the case where the particles experience many scattering events in the time frame of interest. Later we study the short-time case and the relationship of the fluctuating components to noise.

This splitting of the carrier velocity into an average component and a fluctuating component is closely related but not identical to the most common model of carrier transport in a semiconductor device. There we subdivide the average current or particle flux into a drift component and a diffusion component; that is,

$$\mathbf{J}_n = -en\mathbf{v}_n - e\mathbf{D}_n \cdot \nabla n, \tag{4.1}$$

where $\mathbf{v}_n$ is the carrier drift-velocity vector and $\mathbf{D}_n$ is the carrier diffusion tensor, a quantity usually represented as a simple scalar in device modeling. Another common variant would place the diffusion coefficient inside the spatial gradient, and, indeed, at times arguments break out over where the diffusion coefficient should go. Here, however, we view (4.1) (or the alternative form with $\nabla(D_n n)$) as a phenomenological equation. While it is possible to derive an equation of this form from more fundamental equations, such as the Boltzmann equation (a derivation we discuss later), some simplifying assumptions must be made. The difficulty is that different sets of assumptions can be made and used to yield either (4.1) or the alternative (or even other related forms). Hence, we view this as an equation which provides an excellent phenomenological approach for device modeling under appropriate circumstances. We choose to place the diffusion coefficient outside the spatial gradient because it allows us to use a diffusion tensor in anisotropic situations, whereas the other variant cannot use a tensor form for D_n. For cases where the diffusion process is isotropic, a scalar diffusion coefficient is used. Then there effectively will be no difference between the two forms unless the gradient of the diffusion coefficient is a sizable fraction of v_n.

We would like to explore the relationship between (4.1) and a subdivision of the carrier's Brownian motion into average and fluctuating components. We start by dividing $\mathbf{J}_n$ by the mobile charge density $(-en)$ to obtain the mean local velocity

$$\mathbf{v}_{n,\text{ave}} = \mathbf{v}_n + \frac{\mathbf{D}_n \cdot \nabla n}{n}. \tag{4.2}$$

The result will not equal the drift velocity unless ∇n is zero. The diffusion current term of (4.1) therefore does not represent a fluctuating velocity, but instead shows that in the presence of a carrier concentration gradient we will have a local contribution to the average velocity. This contribution is driven by the velocity fluctuations. This point is sometimes forgotten and plays an important role in comparing drift-diffusion models with other approaches. In particular, one cannot use a different transport model and then compare the mean velocity associated with the mean current in that model with the drift velocity of a drift-diffusion model.

From (4.2) we can define an effective diffusion velocity as

$$\mathbf{v}_{\text{diff}} = \frac{\mathbf{D}_n \cdot \nabla n}{n} = \frac{\mathbf{D}_n \cdot \nabla(n/n_i)}{n/n_i}, \tag{4.3}$$

where we have assumed that the intrinsic carrier density, n_i, is not a function of position. This diffusion velocity illustrates that (4.1) can lead to unphysical results in the presence of a sufficiently large carrier density gradient even if all of the normal arguments concerning time scales related to scattering processes are satisfied. The velocity $\mathbf{v}_{\text{diff}}$ is not bounded and can exceed the band structure limit on group velocity (or the speed of light for that matter) when the ratio $\nabla n/n$ becomes large. The intrinsic carrier density was introduced to provide a more transparent comparison of (4.1) with another widely used formulation of carrier transport in which quasi-Fermi levels are used. The electron quasi-Fermi level φ_n is defined by[2]

$$\varphi_n = \Phi - \frac{k_B T_L}{e} \ln \frac{n}{n_i}, \tag{4.4}$$

where k_B is Boltzmann's constant and T_L is the lattice temperature. In (4.4), $\Phi = -E_i/e$ is the total electric potential, where E_i is the intrinsic Fermi level. Using this definition, we then express the carrier density as

$$n = n_i \exp\left[\frac{e(\Phi - \varphi_n)}{k_B T_L}\right]. \tag{4.5}$$

In the quasi-Fermi level formulation the drift-diffusion law (4.1) is replaced by

$$\mathbf{J}_n = -e\mu_n n \nabla \varphi_n, \tag{4.6}$$

where μ_n is the electron chordal mobility. Diffusion currents are still retained here, although the diffusion coefficient has disappeared. If we substitute (4.4) into (4.6), we obtain a gradient of the natural logarithm of the normalized carrier density n/n_i. This gradient can be replaced through use of (4.3). Then using the Einstein relation, which is normally assumed in the development of the quasi-Fermi levels, we obtain

$$\mathbf{J}_n = e\mu_n n \mathbf{F} + e n \mathbf{v}_{\text{diff}}. \tag{4.7}$$

The quasi-Fermi level therefore describes a total potential function for the electron current flow, which includes the normal electric potential and a "diffusive" potential as well. The role of the Einstein relation here is crucial. When quasi-Fermi levels are used, one is assuming that while the electron and hole densities may be out of equilibrium with respect to each other, the electrons (or holes) remain in equilibrium with the lattice or the phonon bath.[3,4] The extension of the quasi-Fermi level concept to the realm of hot-carrier phenomena where this equilibrium is destroyed is not straightforward, although many attempt it by simply using field-dependent mobilities. The central conceptual dilemma, however, is not the field dependency of the mobility but the utilization of the Einstein relation (which applies only in equilibrium) in the treatment of the diffusive contributions to carrier transport in a nonequilibrium situation.

What the diffusion term of (4.1) most directly describes is how velocity fluctuations are translated into position fluctuations over a period of time. This translation in fact plays a crucial role in the use of time-of-flight measurements of the spread of a carrier pulse in the measurement of the diffusion coefficient. The analysis which underlies this experimental technique proceeds by substituting (4.1) into the carrier continuity equation and solving the resulting diffusion equation:

$$\frac{\partial n}{\partial t} = \frac{1}{e} \nabla \cdot (-e n \mathbf{v}_n - e \mathbf{D}_n \cdot \nabla n). \tag{4.8}$$

For a delta function initial condition, the diffusion equation (4.8) has the solution

$$n(x, y, z, t) = \frac{n_0}{8(\pi t)^{3/2} D_{n,\text{per}} \sqrt{D_{n,\text{par}}}} \exp\left\{-\left[\frac{x^2 + y^2}{4 D_{n,\text{per}} t} + \frac{(z - v_n t)^2}{4 D_{n,\text{par}} t}\right]\right\}, \tag{4.9}$$

where $D_{n,\text{per}}$ and $D_{n,\text{par}}$ are the components of the diffusion tensor for diffusion perpendicular and parallel to the direction of the field, and n_0 is the actual number of electrons found in the initial delta function. We can view a carrier density, such as the Gaussian function of (4.9), as a strangely normalized probability function for the location of a single carrier in real space. Using this viewpoint, we can calculate the mean square position fluctuations about directions parallel and perpendicular to the direction of the drift term. These are (in the long-time limit)

$$\langle \Delta x^2 \rangle = 2D_{n,\text{per}}t \qquad (4.10)$$

and

$$\langle \Delta z^2 \rangle = \langle (z - v_n t)^2 \rangle = 2D_{n,\text{par}}t. \qquad (4.11)$$

These formulas are sometimes referred to as the Einstein formulas, because they constitute one of his more important observations about Brownian motion. *These results are of limited validity for short times,* a fact which is developed only by exploring the connection of macroscopic diffusion with microscopic transport.

While the connection between the drift velocity and a more microscopic view is relatively simple, respresenting the diffusion coefficient in terms of some more microscopic model is more difficult. The reason for this difficulty is that, while the diffusion coefficient describes position fluctuations, the microscopic models tend to ignore position and deal with velocity fluctuations. However, we can close this gap by using a subdivision of the individual microscopic carrier velocities into average and fluctuating velocity components; that is,

$$v_i(t) = v_n(t) + \Delta v_i(t), \qquad (4.12)$$

where $v_i(t)$ is the velocity of carrier i at time t, v_n is the time-dependent ensemble average velocity and Δv_i is the time-dependent fluctuation of v_i away from v_n. In this process we gain some insight into the difficulties associated with applying (4.9)–(4.11) on very short time scales. At the end of the chapter, we explore the real problem, which lies not so much with these equations but with the use of the diffusion equation (4.8) on very short time scales.

We begin to connect (4.12) with diffusion by integrating this equation to obtain the time-dependent position of a single carrier,

$$x_i(t) = x_n(t) + \Delta x_i(t), \qquad (4.13)$$

where x_n is the integral of the average velocity and Δx_i is the integral of the velocity fluctuation. The form of the Einstein formulas (4.10) and (4.11) suggests that we can obtain the diffusion coefficient from this approach by examining the mean square position fluctuations. These are expressed as

$$\langle \Delta x^2 \rangle = \left\langle \int_0^t \Delta v_i(u)\, du \int_0^t \Delta v_i(w)\, dw \right\rangle, \qquad (4.14)$$

where $\langle \ \rangle$ represents an ensemble average. We can rewrite (4.14) as

$$\langle \Delta x^2 \rangle = \int_0^t \int_0^t \langle \Delta v_i(u) \, \Delta v_i(w) \rangle \, du \, dw. \tag{4.15}$$

The integrand of our double integral now is the two-time velocity fluctuation autocorrelation function

$$\varphi_{\Delta v}(u, w) = \langle \Delta v_i(u) \, \Delta v_i(w) \rangle. \tag{4.16}$$

We now make several assumptions, all of which are applicable to the steady-state case. First, we assume that $\varphi_{\Delta v}(u, w)$ depends only on the difference $\theta = u - w$ and not on the actual values of u or w. We then rewrite our double integral as one integral over u and a second integral over θ:

$$\langle \Delta x(t)^2 \rangle = \int_0^t du \int_{u-t}^u \varphi_{\Delta v}(u, u - \theta) \, d\theta = \int_0^t du \int_{u-t}^u \varphi_{\Delta v}(\theta) \, d\theta. \tag{4.17}$$

The second assumption is that the autocorrelation function decays on a time scale which is small with respect to t. We can then extend the limits of the θ integration to infinity. This is where the restriction of our analysis to the long-time case most visibly enters. The result is $[\varphi_{\Delta v}(\theta) = 0$ for $\theta < 0]$

$$\langle \Delta x^2(t) \rangle = t \int_{-\infty}^{+\infty} \varphi_{\Delta v}(\theta) \, d\theta. \tag{4.18}$$

By comparing this with our earlier results (4.10) and (4.11), we can now express one transport coefficient, the diffusion coefficient, in terms of an autocorrelation function. That is,

$$D_n = \tfrac{1}{2} \int_{-\infty}^{+\infty} \varphi_{\Delta v}(\tau) \, d\tau. \tag{4.19}$$

In Chapter 10 we note that these velocity fluctuations are an important source of measurable noise in a semiconductor sample. In particular, we can measure noise power spectra which are related to the Fourier transform of $\varphi_{\Delta v}(\tau)$. The integral in (4.19) is the dc term of this transform. Noise measurements therefore are one vehicle for the experimental determination of the diffusion coefficient.[5] This in fact is the contemporary realization of the recommendation made by Einstein in 1906. Here we later explore how mobility can also be related to an integral of an autocorrelation function.

We make one last comment about the diffusion coefficient. The process of integration usually is associated with a "memory"-like behavior, and what is occurring here is that the position fluctuations retain a memory of past velocity fluctuations. This conclusion is reinforced by the connection with the velocity fluctuation autocorrelation function. This function measures the rate at which an individual carrier loses its memory of its initial velocity fluctuation. The initial velocity fluctuations determine the position fluctuations until they have decayed, and that is why the infinite time integration of the velocity autocorrelation function determines the diffusion coefficient. We return to such considerations in Section 4.10.

4.3. DISTRIBUTION FUNCTIONS AND THEIR USE

In the above discussion we focused on the simplest models in which we retained only average information about the carrier density in real space and described the evolution of this quantity in terms of a small set of transport coefficients. We now turn our attention to an approach in which we retain a far more complete description of the underlying microscopic phenomena. It is obvious that our underlying description of the carrier transport process is inherently statistical and that the most information we could hope to have is a description of every carrier velocity (i.e., wave vector $\mathbf{k}$) and every carrier location at a set of M times of interest $\{t_j\}$. If we had N carriers present and labeled each carrier's wave vector $\mathbf{k}_i$ and each position $\mathbf{r}_i$, then we would look for a function

$$f_N[\mathbf{r}_1(t_1), \mathbf{r}_2(t_1) \cdots \mathbf{r}_N(t_1); \mathbf{k}_1(t_1) \cdots \mathbf{k}_N(t_1):$$
$$\mathbf{r}_1(t_2) \cdots \mathbf{r}_N(t_2); \mathbf{k}_1(t_2) \cdots \mathbf{k}_N(t_2):$$
$$\vdots \qquad\qquad \vdots$$
$$\mathbf{r}_1(t_M) \cdots \mathbf{r}_N(t_M); \mathbf{k}_1(t_M) \cdots \mathbf{k}_N(t_M)].$$

This function could be viewed as a joint probability density function for the $2NM$ random variables of interest.

The central difficulty in developing an analytical transport theory is that not only do we not have all of this information, but, in general, we could not solve the equations of motion even if we did. Instead, we commonly replace this complete description by a simpler function, generally called the distribution function, $f(\mathbf{r}, \mathbf{k}, t)$, which describes the probability of finding a carrier at position $\mathbf{r}$ with wave vector $\mathbf{k}$ at time t. A great deal of information is lost in this simplifying step, information which describes various correlations. We have lost all information concerning particle–particle correlations, a topic of central concern to Chapter 7. We also have thrown away the information needed in the computation of the velocity fluctuation autocorrelation function. If we were to use a two-time distribution function, where we preserved the information needed to predict the probability of having velocity v' at time t' and velocity v'' at time t'', we could calculate this function. However, we cannot extract that information from the single time distribution except in the uninteresting case where there is no correlation between the velocity at any two times; i.e., $\varphi_{\Delta v}(t', t'')$ would be a delta function. We return to this point later, as one of the central differences between Monte Carlo methods and Boltzmann transport methods is that the Monte Carlo method can preserve microscopic correlations, whereas the Boltzmann transport equation does not. The suppression of information concerning the number of carriers present also introduces potential confusion concerning the normalization of this function. If it is interpreted as a probability density function, then it is normalized such that, when integrated over all of phase space, the result is unity. We, however, use an interpretation of it as a carrier density function in phase space. Then, when integrated over all of phase space, it will produce N, the number of carriers present in the system.

Our goal will be to predict and understand the flow of current through a semiconductor device. We additionally want to know how the carrier density affects the solution of Poisson's equation. Therefore we need to know how we produce $n(\mathbf{r})$ and $\mathbf{J}(\mathbf{r})$ from $f(\mathbf{k}, \mathbf{r})$. Note that the central step here will be a suppression of information about $\mathbf{k}$. Using our interpretation of the distribution function as a carrier density in phase space, we obtain the desired functions

$$n(\mathbf{r}, t) = \frac{1}{4\pi^3} \int f(\mathbf{k}, \mathbf{r}, t) \, d\mathbf{k} \tag{4.20}$$

and

$$\mathbf{J}(\mathbf{r}, t) = -\frac{e}{4\pi^3} \int f(\mathbf{k}, \mathbf{r}, t) \mathbf{v}(\mathbf{k}) \, d\mathbf{k}. \tag{4.21}$$

The prefactors are obtained by noting that there are $(4\pi^3)^{-1}$ states per unit volume of k-space per unit volume of real space.

It is worthwhile to briefly review one approach to determining the distribution function in equilibrium. (A more sophisticated approach is found in Landsberg.[3]) Consider a system of n_L energy levels. The ith level has energy E_i and has g_i independent quantum states associated with it. We have N fermions which we must distribute over this set of levels in some distribution $\{N_i\}$, where N_i is the number of fermions on the ith level. By combinatorial methods we can compute a function $Q(N_1, N_2, \ldots, N_{n_1})$ which counts the number of statistically independent ways in which we can obtain the distribution $\{N_i\}$. Our goal is to maximize this function subject to the constraints of particle conservation,

$$\sum_{i=1}^{n_L} N_i = N = \text{constant}, \tag{4.22}$$

and energy conservation,

$$\sum_{i=1}^{n_L} E_i N_i = U = \text{constant}. \tag{4.23}$$

At this point we have more equations than unknowns. The condition of maximizing Q yields n_L independent equations in addition to (4.22) and (4.23). We have only n_L unknowns, however. Therefore we define two new unknowns, β and μ_c, and use the method of Lagrangian multipliers. The solution then is the Fermi-Dirac function

$$f_0(E) = \frac{1}{\exp\{\beta(E - \mu_c)\} + 1}, \tag{4.24}$$

which describes the probability that a state of energy E will be occupied by a fermion and therefore differs from the distribution used in (4.20) and (4.21) by a normalization. The quantity β is quite commonly written as

$$\beta = \frac{1}{k_B T}. \tag{4.25}$$

β is connected with energy conservation while the second multiplier μ_c is connected with particle conservation. The multiplier μ_c is often called the chemical potential, and here it can be related to the Fermi energy or level by

$$\mu_c = E_F. \tag{4.26}$$

The Fermi-Dirac function is an even function of energy and also will be an even function of $\mathbf{k}$ as well. The velocity found in (4.21), however, is an odd function of the wave vector, and therefore if we substitute the Fermi-Dirac function into (4.21) we find that the equilibrium current is zero. In order to have current flow, it is necessary to somehow produce a distribution function which is not an even function of $\mathbf{k}$.

One of the factors that maintains this symmetry in equilibrium is detailed balance, the equality of the rate of transition from a state $\mathbf{k}$ to a state $\mathbf{k}'$ with the rate of transition from $\mathbf{k}'$ to $\mathbf{k}$. In order to break the symmetry in the distribution, it will be necessary to also break detailed balance. This will happen when an electric field is applied. What we can retain is a conservation of average quantities, and, in particular, we always retain particle conservation. Energy conservation also can be retained, but, strictly speaking, this happens in the nonequilibrium situation of a flowing current only when we explicitly solve for T simultaneously with our determination of the particle and current densities.

4.4. THE BOLTZMANN TRANSPORT EQUATION

The Fermi-Dirac function was obtained from arguments which contain no dynamic information whatsoever. The situations of interest in submicron semiconductor devices, however, are primarily dynamic in nature. We therefore need an equation which describes the dynamical evolution of the distribution function, and we expect to see potentially large deviations of our distribution function away from the equilibrium Fermi-Dirac function f_0. The Boltzmann transport equation is the most famous example of such an equation. The Boltzmann transport equation is usually obtained in one of two ways: through long rigorous derivations or through short heuristic arguments. We present a short heuristic argument, which is essentially a continuity argument. A more rigorous discussion can be found in Ref. 6.

We start with the basic dynamical laws followed by an electron located inside of our energy bands. These are

$$\frac{d\mathbf{r}}{dt} = \mathbf{v} = \frac{1}{\hbar} \nabla_k E \tag{4.27}$$

and

$$\hbar \frac{d\mathbf{k}}{dt} = \mathbf{F}_L, \tag{4.28}$$

where $\mathbf{F}_L$ is the Lorentz force

$$\mathbf{F}_L = -e(\mathbf{F} + \mathbf{v} \times \mathbf{B}). \tag{4.29}$$

We now write down all of the processes by which the distribution function for some differential volume element $(\mathbf{k}, \mathbf{k} + \Delta\mathbf{k}, \mathbf{r}, \mathbf{r} + \Delta\mathbf{r})$ of our six-dimensional phase-space changes. There are three processes. First, the Lorentz force creates a "velocity" current which moves carriers in and out of our velocity range. This is described by a term of the form

$$\left(\frac{\partial f}{\partial t}\right)_F = \nabla_k \cdot \left(f \frac{d\mathbf{k}}{dt}\right). \tag{4.30}$$

The second process by which the distribution inside our differential phase-space element may change is the movement of carriers in and out of the real-space range of interest; that is,

$$\left(\frac{\partial f}{\partial t}\right)_v = \nabla \cdot \left(f \frac{d\mathbf{r}}{dt}\right). \tag{4.31}$$

Lastly, the carriers may move in and out of the velocity range of interest through scattering events. This is represented by the collision integral

$$\left(\frac{\partial f}{\partial t}\right)_c = \int \{f(\mathbf{k}')[1 - f(\mathbf{k})]S(\mathbf{k}', \mathbf{k}) - f(\mathbf{k})[1 - f(\mathbf{k}')]S(\mathbf{k}, \mathbf{k}')\} \, d\mathbf{k}'. \tag{4.32}$$

Adding all of these terms, we obtain the Boltzmann transport equation

$$\frac{\partial f}{\partial t} = -\left[\nabla \cdot \left(f\frac{d\mathbf{r}}{dt}\right) + \nabla_k \cdot \left(f\frac{d\mathbf{k}}{dt}\right)\right] + \left(\frac{\partial f}{\partial t}\right)_c. \tag{4.33}$$

A convenient viewpoint to adopt here is that the Boltzmann transport equation is a continuity equation for the distribution function in phase space. A more common form of this equation can be obtained if we assume that our force fields do not depend on $\mathbf{k}$, a valid assumption, and by assuming that the velocity is not functionally dependent on r. This last assumption bears a little discussion. $\nabla_k E$ must not be a function of real space if this assumption is to hold. In Chapter 6 we discuss graded band-gap structures in which the energy bands are an explicit function of real-space position and where therefore this particular assumption *is not true*. We will use it here, however, and obtain the most common form for the Boltzmann equation. Substituting (4.27) and (4.28) into (4.33) and utilizing a simple vector identity yields

$$\frac{\partial f}{\partial t} + \mathbf{v} \cdot \nabla f + \frac{\mathbf{F}_L}{\hbar} \cdot \nabla_k f = \left(\frac{\partial f}{\partial t}\right)_c. \tag{4.34}$$

4.5. LINEARIZING THE BOLTZMANN EQUATION

The Boltzmann equation is a nonlinear integrodifferential equation and is difficult to solve. One of the commonly used techniques is to linearize this equation. This is done by representing the distribution function by

$$f(\mathbf{k}, \mathbf{r}) = f_0 + f_1(\mathbf{k}, \mathbf{r}), \tag{4.35}$$

where f_0 is the equilibrium or Fermi-Dirac function and f_1 is a small perturbation produced by the application of a small electric field. The application of a magnetic field does not alter the distribution away from equilibrium and it can easily be shown that[7]

$$-\frac{e}{\hbar}(\mathbf{v} \times \mathbf{B}) \cdot \nabla_k f_0 = \left(\frac{\partial f_0}{\partial t}\right)_c. \tag{4.36}$$

The collision integral can be rewritten as

$$\left(\frac{\partial f}{\partial t}\right)_c = \left(\frac{\partial f_0}{\partial t}\right)_c + \left(\frac{\partial f_1}{\partial t}\right)_c$$

$$= \left(\frac{\partial f_0}{\partial t}\right)_c + \int \{f_1(\mathbf{k}')[[1 - f_0(k)]S(\mathbf{k}', \mathbf{k}) + f_0(k)S(\mathbf{k}, \mathbf{k}')]$$

$$-f_1(\mathbf{k})[f_0(\mathbf{k}')S(\mathbf{k}', \mathbf{k}) + [1 - f_0(\mathbf{k}')]S(\mathbf{k}, \mathbf{k}')]\} \, d\mathbf{k}', \tag{4.37}$$

and since we have assumed a small electric field we have an additional relation

$$\mathbf{F} \cdot \nabla_k f_0 \gg \mathbf{F} \cdot \nabla_k f_1. \tag{4.38}$$

Substituting all of these into our Boltzmann equation yields the linearized Boltzmann equation

$$\left(\frac{\partial f_1}{\partial t}\right)_c + \frac{e}{\hbar}(\mathbf{v} \times \mathbf{B}) \cdot \nabla_k f_1 - \frac{\partial f_1}{\partial t} = \mathbf{v} \cdot \nabla f_0 - \frac{e}{\hbar}\mathbf{F} \cdot \nabla_k f_0. \tag{4.39}$$

For now, we will consider the steady-state case. Since our f_1 is still only an arbitrary function of $\mathbf{k}$ and $\mathbf{r}$, we can replace it by a product of another arbitrary function and the derivative of f_0 with respect to energy; that is,

$$f_1(\mathbf{k}, \mathbf{r}) = -\varphi(\mathbf{k}, \mathbf{r})\frac{df_0}{dE}. \tag{4.40}$$

After some algebra we obtain a new linearized Boltzmann equation

$$(C + \Omega')\varphi(\mathbf{k}, \mathbf{r}) = -\mathbf{X} \cdot \mathbf{v}\frac{df_0}{dE}, \tag{4.41}$$

where the collision operator is

$$\left(\frac{\partial f_1}{\partial t}\right)_c = C\varphi(\mathbf{k}, \mathbf{r}) = \beta \int V(\mathbf{k}, \mathbf{k}')[\varphi(\mathbf{k}') - \varphi(\mathbf{k})]\, d\mathbf{k}' \tag{4.42}$$

and $V(k, k')$ is

$$V(\mathbf{k}, \mathbf{k}') = f_0(\mathbf{k})[1 - f_0(\mathbf{k}')]S(\mathbf{k}, \mathbf{k}'). \tag{4.43}$$

A consequence of detailed balance is

$$V(k, k') = V(k', k). \tag{4.44}$$

The magnetic operator is defined as

$$\Omega'\varphi(\mathbf{k}, \mathbf{r}) = -\frac{df_0}{dE}\frac{e}{\hbar}(\mathbf{v} \times \mathbf{B}) \cdot \nabla_k\varphi(\mathbf{k}, \mathbf{r}), \tag{4.45}$$

and the last remaining operator describes the electromotive forcing functions. It is

$$\mathbf{X} = e\mathbf{F}' + (\nabla T_L)\frac{E - \mu_c}{T_L}, \tag{4.46}$$

where

$$\mathbf{F}' = \mathbf{F} + \frac{\nabla \mu_c}{e} \qquad (4.47)$$

is the electromotive force.

4.6. LINEAR TRANSPORT THEORY IN THE RELAXATION-TIME REGIME

Our goal will be to explore the relationships between the linear Boltzmann equation, the drift-diffusion equation, the quasi-Fermi level formulation and the generalized electromotive force $\mathbf{F}'$. In the process we will use a very common approximation to the collision integral or operator. The integral will be replaced by a relaxation of the perturbed distribution back to the equilibrium form with a characteristic time constant called the relaxation time. We start with the formal solution to our linearized Boltzmann equation (4.41) which is

$$\varphi(\mathbf{k}, \mathbf{r}) = -(C + \Omega')^{-1}\mathbf{X} \cdot \mathbf{v}\frac{df_0}{dE}. \qquad (4.48)$$

If we substitute (4.48) into (4.40) and use (4.47), we then can calculate the electric current density. Remembering that we must integrate $f_0 + f_1$ in (4.21) and that the contribution of the f_0 term will be zero, one can show[7] that

$$\mathbf{J} = \boldsymbol{\sigma} \cdot \mathbf{F}' + \mathbf{L} \cdot \nabla T, \qquad (4.49)$$

where $\boldsymbol{\sigma}$ is the electrical conductivity tensor, which has the elements

$$\sigma_{ij}(\mathbf{B}) = -\frac{e^2}{4\pi^3} \int \frac{df_0}{dE} v_i (C + \Omega')^{-1} v_j \frac{df_0}{dE} \, d\mathbf{k}, \qquad (4.50)$$

and the tensor $\mathbf{L}$, which describes some of the thermoelectric properties of the system (it does not describe the contribution to the heat current produced by the electromotive force), has the elements

$$L_{ij}(\mathbf{B}) = -\frac{e}{4\pi^3 T} \int \frac{df_0}{dE} v_i (C + \Omega')^{-1} v_j (E - \mu_c) \frac{df_0}{dE} \, d\mathbf{k}. \qquad (4.51)$$

The Onsager relations can be obtained from the above by using various symmetry properties of the $C + \Omega$ operators.[7] The most pertinent for our purposes is the relation $\sigma_{ij}(\mathbf{B}) = \sigma_{ji}(-\mathbf{B})$. Since one of our goals is to relate this approach to the drift-diffusion law, it should be noted that the thermoelectric term in (4.49) is missing in the drift-diffusion law.

The electromotive force $\mathbf{F}'$ is generally expressed in terms of a scalar electrochemical potential φ_{ec} by

$$\mathbf{F}' = \frac{1}{e} \nabla \varphi_{ec}. \tag{4.52}$$

The term φ_{ec} can be expressed as

$$\varphi_{ec} = -e\Phi + \mu_c \tag{4.53}$$

and used to recover (4.47) from (4.52). If we have no temperature gradient, equations (4.49), (4.52), and (4.53) are reminiscent of the quasi-Fermi-level expressions (4.4) and (4.6). Indeed, the two would be identical if σ is a simple constant equal to $e\mu_n n$ and if

$$\mu_c = k_B T_L \ln \frac{n}{n_i}. \tag{4.54}$$

There is only one condition where this holds, and that is thermal equilibrium. This actually is related to the role played by the Einstein relation, as (4.54) is used in some derivations of the Einstein relation.[4] We again are in a position of noting that the quasi-Fermi levels are meant to handle a very special type of nonequilibrium situation. They handle the case of two gases, electrons and holes, each of which is in equilibrium with the lattice but where the population densities of these two gases are not in equilibrium with the generation-recombination processes of the system. The chemical and electrochemical potentials, on the other hand, are excellent tools for handling the thermodynamics of near-equilibrium linear transport.

In the previous section we linearized the Boltzmann equation. Yet, even though we attempted to conceal this by using operator notation, the result is still an integrodifferential equation. We now reduce this equation to a differential equation by replacing the collision integral by

$$\left(\frac{\partial f_1}{\partial t}\right)_c = C\varphi(\mathbf{k}, \mathbf{r}) = -\frac{f_1}{\tau}, \tag{4.55}$$

where τ is the relaxation time. We have also set the magnetic field to zero. Justifying the use of a relaxation time of this sort is a question which has been extensively explored. The main features are discussed by Butcher[7] and Conwell.[8] Here we merely note that generally (4.55) can be exact only if the scattering processes are elastic or if they are velocity-randomizing; i.e., all final states on the constant-energy surface corresponding to the final energy are equally probable, in which case $V(\mathbf{k}, \mathbf{k}') = V(\mathbf{k}, -\mathbf{k}')$. For a wide variety of other situations it is possible to define some form of approximate relaxation time, but the situations tend to become increasingly complex, with relaxation time tensors being used on occasion. Since there are alternative techniques of similar complexity but greater accuracy for hot-carrier problems, this will be the last time that we use the relaxation time approximation for the Boltzmann equation.

Using this approximation, we may write the distribution function as

$$f(\mathbf{k}, \mathbf{r}) = f_0 - \tau\left(\mathbf{v} \cdot \nabla f_0 - \frac{e}{\hbar} \mathbf{F} \cdot \nabla_k f_0\right). \tag{4.56}$$

When this solution is inserted into (4.21), the current is found to be (for one-dimensional situations)

$$J = -\frac{e^2 F}{4\pi^3 \hbar} \int \tau \frac{\partial f_0}{\partial k} v(k) \, dk + \frac{e}{4\pi^3} \int \tau v \frac{\partial f_0}{\partial x} v \, dk. \tag{4.57}$$

The first term of (4.57) can be rewritten as $J_n = en\mu_n F$, where the mobility is defined by

$$\mu_n = -\frac{e}{\hbar} \frac{\int \tau(\partial f_0/\partial k)v \, dk}{\int f_0 \, dk}. \tag{4.58}$$

Rewriting the second term as a diffusion term is also possible. There, as Stratton discusses in more detail,[9] one can either obtain (4.1) where the diffusion coefficient is located outside the gradient operator, or a similar equation with the diffusion coefficient located inside the gradient operator, or one can obtain various other equations with various thermoelectric terms included as well. Therefore the drift-diffusion law (4.1) is best viewed as a phenomenological equation whose range of validity is restricted to situations in which thermoelectric effects are unimportant.

On the surface it may appear strange to first argue, as we did, that the diffusion coefficient is intimately related to microscopic correlations, and to then argue that some form of a diffusion coefficient can be obtained from Boltzmann transport theory in which we do not have the capability of constructing these microscopic correlation functions. No contradiction actually occurs. First, while it is possible to construct D_n from microscopic correlations, diffusion is a macroscopic phenomena. The Boltzmann picture is more than adequate for describing the same macroscopic phenomena. Second, we obtain the diffusion coefficient from the microscopic correlation function only in situations where there is a well-defined steady state and where the time of interest is long with respect to the decay time of the correlation function. In other words, we require that the initial and final microstates of the system be uncorrelated.

4.7. NONLINEAR ENERGY-MOMENTUM-CONSERVING MODELS

In the above discussion we greatly simplified the Boltzmann equation by first linearizing the equation and then replacing the collision integral by a simple relaxation time. While this approach does yield analytical equations which can be straightforwardly solved, a great deal of information has been thrown away. Of particular importance in the context of submicron devices is the linearization step, because in this step all nonlinear hot-carrier effects are thrown away. It is possible, however, to develop a system of equations from the Boltzmann equation in which the collision integral is replaced by relaxation times without first linearizing the equation. This is done by employing the method of moments. We calculate the nth moment by multiplying the Boltzmann equation by k^n and then integrating over k. The equation for the nth moment then is

$$\int \mathbf{k} \left\{ \frac{\partial f}{\partial t} + \mathbf{v} \cdot \nabla f + \frac{\mathbf{F}}{\hbar} \cdot \nabla_k f = \left(\frac{\partial f}{\partial t} \right)_c \right\} d\mathbf{k}. \tag{4.59}$$

The next step is to replace the somewhat cumbersome collision terms by relaxation terms. Actual justification of this rests on certain assumptions concerning the distribution function. One often finds this approach therefore coupled with the explicit assumption of a distribution function which satisfies these properties. The most common form of such a function is the displaced Maxwellian

$$f(E) = A \exp \left[-\frac{m^*(v - v_d)^2}{k_B T_e} \right]. \tag{4.60}$$

In (4.60) v_d is the drift velocity and T_e is a carrier temperature, which, in general, will be larger than T_L. It is usually argued that when carrier–carrier interactions are strong, these interactions drive the distribution to a displaced Maxwellian form, as this form is indeed the solution to the Boltzmann equation when only these interactions are present. Carrier–carrier interactions only redistribute energy and momentum from one carrier to another and leave the average quantities unchanged. The carrier–lattice interactions, however, do change the average carrier momentum and energy. Therefore when the rate of energy and momentum exchange through carrier–carrier interactions is large when compared with the corresponding carrier lattice and carrier–field rates, the following approach can be justified. Model the carrier–carrier interactions by using a Maxwellian, and model the carrier–lattice and carrier–field interactions by developing a set of equations which describe the temporal evolution of v_d and T_e.

Another line of attack is the use of a phenomenological justification for the approach. Instead of assuming a distribution function, one instead assumes that the system of equations developed by the method of moments applies to one's problem and seeks out a set of relaxation times which provide the best fit of the results to the situation of interest. This is usually done by fitting the results to experimentally determined velocity-field data, diffusion-field data, and impact ionization data. Monte Carlo methods have been used to estimate these times as well.

To illustrate how this analysis works, we review one such model, that of Mains et al.,[10] in some detail. Mains et al. do not use a displaced Maxwellian, but do develop a set of partial differential equations which conserve carrier density, momentum, and energy by applying the method of moments. The model is a bipolar model in which one valence-band and two conduction-band valleys are modeled. Impact ionization appears as a generation term in the continuity equation, which is the zeroth moment. A momentum conservation equation is obtained from the first moment, and the second moment yields an energy conservation equation. The carrier continuity equation for the ith valley is

$$\frac{\partial n_i}{\partial t} = -\frac{\partial(v_i n_i)}{\partial x} + \left(\frac{\partial n_i}{\partial t}\right)_c. \tag{4.61}$$

Note that the equation for the zeroth momentum contains a term dependent on the first moment v_i. Generally the $(m+1)$st moment appears in the equation for the mth moment. Eventually this hierarchy must be truncated, generally by simply assuming that all higher moments are negligible. Mains et $al.$ ignored all moments higher than the second. In (4.61) the subscript i refers to one of a set of nonequivalent valleys, n_i is the electron concentration in the ith valley, while v_i is the average velocity of the carriers in the ith valley under all conditions (and therefore includes diffusive transport in the presence of concentration gradients). The momentum balance equation obtained for the first moment is

$$\frac{\partial v_i}{\partial t} = -\frac{1}{2}\frac{\partial v_i^2}{\partial x} + \frac{qF}{m^*} - \frac{2}{3m_i^* n_i}\frac{\partial}{\partial x}[n_i(w_i - \tfrac{1}{2}m_i^* v_i^2)] + \left(\frac{\partial v_i}{\partial t}\right)_c. \tag{4.62}$$

Here we are assuming that q is the actual charge of the carrier and equals e for electrons and $-e$ for holes; m_i^* is the conductivity effective mass of the ith valley; and w_i, the average total energy of a carrier in the ith valley, is

$$w_i = \frac{1}{2n_i} m_i^* \int v^2 f d^3 v. \tag{4.63}$$

The energy conservation equation obtained from the second moment is

$$\frac{\partial w_i}{\partial t} = -v_i \frac{\partial w_i}{\partial x} + qFv_i - \frac{2}{3m_i^* n_i}\frac{\partial}{\partial x}[n_i v_i(w_i - \tfrac{1}{2}m_i^* v_i^2)] - \frac{1}{n_i}\frac{\partial h_i}{\partial x} + \left(\frac{\partial v_i}{\partial t}\right)_c, \tag{4.64}$$

where

$$h_i = \tfrac{1}{2}m_i^* \int (v_x - v_i)[(v_x - v_i)^2 + v_y^2 + v_z^2]f d^3 v, \tag{4.65}$$

a heat flow term, is the neglected third moment.

In the above equations, the terms containing a subscript c describe the effects of collisions. They all result from the collision integral of the Boltzmann equation and usually have the form of a relaxation process controlled by an energy-dependent relaxation time. These integrals could be analytically evaluated if the displaced Maxwellian form is retained. However, better fits to experimental steady-state transport data can be obtained by selecting relaxation times which provide a good agreement with experimental or Monte Carlo data. Curves of these various parameters are provided by Mains et al.[10]

The various collision terms in the continuity equations are (for a two valley model)

$$\left(\frac{\partial n_1}{\partial t}\right)_c = -\frac{n_1}{\tau_{n1}(w_1)} + \frac{n_2}{\tau_{n2}(w_2)} \tag{4.66}$$

and

$$\left(\frac{\partial n_2}{\partial t}\right)_c = -\frac{n_2}{\tau_{n2}(w_2)} + \frac{n_1}{\tau_{n1}(w_1)} + \alpha_n(w_2)n_2 + \alpha_p(w_p)p. \tag{4.67}$$

The τ_{n1} and τ_{n2} terms describe the process of intervalley transfer: τ_{n1} describes transfer from valley 1 to valley 2, while τ_{n2} describes transfer from valley 2 to valley 1. (Mains et al. assumed that only electrons in valley 2, the satellite valley, undergo impact ionization with both resultant electrons appearing in valley 2.) The hole continuity equation is similar to the electron continuity equations, but utilizes the generation term

$$\left(\frac{\partial p}{\partial t}\right)_c = \alpha_n(w_2)n_2 + \alpha_p(w_p)p. \tag{4.68}$$

The collision terms encountered in the momentum balance equations are[10]

$$\left(\frac{\partial v_1}{\partial t}\right)_c = -\frac{v_1}{\tau_{m1}(w_1)}, \tag{4.69}$$

$$\left(\frac{\partial v_2}{\partial t}\right)_c = -\frac{v_2}{\tau_{m2}(w_2)} - v_2\frac{p\alpha_p(w_p)}{n_2}, \tag{4.70}$$

and

$$\left(\frac{\partial v_p}{\partial t}\right)_c = -\frac{v_p}{\tau_{mp}(w_p)} - v_p\frac{n_2\alpha_n(w_2)}{p}. \tag{4.71}$$

The terms involving τ_{m1}, τ_{m2}, and τ_{mp} represent momentum loss due to various scattering events other than the creation of a new carrier. The second terms in (4.70) and (4.71) represent the change induced by the contribution of the newly created carrier to the average momentum under the assumption that the newly created carrier has zero momentum. Therefore, if we consider electrons in valley two as an example, the contribution made to the total momentum by the newly created carrier is

$$\left(\frac{\partial(n_2 v_2)}{\partial t}\right)_{new} = \left(v_2 \frac{\partial n_2}{\partial t}\right)_{new} + n_2 \frac{\partial v_2}{\partial t} = 0. \tag{4.72}$$

However,

$$\left(\frac{\partial n_2}{\partial t}\right)_{new} = \alpha_p p, \tag{4.73}$$

and therefore

$$\left(\frac{\partial v_2}{\partial t}\right)_{new} = -v_2 \frac{p\alpha_p(w_p)}{n_2}. \tag{4.74}$$

The collision terms involved in the energy balance equations are

$$\left(\frac{\partial w_1}{\partial t}\right)_c = -\frac{w_1 - w_{th}}{\tau_{e1}(w_1)} - \frac{w_1}{\tau_{e(1-2)}(w_1)} + \frac{w_2}{\tau_{e(2-1)}^*(w_2)} \frac{n_2}{n_1}, \tag{4.75}$$

$$\left(\frac{\partial w_2}{\partial t}\right)_c = -(w_2 - w_{th})\left(\frac{1}{\tau_{e2}(w_2)} + \frac{p\alpha_p(w_p)}{n_2}\right) - \frac{w_2}{\tau_{e(2-1)}(w_2)}$$

$$+ \frac{w_1}{\tau_{e(1-2)}^*(w_1)} \frac{n_1}{n_2}, \tag{4.76}$$

and

$$\left(\frac{\partial w_p}{\partial t}\right)_c = -(w_p - w_{th})\left(\frac{1}{\tau_{ep}(w_p)} + \frac{n_2\alpha_n(w_2)}{p}\right), \tag{4.77}$$

where

$$w_{th} = \tfrac{3}{2}k_B T_L \tag{4.78}$$

is the thermal energy of the semiconductor lattice which is at temperature T_L. The terms τ_{e1}, τ_{e2}, and τ_{ep} are energy relaxation rates for intravalley scattering processes, although τ_{e2} also includes the contribution of equivalent intervalley scattering. The terms involving $\tau_{e(i-j)}$ represent an energy change for electrons in valley i due to transfers of electrons from valley i to valley j, while those involving $\tau_{e(i-j)}^*$ represent the energy change due to transfer from valley j to valley i.

The central difficulty is not deriving the equations but obtaining the various relaxation times as a function of energy. One approach uses a Monte Carlo simulation. A field is applied, and one electron is followed through many scattering events. The various energy and momentum changes experienced by this electron due to scattering events are stored and used at the end to estimate the average rate of energy and momentum change due to scattering. This is assumed to provide good estimates of these terms for the time average of the electron energy for the given applied field. One further demands that the solution be reduced to well-established values for steady-state transport coefficients. For a single-valley system, this reduction yields field-dependent mobilities and diffusion coefficients, which are

$$\mu = \frac{q\tau_{mi}}{m_i^*} \qquad (4.79)$$

and

$$D = \frac{\tau_{mi} k_B T_{ei}}{m_i^*}, \qquad (4.80)$$

where the ith valley carrier temperature is T_{ei}. These two coefficients satisfy an Einstein relation in which the temperature used is T_{ei}. Equation (4.79) is equivalent to (4.58) with the integrals evaluated for $f(E) \neq f_0(E)$. Equation (4.80) is related to integral expressions found in Stratton.[9] Similar results have been developed for a three-valley model by Woolard et al.[11]

This type of approach implicitly assumes that the carrier distribution will always closely resemble that of the steady-state case with corresponding average carrier energy and momentum. It therefore has difficulties modeling situations in which the dominant factor is a warping of the distribution function. One such case is the Jones–Rees[12] effect, which occurs in systems with two nonequivalent valleys. It is illustrated in Figure 4.1. The important point here is that the intervalley

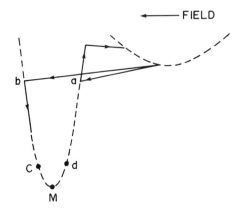

FIELD

FIGURE 4.1. The Jones–Rees effect. A carrier scattering from the higher valley to the lower valley can enter the lower valley either traveling with the field or against it. The field delivers energy to electrons traveling with it while electrons traveling opposite to the field deliver energy to it. Therefore electrons entering with velocity opposite to the field are more likely to remain in the lower energy valley.

scattering of carriers from valley 2 to valley 1 introduces them with equal probability along a constant-energy circle which lies in the plane perpendicular to the paper. (Remember that the central valley is a conical solid in k-space.) There is a field present though, and it tends to keep electrons which enter on the right half of this circle above the threshold for intervalley scattering. Such electrons will therefore rapidly scatter back to valley 2. Electrons which enter on the left-hand side of this circle, however, have a negative group velocity, and as they move opposite to the field in real space, they lose energy to the field and fall below the threshold for intervalley scattering. The overall effect therefore is to select electrons with negative velocities as central valley electrons.

One last point should be made about the form of these balance equations. Note that for the case of zero field, the average velocities will be 0 and the average carrier energy will be w_{th}.

When one compares the conventional drift-diffusion (DD) model with this energy- and momentum-conserving (EM) model, it is obvious that the EM model is much more complicated. A complete DD model in which the electric field is simultaneously solved along with the various carrier densities and currents utilizes five partial differential equations and one algebraic equation for the six unknowns F, G, n, p, J_n, and J_p. Six field-dependent parameters are used, and the numerical methods and boundary conditions are relatively well understood. The corresponding EM model uses 10 partial differential equations for 10 unknowns: F, three carrier densities, three carrier velocities, and three carrier energies. Fourteen energy-dependent parameters are used, and the appropriate numerical techniques and boundary conditions are not well developed. Stability problems in fact prevented Mains et al.[13] from performing EM simulations at high drive levels for some structures. The complexity of this model, a one-dimensional model at that, is well worth noting. There is a common bias that somehow a model based on partial differential equations must be simpler than the particle-based models which we begin to discuss later in this chapter. However, it is not clear that the overall trade-off between programming complexity, physical accuracy, and computational stability obtained when applying an EM model to a complex submicron device is significantly better than that of a particle-based model in which comparable assumptions are made concerning energy-band structure and scattering processes.

The central difference between these EM models and the more conventional DD model is that we abandon the concept that all transport and ionization parameters are uniquely determined by the local instantaneous field. Instead they are uniquely determined by the local instantaneous average carrier energy. There are simpler variations on this theme. The simplest approach is to utilize the same models for transport and ionization as are conventionally used but to add an additional equation which models the dynamical evolution of the average carrier energy. All transport and ionization parameters are then re-expressed as functions of the local, instantaneous energy. Certain transient phenomena, however, most notably the velocity overshoots[14] discussed in Chapter 5, are not included in models which do not allow the carrier velocity to move away from the value predicted by the steady-state velocity associated with the average energy. The simplest and, therefore, most widely used energy balance equation is an old one which was popularized for simple device models by Shur.[15] It is

$$\frac{\partial w}{\partial t} = qFv_m - \frac{w - w_{\text{th}}}{\tau_E(w)},$$ (4.81)

where v_m is the average carrier velocity, w_{th} is the equilibrium carrier energy, and τ_E is an energy relaxation time. This equation completely neglects any effects associated with spatial inhomogeneity and therefore cannot quantitatively model any submicron device. An example of another and more sophisticated approach in which only energy transients were followed is that of Kafka and Hess,[16] who used the following energy equation:

$$\frac{\partial}{\partial t}[(n + p)w] = (J_n + J_p)F + n\left(\frac{dw}{dt}\right)_{n,c} + p\left(\frac{dw}{dt}\right)_{p,c}$$

$$+ \frac{\partial}{\partial x}\left[2k_B(D_n n + D_p p)\frac{2}{3k_B}\frac{\partial w}{\partial x} + \frac{1}{q}(J_n + J_p)\frac{\langle \tau E^2 \rangle}{\langle \tau E \rangle}\right],$$ (4.82)

where E is the energy of a single carrier and $\langle \ \rangle$ denotes an average over the distribution. The terms $(dw/dt)_{n,p,c}$ denote collision terms for electrons and holes. Unlike the simple Shur balance equation, an electronic heat conduction term which is needed for spatially inhomogenous device modeling is included by Kafka and Hess. They asserted that this term is important and demonstrated that they obtained different results when it was included.

While there is no reason to doubt the validity of this conclusion, doubts have been raised concerning their methodology.[17] Since these doubts deal with an important but subtle point, it is worth repeating the discussion. The only independently known parameters are the steady-state field-dependent transport and ionization parameters. These parameters are obtained by fitting experimental data to a drift-diffusion model. All other models obtain their parameters by adjusting the model to allow it to reproduce these measured parameters. Kafka and Hess perform such a parameter adjustment with one form of their model in which one term of one equation is arbitrarily set to zero and then obtain different results when the full model is run with the same parameters. The question which has been raised concerns the possibility that if they used the full model with these parameters and ran it under conditions where the steady-state ionization data used in the parameter fit should be reobtained, that this data would not be reobtained. In short, by not recalibrating the model with all terms present, they may have not only changed the model but also the material. The important but subtle point is that the parameters obtained in the fitting process are not independent of the equations used in the model. Instead the parameters and the equations taken as a self-consistent whole constitute the model of a specified material.

As we noted, however, there is no reason to doubt the validity of their conclusion that heat conduction effects are important. This is significant in that such effects are neglected in the model of Mains et al. and also that of Froelich,[17] both of whom truncate their moment hierarchy at the second moment. This involved the arbitrary assumption that h_i is zero in (4.64) and (4.65). We therefore have models in which the first and second moments are fully retained,[10,17] models in which the second and third moments are retained,[16] and models in which all three moments are retained.[11]

The simplest model which includes the possibility of a velocity overshoot uses the momentum balance equation

$$\frac{\partial}{\partial t} v = \frac{qF}{m^*} - \frac{v}{\tau_m(w)} \tag{4.83}$$

in conjunction with equation (4.81). This particular set of equations is the set popularized by Shur, and equation (4.83) shares the same limitations as equation (4.81). The model can be, and has been, used since the 1950s to provide simple estimates of the possible importance of energy and momentum relaxation.

4.8. THE MONTE CARLO TECHNIQUE

In all of the models discussed, the carrier density is a continuous parameter whose evolution in time and space is described by a set of partial differential equations which include a continuity equation and a set of transport laws. While this approach has been the backbone of device modeling, we have seen that attempts at developing a variation in which transport and generation rates are not determined by a quasi-static approximation yield a complicated set of equations whose boundary conditions are not well understood and for which solution procedures are also poorly developed. There is an alternative which over the past 20 years has been widely used in more basic studies of the physics of hot-carrier transport. In this approach we do not treat the mobile charge density as a continuous valued function or fluid, but instead we track individual point charges which are accelerated by the applied fields and experience various scattering or collision events. In such models we need to directly incorporate a statistical model of at least the scattering events, and this is done by the use of Monte Carlo techniques. No partial differential equations are used in determining the location of the charge in the system. We still, however, will need to solve Gauss's law or Poisson's equation in order to relate the local field or potential to this charge distribution for a specified set of boundary conditions.

In the Monte Carlo technique we simply use random number generation to directly simulate the underlying statistical nature of carrier transport. There are three basic steps in this procedure.

1. Random number generation is used to produce a time-of-flight to the next scattering event.

2. The carrier is accelerated through k-space by the field for this time-of-flight.
3. More random numbers are used to simulate the statistics of the scattering process.

In the Monte Carlo technique we therefore deal directly with the band structure, the total scattering rate $\lambda(\mathbf{k})$, and the transition rate $S(\mathbf{k}, \mathbf{k}')$.

The difficulty in step 1 is that the electronic wave vector $\mathbf{k}$ changes during the time-of-flight. Since $\lambda(\mathbf{k})$ is not constant, the scattering rate also is changing during the time-of-flight. These two interact to produce the following result. The probability that an electron which was scattered at time t_0 has not been scattered since (i.e., has experienced a free flight through k-space under the influence of the field until time t) is $\exp\{-\int_{t_0}^{t} \lambda[\mathbf{k}(s)] \, ds\}$. Now the probability that this same electron experiences a scattering event at $t = 0$ and its next scattering event in the differential time interval $(t, t + dt)$ is

$$P(t) \, dt = \lambda[\mathbf{k}(t)] \exp\left\{-\int \lambda[\mathbf{k}(u)] \, du\right\}. \qquad (4.84)$$

As we have seen, $\lambda(\mathbf{k})$ is a nasty-looking nonlinear function. Here it appears in an integral which is the argument of an exponential and generally messes up the whole problem. We can greatly simplify the mathematics here without introducing any physical simplifications by using an elegant idea called self-scattering. If $\lambda(\mathbf{k}) = \Gamma$, where Γ is constant, then

$$P(t) \, dt = \Gamma \exp[-\Gamma t]. \qquad (4.85)$$

We then could generate our time-of-flight by generating a random number r_1, evenly distributed between 0 and 1, and set the time-of-flight to be

$$t_r = \frac{-\ln(r_1)}{\Gamma}. \qquad (4.86)$$

This last equation is obtained from the identity $P(t_r) = r$, remembering that Γ is a constant. Self-scattering[18] allows us to effectively use a constant scattering rate $\lambda = \Gamma$ and the simple flight time estimate of equation (4.86), even though the physical scattering processes are quite complex. We introduce a fictitious scattering process which we call self-scattering. The transition rate for this self-scattering process is $S(\mathbf{k}, \mathbf{k}') = \lambda_{ss}(\mathbf{k})\delta(\mathbf{k} - \mathbf{k}')$. Therefore in a self-scattering event, nothing happens. The carrier energy and wave vector are left unchanged. The total scattering rate for the self-scattering process is set to the value $\lambda_{ss}(\mathbf{k}) = \Gamma - \lambda(\mathbf{k})$, where Γ is an arbitrary constant. The constant Γ is generally chosen to equal the maximum value of the total physical $\lambda(\mathbf{k})$. When self-scattering is added to our Monte Carlo simulation, we obtain a total (real plus self-) scattering rate equal to Γ and we can then use (4.86) to generate flight times. The result is a great simplification of the first step at the cost of a relatively minor complication of the third step.

Step 2 of the Monte Carlo process is very simple. We solve the equation

$$\mathbf{k}(t_0 + t_r) = \mathbf{k}(t_0) + \frac{e\mathbf{F}}{\hbar},$$

(4.87)

where $\mathbf{k}(t_0)$ is the value of the k-vector after the scattering event which occurred at time t_p and $\hbar$ is the reduced Planck's constant.

Once self-scattering is introduced, steps 1 and 2 become very easy. It is step 3 that poses most of the computational effort in a Monte Carlo simulation. Note in the following discussion that the use of self-scattering clearly does not greatly affect the complexity of this step. Step 3 starts by selecting which of the scattering events occurs. In Figure 4.2 we illustrate how this is done. Here the subscripts on the scattering rates λ_i denote either a specific physical scattering process, if i is a numeral, or self-scattering in the case of λ_{ss}. The ith curve is constructed by simply adding the rates for processes 1 through i and can be normalized by dividing the total scattering rate by Γ. The vertical line represents the energy E_1 of the carrier at the end of the free flight immediately before it scatters. In normalized scattering rates, this line is the unit interval, and each of the subintervals into which it is divided by the various curves shown corresponds to one of the scattering processes, either physical or self-scattering. The length of each subinterval is the relative probability that that particular scattering process occurs, given that a carrier of energy E_1 is scattering. A random number r_2, uniformly distributed from 0 to 1, is generated. This random number will fall in one of the subintervals of Figure 4.2, and the corresponding scattering process is the process used to terminate the flight in question.

The above scheme ensures that we will accurately simulate the statistics describing the relative scattering rates $\lambda(\mathbf{k})$. The next challenge is to accurately simulate the statistics that describe the change in energy and momentum which

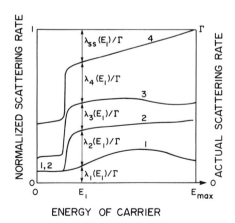

FIGURE 4.2.   Selection of scattering process in Monte Carlo simulations.

occurs in each of the scattering processes $S(\mathbf{k}, \mathbf{k}')$. The easiest is self-scattering. If the process selected is self-scattering, the electronic state is left unchanged and the simulation returns to step 1 and generates the next time-of-flight. The various physical processes can be divided into several categories defined by the choices between elastic and inelastic processes, intravalley and intervalley processes, and momentum-randomizing and -nonrandomizing processes. In all cases we need to choose a final state which lies on the appropriate constant-energy surface in the appropriate valley. If the process is inelastic, it is necessary to shift the energy of the final state away from that of the initial state by the appropriate amount. Similarly, if the process is an intervalley process, it is necessary to update the valley indicator. No additional random numbers are generated in determining the final energy and valley.

Additional random numbers, however, are always needed to determine the final momentum. If the process is momentum-randomizing, then all k-vectors which terminate on the final constant-energy surface are equally probable. Generally, the simulation first uses random numbers to generate an angle between the final-state wave vector $\mathbf{k}'$ and the applied field vector. The set of k-vectors having this angular relation with the field defines a closed path on the constant-energy surface. Any such k-vector can be described in terms of a component parallel to the field, k'_z, and a component perpendicular to the field, k'_p. While quite commonly the symmetry of the problem allows us to stop here, in some situations we must actually specify which of these k-vectors has been chosen. This can be done by splitting k'_p into two components which specify a given location on the closed path along the constant-energy surface.

The simplest case is shown in Figure 4.3. Here the constant-energy surface is spherical, and the closed path is a circle. The term k'_z will be $k'_{\text{mag}} \cos \theta$, and k'_p will be $k'_{\text{mag}} \sin \theta$, where k'_{mag} is the magnitude of $\mathbf{k}'$ and θ is the angle between it and the field. It will always be necessary to specify θ. Since the scattering

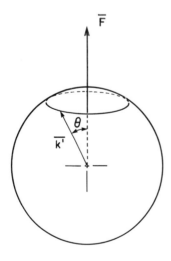

FIGURE 4.3. Selection of angle θ between field vector $\mathbf{F}$ and final state vector $\mathbf{k}'$ for spherical constant-energy surfaces. All states on circle shown are equally probable, but probability of θ is proportional to circumference of associated circle.

process does not favor any of the states on the surface, the probability of scattering to a $\mathbf{k}'$ which yields angle θ is proportional to the number of states available which yield this angle. Since the density of states is uniform along the sphere, the number of states available which yield an angle θ is proportional to the circumference of the circle characterized by this angle. This circumference in turn is proportional to the circle's radius $k'_{mag} \sin \theta$. In fact, the probability density function for scattering into the interval $(\theta, \theta + d\theta)$, $P(\theta)\, d\theta$, is $\sin \theta\, d\theta / 2$. The probability distribution function is a function which describes the probability that the scattering angle is less than θ. This function is

$$F(\theta) = \int_0^\theta P(\theta)\, d\theta = \tfrac{1}{2}(1 - \cos \theta), \tag{4.88}$$

and it has the useful property that it runs from 0 to 1 in a monotonically increasing fashion. Therefore it can be used in conjunction with a random number generator which generates numbers uniformly between 0 and 1 to generate the angle θ. This is illustrated graphically in Figure 4.4, but in the simulation one simply generates a number r_θ and then transforms it into θ by the rule

$$\theta = \cos^{-1}(1 - 2r_\theta). \tag{4.89}$$

If it is necessary to decompose k'_p, this can be done by generating another random number between 0 and 2π. This random number is the angle φ, and the two components of k'_p are then $k'_p \cos \varphi$ and $k'_p \sin \varphi$. Having done this, we are now ready to apply step 1 again.

  If the process is non-momentum-randomizing, a different procedure must be followed in order to choose $\mathbf{k}'$. In this case the angle of interest β is the angle between the initial vector $\mathbf{k}$ and the final vector $\mathbf{k}'$. Proceeding along lines similar to that above, our first task is to find $P(\beta)$, the probability density function for β. The integral of this function is the probability distribution function $F(\beta)$ for β, and it generally behaves similarly to our previous probability distribution

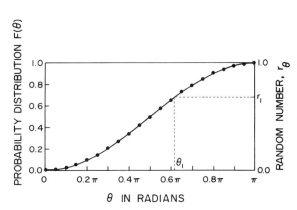

FIGURE 4.4. The probability distribution $F(\theta)$ for situation of Figure 4.3. If random number r_θ is generated, then angle θ is selected.

$F(\theta)$. Unfortunately, it is not always possible to find a closed-form expression for $F(\beta)$. Under these circumstances one can either numerically integrate the probability density $P(\beta)$ or employ the rejection technique. In the rejection technique one generates two random numbers: r and β_r. This pair of numbers describes the coordinates of a point in a plane similar to that shown in Figure 4.5. The horizontal axis is β, and the function plotted is $P(\beta)$. The number r is uniformly distributed from 0 to 1 and describes the vertical coordinate, while β_r is uniformly distributed from 0 to π and describes the horizontal coordinate. If the point selected lies below the $P(\beta)$ curve, then the angle β_r is accepted. If the point lies above the $P(\beta)$ curve, then the angle is rejected and a new random pair of coordinates is selected. This new pair will then be tested in exactly the same fashion as was the previous pair. This process is repeated until an acceptable pair is generated. The process just described, when repeated many times in the course of the simulation, is equivalent to performing a Monte Carlo integration of $P(\beta)$ by the "hit-or-miss" technique.[19]

The results of applying this three-step Monte Carlo process many times are shown in Figure 4.6. Notice that shifts in directions perpendicular to the applied field occur only as a result of the scattering events, while changes in the k-direction which is parallel to the field occur as a result of both scattering and acceleration by the electric field. Here we also show a real-space trajectory which similarly has a field accelerating the carrier in the z-direction. The real-space trajectory is continuous and is a realization of a Brownian path. A Brownian path is described by a random process which is a continuous but nondifferentiable function of time.[20]

In the early applications of the Monte Carlo method, it was used to estimate parameters which appeared in other models, most commonly the nonohmic drift-diffusion model. (In recent years it has been increasingly used in the direct simulation of small semiconductor devices. Such techniques will be discussed in Chapter 5.) For example, a field-dependent drift velocity could be estimated by applying N iterations of the above three-step procedure for a constant field F.

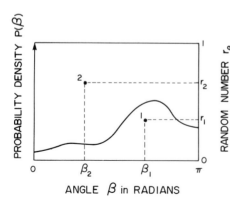

FIGURE 4.5. Use of rejection technique for selecting scattering angle β. The pair of random numbers r_1 and β_1 are selected, and the angle β_1 is used while the pair of random numbers r_2 and β_2 would be rejected.

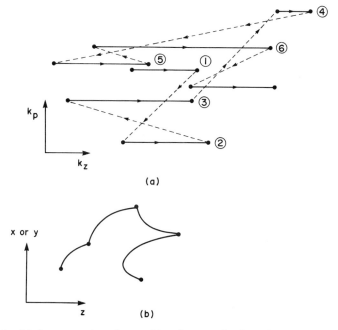

FIGURE 4.6. (a) k-space trajectories resulting from application of Monte Carlo method. (b) Real-space trajectories resulting from application of Monte Carlo method.

The drift velocity then is estimated by

$$V_d(E) = \frac{1}{N} \sum_{i=1}^{N} v(\mathbf{k} \text{ at end of first step}). \tag{4.90}$$

Diffusion coefficients[21] could be estimated by evaluating the average velocities v_z and v_x of an electron during each of many time intervals, each of duration Δt, squaring each of these values v_z and v_x, and taking the average of these squares. Here we have assumed that the field is applied in the z-direction, and we obtain the diffusion coefficients

$$D_{\text{par}} = \tfrac{1}{2}(\langle v_z^2 \rangle - \langle v_z \rangle^2)\, \Delta t \tag{4.91}$$

and

$$D_{\text{per}} = \tfrac{1}{2}\langle v_x^2 \rangle\, \Delta t, \tag{4.92}$$

where $\langle\ \rangle$ denotes the averages just described.

4.9. BROWNIAN DYNAMICS

At this point we seem to always have a choice between models which treat electrons and holes as fluids or models which treat electrons, holes, *and the scattering processes* as particulate. There is a middle ground, however, which has been partially explored for systems of chemically interacting molecules in a fluid medium,[22,23] and has been used by Lippens *et al.*[24,25] to simulate an important millimeter-wave semiconductor source, the IMPATT. In this middle ground the electrons and holes are treated as particles undergoing a Brownian motion (chemical physicists call this approach Brownian dynamics). This means that while the charge carriers are treated as individual particles, the scattering processes are not treated as a sequence of individual scattering events. Instead, each particle follows a drift-diffusion type of behavior in which the role of scattering is treated in an average fashion using a set of transport and ionization coefficients.

This model, like other simulation tools, divides space-time into space steps and time steps. In an individual time step, a carrier can change its position. The position change is computed by first drifting the carrier a distance $\mu F \Delta t$, where μ is a mobility which is a function of the particle energy. Following this drift, a random position change is introduced, using random number generation guided by the Einstein formula (4.10). The diffusion coefficient used in (4.10) during this step is determined by the particle energy. Impact ionization is treated in a random fashion by using random number generation again, this time guided by an ionization probability which is determined by the carrier energy. This allows ionization thresholds to automatically be introduced by assigning an ionization probability of zero to a carrier whose energy is below threshold. Since all of the parameters are functions of particle energy, this quantity needs to be calculated as well. Energy is calculated for each particle by first adding the energy gained by the particle in moving a distance Δx through the field (this Δx includes both the drift and diffusion position shifts) to the energy of the particle at the beginning of the time step and then subtracting the energy lost to the lattice during the time step. This energy loss is $(w - w_{th})\Delta t / \tau$, where τ is an energy relaxation time which also is determined by the carrier energy.

There is an important feature, central to the Brownian dynamic approach, which can escape casual notice. The relations used in the above paragraph look suspiciously like energy balance relations. The energy balance relations, however, were not interpreted as describing the evolution of the energy of a single carrier but as describing the evolution of the average carrier energy. In the Brownian dynamic approach, however, we are using these relations to describe the dynamics of a single carrier. Formally justifying this approach is difficult. A more sophisticated variant would be to replace the energy relaxation time by an energy relaxation rate

$$R_E(\mathbf{k}) = \left\{ \int S(\mathbf{k}, \mathbf{k}')[E(\mathbf{k}) - E(\mathbf{k}')] \, d\mathbf{k}' \right\}^{-1}. \qquad (4.93)$$

A similar equation could be used to describe the momentum relaxation of a single carrier. The crucial difference between (4.93) and the usual energy relaxation time form is that the distribution function appears in the normal form but is missing in (4.93). It is not needed there because $R_E(\mathbf{k})$ will only be used when the carrier is known to be in state **k**. An expanded discussion of these considerations is found in the next section.

4.10. RETARDED LANGEVIN EQUATIONS

We now consider another Brownian motion model for carrier transport in a semiconductor. While one can attempt to derive a Langevin equation, which is a differential equation with a random force term, for carrier transport, the derivation is long and well beyond the limits of our present discussion. Instead we provide a heuristic argument. We are interested in calculating the time derivative of the velocity of a single carrier. There are three sources for this time variation. First, the carrier can be accelerated by applied fields. Second, there will be random changes in carrier velocity due to scattering, which we can represent by a random force $R(t)$. Last, there will be a net momentum relaxation to the lattice through scattering, which in the steady-state counteracts the net momentum gain from the field. Note that we have basically broken the scattering contribution into two parts: an average contribution, which we represent as a frictional damping, and a fluctuating force. Our Langevin equation for a single electron then is[26]

$$m^* \frac{dv}{dt} = -m^* \int_0^t \gamma(t-u)v(u)\, du + R(t) + eF(t)H(t), \qquad (4.94)$$

where $\gamma(t-u)$ is called the memory function. It is the use of this convolution integral, as opposed to relaxation times, that causes this approach to be referred to as a retarded Langevin equation. We have assumed that the field is zero for negative time through the use of the Heaviside step function

$$H(t) = \begin{cases} 0 & \text{if } t < 0, \\ 1 & \text{if } t \geq 0. \end{cases} \qquad (4.95)$$

Therefore, we start with an initial ensemble of carriers, each of which obeys (4.94), which is in thermal equilibrium with the bath. This means that our initial conditions on ensemble averages are

$$\langle v(0) \rangle = 0 \qquad (4.96)$$

and

$$m^* \langle v(0)^2 \rangle = k_B T_L. \qquad (4.97)$$

The solution proceeds by taking the Laplace transform of (4.94):

$$m^* v(s)s - m^* v(0) = \frac{e}{s} F(s) - m^* v(s)\gamma(s) + R(s), \qquad (4.98)$$

which can be rewritten as

$$v(s) = X(s)\left[v(0) + \frac{eF(s)}{m^*s} + \frac{R(s)}{m^*}\right], \tag{4.99}$$

where

$$X(s) = \frac{1}{s + \gamma(s)}. \tag{4.100}$$

Returning to the time domain, we find

$$v(t) = v(0)X(t) + \frac{e}{m^*}\int_0^t F(u)X(t-u)\,du$$

$$+ \frac{1}{m^*}\int_0^t R(u)X(t-u)\,du. \tag{4.101}$$

We now take the ensemble average of $v(t)$, noting that $\langle R(t)\rangle = 0$ and applying (4.97). The result is

$$\langle v(t)\rangle = v_d(t) = \frac{e}{m^*}\int_0^t F(u)X(t-u)\,du. \tag{4.102}$$

In setting the ensemble average equal to a drift velocity, we are implicitly assuming, as we have throughout this section, a spatially uniform system. Our task now is to find expressions for $X(t)$. Since we have expressed the diffusion coefficient in terms of an integral of the velocity autocorrelation function, and we know that the drift velocity in the steady state will be described by a mobility that is connected with the diffusion coefficient at equilibrium by the Einstein relation, we will try to express $X(t)$ in terms of an integral of the velocity autocorrelation as well. To do this, we multiply (4.101) by $v(0)$ and take another ensemble average. The result is a two-time velocity autocorrelation function

$$\langle v(0)v(t)\rangle = \langle v(0)^2\rangle X(t) = \varphi_v(0, t), \tag{4.103}$$

where we have assumed that

$$\langle v(0)R(t)\rangle = 0. \tag{4.104}$$

Therefore, as $v_d(0) = 0$,

$$X(t) = \frac{\varphi_v(0, t)}{\langle v(0)^2\rangle} \tag{4.105}$$

and

$$v_d(t) = \frac{e}{k_B T_L} \int_0^t F(u)\varphi_v(0, t - u) \, du. \tag{4.106}$$

Now, do we get the Einstein relation? We have a steady-state chordal mobility defined by [$F(u)$ constant]

$$\mu = \lim_{t \to \infty} \frac{e}{k_B T_L} \int_0^t \varphi_v(0, t - u) \, du, \tag{4.107}$$

and a steady-state diffusion coefficient defined by (4.19). We will have the Einstein relation if

$$\int_0^\infty \varphi_v(0, t - u) \, du = \tfrac{1}{2} \int_{-\infty}^{+\infty} \varphi_{\Delta v}(\tau) \, d\tau. \tag{4.108}$$

Equality here will occur if $\varphi_{\Delta v}(t_1, t_2) = \varphi_v(t_1, t_2) = \varphi_v(t_2 - t_1) = \varphi_v(t_1 - t_2)$. The equality of φ_v and $\varphi_{\Delta v}$ requires a zero drift velocity and the time symmetry is needed to obtain the proper relationship between the "single-sided" and "double-sided" time integrals. These conditions require that $F(t)$ be 0 in (4.95), and we therefore recover the Einstein relation in equilibrium.

We now consider the simple special case of a small ac field. We let

$$F(t) = F_{ac} \exp(j\omega t) \tag{4.109}$$

and calculate the corresponding ac velocity

$$v(t) = v_{ac} \exp(j\omega t), \tag{4.110}$$

where both F_{ac} and v_{ac} are complex-valued. If $\varphi_v(t)$ were a simple exponential decay with time constant τ, then

$$v(t) = \frac{e}{k_B T_L} \int_0^t F_{ac} \exp(j\omega u)\langle v(0)^2\rangle \exp\left[\frac{-(t - u)}{\tau}\right] du, \tag{4.111}$$

which yields an ac term

$$v_{ac} = \frac{e\tau}{m^*} \frac{F_{ac}}{1 + j\omega\tau}. \tag{4.112}$$

We therefore have the ac mobility

$$\mu_{ac} = \frac{e\tau}{m^*} \frac{1}{1 + j\omega\tau}. \tag{4.113}$$

For the same autocorrelation function the dc chordal mobility is

$$\mu = \frac{e\tau}{m^*}. \tag{4.114}$$

We have this type of simple exponential decaying autocorrelation function only near equilibrium. This also is the linear transport regime in which the differential mobility equals the chordal mobility. In the present analysis any field dependence of the mobility leading to nonlinear responses has been buried in φ_v. The τ used in the equations above corresponds to the momentum relaxation time for the system in the linear transport regime, and the ac mobility is closely approximated by the differential mobility for frequencies which are small when compared with the inverse of this relaxation time.

Philosophically, models of this sort for high-field transport are troublesome. The individual carrier's real microscopic transport process is determined by processes which depend only on the present state of the carrier and not on how it arrived in that state. On the other hand, these processes strongly depend in a nonlinear fashion on that state. In other words, individual carrier transport is Markovian but nonlinear. The retarded Langevin analysis above, however, is non-Markovian, as a result of the memory effects contained in the convolution integral, but linear in the sense that no explicit dependency of the memory function on the velocity is included. We have also not dealt with the relationship which exists in the near-equilibrium setting between the memory function and the random force term. We return to this question in Chapter 10.

REFERENCES

1. A. Einstein, *Investigations on the Theory of the Brownian Movement*, Dover, New York (1956). This reprints Einstein's papers from *Ann. Phys.* of 1905 and 1906.
2. S. M. Sze, *Physics of Semiconductor Devices*, 2nd ed., Wiley, New York (1981).
3. P. T. Landsberg, in: *Handbook on Semiconductors* (William Paul, ed.), North-Holland, Amsterdam (1982).
4. C. Kittel and H. Kroemer, *Thermal Physics*, 2nd ed., W. H. Freeman, San Francisco (1980).
5. J. P. Nougier, in: *Physics of Nonlinear Transport in Semiconductors* (D. K. Ferry, J. R. Barker, and C. Jacoboni, eds.), Plenum Press, New York (1980).
6. O. E. Lanford III, in: *Nonequilibrium Phenomena. I: The Boltzmann Equation* (J. L. Lebowitz and E. W. Montroll, eds.), North-Holland, Amsterdam (1983).
7. P. N. Butcher, in: *Electrons in Crystalline Solids*, International Atomic Energy Agency, Vienna (1973).
8. E. Conwell, in: *Handbook on Semiconductors* (William Paul, ed.), North-Holland, Amsterdam (1982).
9. R. Stratton, *IEEE Trans. Electron Dev.* **ED-19**, 1288 (1972).
10. R. K. Mains, G. I. Haddad, and P. A. Blakey, *IEEE Trans. Electron Dev.* **ED-30**, 1327 (1983).
11. D. L. Woolard, R. J. Trew, and M. A. Littlejohn, *Fifth Int. Conf. on Hot Carriers in Semiconductors*, Boston, MA (1987).
12. D. Jones and H. D. Rees, *J. Phys. C* **6**, 1781 (1973).
13. R. K. Mains, M. A. El-Gabaly, G. I. Haddad, and J. P. Sun, *IEEE Trans. Electron Dev.* **ED-31**, 1273 (1984).

14. J. G. Ruch, *IEEE Trans. Electron Dev.* **ED-19**, 652 (1972).

15. M. S. Shur, *Electron. Lett.* **12**, 615 (1976).

16. H. J. Kafka and K. Hess, *IEEE Trans. Electron Dev.* **ED-28**, 831 (1981).

17. R. K. Froelich, *Computer Modeling of Millimeter-Wave IMPATT Diodes*, Ph.D. thesis, University of Michigan (1982).

18. H. D. Rees, *J. Phys. Chem. Solids* **30**, 643 (1969).

19. S. J. Yakowitz, *Computational Probability and Simulation*, Addison-Wesley, Reading, MA (1977).

20. S. Karlin and H. M. Taylor, *A First Course in Stochastic Processes*, 2nd ed., Academic Press, New York (1975).

21. W. Fawcett, in: *Electrons in Crystalline Solids*, International Atomic Energy Agency, Vienna (1973).

22. D. L. Emak and J. A. McCammon, *J. Chem. Phys.* **69**, 1352 (1978).

23. W. F. van Gunsteren and H. J. C. Berendsen, *Mol. Phys.* **45**, 637 (1982).

24. D. Lippens and E. Constant, *J. Phys. Coll.* **42**(C7), 207 (1981).

25. D. Lippens, J.-L. Nieruchalski, and E. Constant, *IEEE Trans. Electron Dev.* **ED-32**, 2269 (1985).

26. J. Zimmermann, P. Lugli, and D. K. Ferry, *J. Phys. Coll.* **42**(C7), 95 (1981).

5

Transient Hot-Carrier Transport

5.1. ENSEMBLE MONTE CARLO TECHNIQUES

5.1.1. Direct Comparison with Retarded Langevin Equations

When a system with many time constants is excited by a temporally varying forcing function, the system's response can exhibit a variety of features such as overshoots, undershoots, and phase shifts. When the forcing function varies slowly on the scale of the time constants, it is common to use quasi-static analysis techniques in which the instantaneous time-varying response is modeled by using the steady-state response of the system to a constant forcing function equal in magnitude to the actual instantaneous forcing function. For the transport problems considered here, the system is an ensemble of carriers embedded in a set of energy bands and interacting with a phonon bath. The time constants include energy and momentum relaxation times or various scattering rates, the forcing function is the electric field, and the response of greatest interest is the ensemble average carrier velocity. The quasi-static analysis technique is the use of a steady-state drift-diffusion transport law. The response of the ensemble in situations where this steady-state or quasi-static analysis is inappropriate is called the transient dynamic response (TDR). Describing carrier transport in the TDR regime is the central topic of this chapter.

The details of TDR are exceedingly important for semiconductor devices operating at very high frequencies or for those which exist in a physically small space,[1,2] since for these situations the ensemble may never exist in a steady state. The most commonly studied variation subjects the carriers to a sudden step in field from an initial value of zero to a nonzero value. This temporal step is viewed as representing the field seen by a moving ensemble of carriers leaving a low-field region (such as the source of an FET) and entering a high-field region (such as the channel under the gate of the FET). The ensemble evolves temporally and eventually to a new, steady-state distribution characterized by the magnitude of the electric field (as discussed in Chapter 4). The interesting aspect is that during the transient the average velocity is not only poorly predicted by the

173

steady-state velocity-field characteristic, but actually is significantly underestimated. This is referred to as the velocity overshoot, even though the carrier velocity never does anything improper at all. The only thing improper would be to expect the steady-state velocity to work as a velocity estimator for a non-steady-state situation.

Since the 1970s, Monte Carlo (MC) techniques[3-13] have been the workhorse for theoretical investigations of transport in the TDR for semiconductors. A commonly held belief is that this approach is justified because the MC method provides a solution for the Boltzmann transport equation (BTE). However, the MC technique is a random walk process in momentum space, and the general probability integrals reduce to a path integral of the BTE only in the steady-state limit.[14] This, however, is not a terrible thing because the BTE itself is valid (as a carrier transport model) only in this same limit, and it has been known for some time that more generalized transport equations accede to a path integral solution which agrees with ensemble Monte Carlo techniques on the short time scale.[15] We in fact will discover that the retarded Langevin equations of Chapter 4 are more closely related to Monte Carlo calculations than are the balance equations derived from the BTE by application of the method of moments.

In the previous chapter we discussed both the basics of the Monte Carlo technique and the use of various analytical formulations, such as retarded Langevin equations (RLE) for carrier transport models. Here we start with a discussion of the relationship that exists between Monte Carlo techniques and analytical techniques. The essential tool is the use of an ensemble Monte Carlo (EMC) technique. In Figure 5.1, we show two variations on Monte Carlo techniques which differ only in an interchange of the outer two loops. This simple change, however, is important, as bringing the electron counter inside the time loop allows us to simulate the evolution of an ensemble of carriers and to conduct studies similar to the diffusion studies of Chapter 4. We can additionally introduce electron–electron interactions of the sort described in Chapter 7 as well. This chapter therefore starts with discussions of both the EMC technique and of various methods by which we obtain analytical equations which are in some sense equivalent to the EMC process. These methods are closely related to the view of an EMC program as a model for a random walk in momentum space. We also discuss an iterative technique for solving the Boltzmann equation. The last half of the chapter reviews various studies, mainly experimental, of TDR.

Several of the most important effects and the potential interrelationship of the EMC and RLE approaches were illustrated by Lugli et al.[16] Recall from Chapter 4 that a nonstationary two-time correlation function for the velocity fluctuations can be defined as

$$\phi_{\Delta v}(t, t_0) = \langle [v(t) - \langle v(t) \rangle][v(t_0) - \langle v(t_0) \rangle] \rangle. \tag{5.1}$$

The angular brackets represent an ensemble average. This function can be directly calculated using the EMC technique, allowing us to check an important property

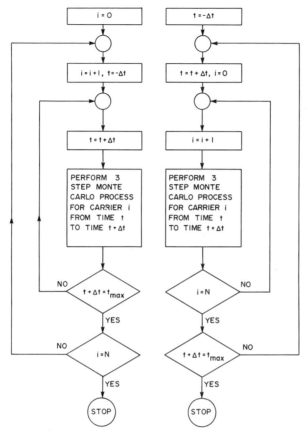

FIGURE 5.1. Flowcharts for the ensemble Monte Carlo technique (charted on right) and the single particle Monte Carlo technique (charted on left).

of the solutions to a retarded Langevin equation, namely that the transient velocity is given by

$$\langle v(t) \rangle = \frac{eF}{m^*} \int_0^t \phi'_{\Delta v}(\tau, 0) \, d\tau, \tag{5.2}$$

where $\phi'_{\Delta v}(\tau, 0)$ is the normalized correlation function

$$\phi'_{\Delta v}(\tau, 0) = \frac{\phi_{\Delta v}(\tau, 0)}{\phi_{\Delta v}(0, 0)}. \tag{5.3}$$

Both sides of (5.2) have been calculated for an overshoot study in Si and are compared in Figure 5.2. They agree within the accuracy of the EMC technique. While this agreement also occurs in similar cases for a multivalley system, such as GaAs, there are experimentally accessible initial conditions for which (5.2)

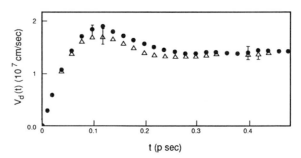

FIGURE 5.2. Comparison of the integral of the velocity autocorrelation function (solid circles) with the actual transient ensemble average velocity. The circles are a solution of equation (5.2), while the triangles are the ensemble average velocity. The material was silicon, and a field of 50 kV/cm was applied along the ⟨100⟩ direction.

fails.[17] The reason for the failure is that in the derivation of (5.2), as reviewed in Chapter 4, we use a linear RLE in which the memory function is not sensitive to the present velocity. In the case for which the failure occurred, there was a significant difference in the scattering environment seen by carriers with an initial negative velocity as opposed to those with an initial positive velocity.

The crucial point here is that the EMC calculation was well modeled in at least certain circumstances by a retarded transport equation. A central message of the following discussion is that this connection between EMC techniques and retarded transport equations is not an accident. In the next section we explore the non-Markovian properties of the transport system. We develop a generalized transport equation by using a path integral formulation of the semiclassical transport problem. The random walk equations of the EMC simulation are shown to produce the same temporal equation. Finally, we develop retarded transport equations which describe the temporal evolution of the moments and explore their consequences.

5.2. PROJECTION OPERATORS AND TRANSPORT EQUATIONS

In statistical physics, the equation of motion of any dynamical variable $\alpha(R, p)$, depending explicitly only on the position R and momentum p (and in particular not on time), can be written as

$$i \frac{\partial \alpha}{\partial t} = -L\alpha, \tag{5.4}$$

where the Liouville operator $L = i\{H, \}$ in classical mechanics and the braces are Poisson brackets, which are defined by

$$\{u, v\} = \frac{\partial u}{\partial q_i} \frac{\partial v}{\partial p_i} - \frac{\partial u}{\partial p_i} \frac{\partial v}{\partial q_i}, \tag{5.5}$$

where q_i are configuration coordinates and p_i are the corresponding conjugate momenta. The Hamiltonian H is, commonly for our problems, of the form

$$H = H_e + H_{e\text{-}F} + H_{e\text{-}L},\tag{5.6}$$

where H_e is the electronic Hamiltonian (sometimes including the Coulombic carrier–carrier interaction), $H_{e\text{-}F}$ is the Hamiltonian describing the field acceleration of the carriers, and $H_{e\text{-}L}$ is a term describing the carrier-lattice interaction. Equation (5.4) stems from the Liouville equation for the time dependence of the phase-space ensemble density function $f(p, R, t)$:

$$i\frac{\partial f}{\partial t} = Lf.\tag{5.7}$$

The sign difference between (5.4) and (5.7) is significant.

For the many particle systems of interest here, the above relations involve a dauntingly large number of variables. Therefore, if we are to develop an analytical formulation which we can potentially solve, or at least intelligently discuss, somehow we must replace this problem by one involving a much smaller number of variables. This formally can be described as a projection of the problem from the large-dimensional space into a space of far fewer dimensions. In the following, we introduce a general projection operator such that[18]

$$Pf = f_1,\tag{5.8}$$

$$P^2 = P,\tag{5.9}$$

$$Q = (1 - P),\tag{5.10}$$

$$PQ = QP = 0,\tag{5.11}$$

$$Q^2 = Q,\tag{5.12}$$

and for any f,

$$f = Pf + Qf,\tag{5.13}$$

all of which are normal properties for a projection operator. Here f_1 can be completely described by a hopefully small number of variables. Equation (5.8) describes the projection P, and (5.10) describes the unprojected portion Q of the function. Once we have projected into the subspace, then a further projection into the subspace will not alter the function (5.9). Similarly, once we have extracted the unprojected portion of f, a similar extraction leaves the result unchanged (5.12). However, as (5.11) and (5.13) note, all portions of the original function f are either projected onto f_1 or left unprojected. In general, operation with P could produce, e.g., the diagonal part of a Green's function, the average value of f ($Pf = \langle f \rangle$), etc. In principle, each f chosen for (5.7) is a sample from the complete phase-space representation, and Pf can represent the average single-particle distribution function.[19]

A kinetic equation for f_1 can be developed. To proceed, we multiply (5.7) first by P and then by $(1-P) = Q$, and obtain the equations

$$iP\frac{\partial f}{\partial t} = i\frac{\partial f_1}{\partial t} = PL(Pf + Qf) = PLf_1 + PLQf \qquad (5.14)$$

and

$$i(1 - P)\frac{\partial f}{\partial t} = i\frac{\partial Qf}{\partial t} = QL(Pf + Qf) = QLf_1 + QLQf. \qquad (5.15)$$

Note that as the projections P and Q project into very different spaces, which in fact contain no information concerning the other, (5.14) and (5.15) are independent of each other. We therefore have two independent equations in the unknowns Pf (or f_1) and Qf. These equations are coupled through the Liouville operator, which can transform unprojectable portions of f into projectable portions, and vice versa. Now, we formally solve the second of these for Qf, which gives

$$Qf(t) = e^{-itQL}Qf(0) - i\int_0^t dt'\, e^{-it'QL}QLf_1(t - t'), \qquad (5.16)$$

where $Qf(0) = f(0) - Pf(0)$. Then (5.16) can be used to replace Qf in (5.14), and

$$\frac{\partial f_1}{\partial t} + iPLf_1 = -\int_0^t dt'\, [PL\, e^{-it'QL}QL]f_1(t - t') - iPL\, e^{-itQL}Qf(0). \qquad (5.17)$$

When L can be represented as having a diagonal part and a weak nondiagonal part, a simple semiclassical perturbation theory is useful, as normally applied to the BTE. Suppose that L can be represented as

$$L = L_0 + L', \qquad (5.18)$$

where L_0 is diagonal and L' is weak with no diagonal elements. The second term on the left is easy, and

$$PLf_1 = L_0 f_1. \qquad (5.19)$$

This term leads generally to the field and spatial gradient streaming terms, of which only the field term is of interest here. The contribution from the integral in (5.17) is somewhat more difficult. However, noting that since diagonal terms are omitted, $QLf_1 = QL'f_1$, and using (5.12) to show that

$$e^{-itQL}QA = \left[1 - itQL + \frac{(it)^2QLQL}{2} + \cdots\right]QA = Q\, e^{-itLQ}QA, \qquad (5.20)$$

we may write

$$PL\, e^{-it'LQ} QL = PLQ\, e^{-it'LQ} QL \simeq PL'\, e^{-it'L_0 Q} QL'. \qquad (5.21)$$

It is not a simple matter to show that this is just the normal scattering operator in the weak coupling theory. However, it can be done in our semiclassical approach,[18,20-24] with the exception that the energy-conserving joint-spectral density function only becomes a δ function in the infinite time limit. Thus, since L_0 includes the field, it is possible to obtain a field-scatterer interaction leading to the intracollisional field effect.[20-23] Then (5.17) can be written in the more usual form

$$\frac{\partial f_1}{\partial t} + iL_0 f_1 = -\int_0^t dt' \sum_{p'} [W(p', p, t') f_1(p', t - t') - W(p, p', t) f_1(p, t - t')]$$

$$- iPL\, e^{-itQL} Qf(0). \qquad (5.22)$$

With the exception of the last term, (5.22) has the form of the kinetic equation used earlier for generalized retarded transport, although this earlier form was obtained on quantum mechanical grounds.[21,25] We will return to a discussion of this last term later.

5.3. PATH INTEGRAL SOLUTIONS TO GENERAL TRANSPORT EQUATIONS

The nonlocal equation (5.22) cannot be immediately written as a Chambers-Rees-type path integral because of the inherent retardation of the scattering term. Therefore, the numerical evaluation of (5.22) remains a formidable challenge. However, by generalizing the concept of the self-scattering process, we can obtain a relatively simple path integral formulation for $f_1(p, t)$ which is similar in form to the more usual path integral formalism for the BTE.[26,27] In the following we ignore, for the moment, the memory terms. We return to them in Section 5.5. To proceed, we add and subtract identical terms to the right-hand side of (5.22), so that we can define[28]

$$W^*(p', p, t, t') = W(p', p, t, t') + [\Gamma(t, t') - \Gamma_{out}(p, t', t)]\delta_{p,p'}, \qquad (5.23)$$

where $\delta_{p,p'}$ is the Kronecker delta function,

$$\Gamma_{out}(p, t', t) = \sum_{p'} W(p', p, t, t'), \qquad (5.24)$$

$$\Gamma(t, t') = \Gamma_0 \delta(t - t'), \qquad (5.25)$$

and $\delta(t - t')$ is the Dirac delta function, with Γ_0 a positive constant which equals the total (real plus self-) scattering rate. The kinetic equation (5.22) may then be written as, with the field term explicitly introduced,

$$\left(\frac{\partial}{\partial t} + e\mathbf{F} \cdot \nabla_p + \Gamma_0\right)f_1 = \sum_{p'} \int_0^t dt' \, [W^*(p', p, t - t')f_1(p', t')], \qquad (5.26)$$

which may now be solved by the method of characteristics.[29] We transform to a coordinate system determined by the collision-free trajectories of the particles as

$$p \to p^*(s), \qquad t \to s, \qquad (5.27)$$

where

$$\frac{dp^*(s)}{ds} = eF, \qquad p^*(s) = p^*(0) + eFs, \qquad p^*(t) = p. \qquad (5.28)$$

Our purpose is to replace the two partial derivatives on the left-hand side of (5.26) by a single total derivative in terms of the path variable. Our function f_1 is transformed by

$$f_1(p, t) \to f_1(p^*(0) + eFs, s). \qquad (5.29)$$

Investigating the total derivative of the transformed f_1 with respect to s, we find

$$\frac{df_1(p^*(0) + eFs, s)}{ds} = \frac{\partial f_1(p^*(0) + eFs, s)}{\partial s}$$

$$+ \frac{\partial f_1(p^*(0) + eFs, s)}{\partial(p^*(0) + eFs)} \frac{\partial(p^*(0) + eFs)}{\partial s} \qquad (5.30)$$

$$= \frac{\partial f_1(p^*(0) + eFs, s)}{\partial s} + \frac{\partial f_1(p^*(0) + eFs, s)}{\partial(p^*(0) + eFs)} eF, \qquad (5.31)$$

or, in terms of the original variables,

$$\frac{df_1(p^*(0) + eFs, s)}{ds} \to \left(\frac{\partial}{\partial t} + e\mathbf{F} \cdot \nabla_p\right)f_1. \qquad (5.32)$$

Then (5.21) becomes

$$\frac{df_1}{ds} + \Gamma_0 f_1 = \sum_{p'} \int_0^s dt' \, [W^*(p', p^*, s - t')f_1(p', t')], \qquad (5.33)$$

and

$$f_1(p, t) = \int_0^t dt' \, e^{-\Gamma_0(t-t')} \sum_{p'} \int_0^{t'} dt'' \, [\, W^*(p', p - eFt'', t'')f_1(p', t' - t'')].$$ (5.34)

In the limit of long times $t \gg 0$, instantaneous collisions, and no quasi-particle effects from the field, (5.34) reduces to the normal path integral form of the BTE.[26,27] However, (5.34) is not a completely general solution to (5.26). The introduction of self-scattering limits the applicability of (5.34) to those situations in which this fictitious scattering process can be incorporated automatically in the ensemble Monte Carlo technique. Additionally we have omitted the initial condition, discussed below.

We have pursued the path integral formulation first so as to introduce the trajectory coordinates of (5.28), as we will have need of these in the next section.

5.4. NONEQUIVALENCE OF BOLTZMANN THEORY AND ENSEMBLE MONTE CARLO

Before we enter this section, please remember that, quite contrary to popular supposition, the Boltzmann theory is only an approximation for the hot-carrier transport problem. The ensemble Monte Carlo technique obviously includes in a direct fashion, single-particle, multiple-time correlations that are thrown away in the Boltzmann equation, which after all only solves for a single-particle, single-time distribution function. In fact, in the context of a semiclassical approximation, the EMC technique is more accurate at short time scales than is the BTE. In this section we will more formally study this nonequivalence, following the general approach of Klafter and Silbey[30] and Kenkre et al.[31]

We begin with the single-particle MC process. This process consists of a sequence of free flights, each of which terminates in a scattering event. We define $S_n(p_0, p, t)$ to be the conditional probability of an electron, initially at p_0 at time $t = 0$, passing through the state p at time t during the course of the nth free flight. The explicit time dependence of this definition must be maintained since the electron can pass through p at any time $0 < t < \infty$. We can, however, develop an iterative relationship for S_n.

In order to pass through p at time t during the nth free flight, the $(n-1)$st flight must have terminated in a scattering event that left the carrier in a momentum state p_{n-1} which lies on the free path passing through state p at time t—that is, a p_{n-1} such that

$$p = p_{n-1} + e \int_{t-\tau'}^t F(u) \, du,$$ (5.35)

where $t - \tau'$ is the time at which the $(n-1)$st flight terminated and $F(u)$ is the electric field seen by the carrier at time u. If we were to restrict ourselves to the case of a constant electric field, then

$$p_{n-1} = p - e[t - (t - \tau')]F = p - eF\tau'. \tag{5.36}$$

The similarity between this result and the path variable formulation of (5.28) is not accidental.

The above two equations tell us how the $(n-1)$st flight must terminate. We can see that the conditional probability S_n must depend on the product of the conditional probability that the $(n-1)$st flight terminates by scattering into the state p_{n-1} lying on the appropriate collision free path, and the probability that there is no scattering event in the time interval $t - \tau'$ to t. If we define $M_{n-1}(p_0, p_{n-1}, t, \tau')$ to be the conditional probability that, given an initial momentum p_0 at time zero, the $(n-1)$st free flight terminated at time $t - \tau'$ in a scattering event which left the particle in momentum state p_{n-1}, then $S_n(p_0, p, t)$ is

$$S_n(p_0, p, t) = \int_0^t d\tau' \, M_{n-1}(p_0, p - eF\tau', t, \tau') \, e^{-\Gamma_0 \tau'}. \tag{5.37}$$

The factor $e^{-\Gamma_0 \tau'}$ is the probability, given a constant scattering rate Γ_0 (which here includes self-scattering), that no scattering event occurs in a time interval of length τ'. The iterative relationship for constructing S_n is obtained by expressing M_{n-1} in terms of S_{n-1} and $W^*(p', p)$, the transition rate from p' to p. This is done by first using S_{n-1} to describe the conditional probability of a carrier passing through p' at some time t'', while using the transition rate W^* to describe the probability that at this time t'' the flight is terminated by a transition to the appropriate path for S_n. The relation is

$$M_{n-1}(p_0, p - eF\tau', t, \tau') = \sum_{p'} S_{n-1}(p_0, p', t - \tau') W^*(p', p - eF\tau'). \tag{5.38}$$

Then the conditional probability function S_n satisfies the iterative relationship[14]

$$S_n(p_0, p, t) = \sum_{p'} \int_0^t dt' \, S_{n-1}(p_0, p', t - t') W^*(p', p - eFt') \, e^{-\Gamma_0 t'}. \tag{5.39}$$

If we formally define the operator $\tilde{W}$ by

$$\tilde{W} S_i = \sum_{p'} \int_0^t dt' \, S_{i-1}(p_0, p', t - t') W^*(p', p - eFt') \, e^{-\Gamma_0 t'}, \tag{5.40}$$

then (5.39) is the nth term in the series

$$S = S_1 + \tilde{W} S_1 + \tilde{W}\tilde{W} S_1 + \tilde{W}\tilde{W}\tilde{W} S_1 + \cdots, \tag{5.41}$$

where $S_1(p_0, p, t)$ is the conditional probability that the carrier started in state p_0 at time zero and traveled through state p at time t during its first free flight. There is only one p_0 for which $S_1(p_0, p, t)$ is not zero, as there is only one ballistic free flight path from the initial momentum state to the final momentum state. This first term therefore equals

$$S_1(p_0, p, t) = e^{-\Gamma_0 t}\delta(p - p_0 - eFt). \tag{5.42}$$

To convert these conditional probabilities to distribution functions, one must weight $S_1(p_0, p, t)$ by the initial distribution function $f(p_0, t = 0)$. When this is done, one must be careful to set p_0 and p appropriately in the first term of the expansion. Each following term in the expansion represents the probability of moving from p_0 to p in time t with a number of scattering events equal to the number of applications of the operator $\tilde{W}$, which automatically places the particle on the appropriate free path. These terms are each a nesting of time-ordered integrations, and one must make sure that deep inside the nested integrals in the integrand we eventually use

$$S_1(p_0, p', t - t') = e^{-\Gamma_0(t-t')}\delta[p' - p_0 - eF(t - t')], \tag{5.43}$$

where t' is the smallest intermediate time. The total expansion therefore includes all possible ways of combining scattering events and free flights to move from p_0 to p in time t. This expansion is the iterative solution of the integral equation[32]

$$S(p_0, p, t) = \sum_{p'} \int_0^t dt'\, W^*(p', p, t')\, e^{-\Gamma_0 t'} S(p_0, p', t - t') \tag{5.44}$$

and corresponds to general expansions found in quantum mechanics when the interaction representation is used. Therefore, (5.34) is the solution of the equation[24,29]

$$\frac{dS}{dt} + \Gamma_0 S = \sum_{p'} W^*(p', p)S(p_0, p', t). \tag{5.45}$$

In (5.45) we recognize that Γ_0 includes all relevant outscattering contributions, including self-scattering. However, the connection to the BTE is recognized only when the total derivative (first term on the left-hand side) is interpreted as a coalescence of the streaming terms into a path variable form.[29] To indicate this, we use the symbol s, following Budd,[29] to be the equivalent "time" along the path and rewrite (5.45) as

$$\frac{dS}{ds} + \Gamma_0 S = \sum_{p'} W^*(p', p)S(p_0, p', s). \tag{5.46}$$

Next, we rewrite (5.46) and introduce the generalized collision operator as

$$\frac{dS}{ds} = \check{C}S = (\tilde{W} - \Gamma_0)S, \tag{5.47}$$

where $\tilde{W}$ is defined in (5.40). In general, we have

$$S(s) = e^{\check{C}s}S(0) = G(s)S(0) \tag{5.48}$$

and

$$\frac{dS(0)}{ds} = \check{C}S(0) = 0 \tag{5.49}$$

in equilibrium.

We can now make a connection to the BTE by defining an initial state distribution function of the N electrons used in the EMC. Thus we can set

$$f(p, 0) = \sum_{p_0} n(p_0)S(p_0, p, 0) \tag{5.50}$$

with $\sum_{p_0} n(p_0) = 1$. A typical projection operation at time t can be defined on any arbitrary G, such as that defined in (5.48), by

$$PG = f_1(p, t) \int dp' \, G(p', t), \tag{5.51}$$

with the integral over p or p' normalized to unity, and with $f_1(p, t)$ parameterized in such a way that all central moments are equal to those in the actual distribution.[33] Such an approach has been used to calculate equations of motion for correlation functions.[18]

We now proceed just as we did going from (5.7) to (5.17). We first combine (5.47) through (5.49) to obtain as our starting point

$$\frac{dS}{ds} = \frac{d}{ds}[G(s)S(0)] = \frac{dG(s)}{ds} S(0). \tag{5.52}$$

Since $S(0)$ is a constant, we focus our attention on the derivative of $G(s)$. We now operate on (5.52) first with P and then with Q. The two resulting independent equations are

$$\frac{d}{ds}(PG) = P\tilde{C}G = P\tilde{C}[PG + QG] = P\tilde{C}PG + P\tilde{C}QG \tag{5.53}$$

and

$$\frac{d}{ds}(QG) = Q\tilde{C}G = Q\tilde{C}[PG + QG] = Q\tilde{C}PG + Q\tilde{C}QG. \tag{5.54}$$

Then by inserting the solution to (5.54) into (5.53), we obtain

$$\frac{d}{ds}(PG) = P\tilde{C}PG + \int_0^s dt'\, P\tilde{C}Q\, e^{Q\tilde{C}(s-t')}Q\tilde{C}PG(t') + P\tilde{C}Q\, e^{Q\tilde{C}s}QS(0). \tag{5.55}$$

Note that this equation has a convolution of the function PG with a memory function, the retarded form similar to the RLE. The form of (5.55) is a standard one in transport theory and was initially obtained by Zwanzig.[18] The N-body problem represented by the reduced (5.55) is not solvable due to the number of variables involved.[19] As we are generally interested in a specific average of the evolving system, we may look for an equation of motion of this average quantity (drift velocity, mean energy, etc.). More generally, a projection operator is useful and leaves us with an equation such as (5.55). When a projection operator of this sort is applied, most of the information contained in the description of the entire system is thrown or projected away. This has often been regarded as introducing irreversibility into the equation of motion for the quantity of interest, as well as retardation and memory terms. This phenomenon was thoroughly studied by Zwanzig.[34] Choosing a finite set of parameters such as momentum and energy, each having a measurable average, allows a microcanonical distribution that is parameterized with this parameter set to be defined on a subspace of phase space. This, in turn, satisfies a generalized Fokker–Planck (F.P.) equation containing memory functions in the convolution integrals. Any equation of motion, obtained from this F.P. equation, for the averaged values of the set of parameters exhibits retardation, and the memory function appearing in each of these equations can be given the meaning of a correlation function.

We now want to examine the propagator for PG and introduce the Laplace transform via

$$G(z) = \int_0^\infty G(s)\, e^{-sz}\, ds \tag{5.56}$$

and

$$[z - \langle\tilde{C}\rangle - \langle(\delta\tilde{C})^2(z - Q\tilde{C})^{-1}\rangle]PG(s) = PG(0), \tag{5.57}$$

where $\delta\tilde{C} = \tilde{C} - \langle\tilde{C}\rangle$, and we ignore the memory term for the moment. Returning to (5.47), we also have

$$G(z) = (z - \tilde{C})^{-1}G(0) \tag{5.58}$$

and

$$\langle (z - \tilde{C})^{-1} \rangle = [z - \langle \tilde{C} \rangle - \langle (\delta \tilde{C})^2 (z - Q\tilde{C})^{-1} \rangle]^{-1} = [z - \tilde{M}(z)]^{-1}. \qquad (5.59)$$

The last term can be expanded as

$$[z - \tilde{M}(z)]^{-1} = \frac{1}{z} \{ 1 + \tilde{M}(z)[z - \tilde{M}(z)]^{-1} \}. \qquad (5.60)$$

Combining (5.48), (5.57), and (5.60), we then finally achieve the following equation for f_1:

$$z f_1(z) = f_1(0) + \tilde{M}(z) f_1(z) + \langle \tilde{C} Q(z - Q\tilde{C})^{-1} Q f(0) \rangle, \qquad (5.61)$$

or, transforming back to the time domain,

$$\frac{df_1(s)}{ds} = \int_0^s dt' \, \tilde{M}(s - t') f_1(t') + P\tilde{C} e^{Q\tilde{C}s} Q f(0), \qquad (5.62)$$

which is in path variable form. This is now equivalent to (5.22) or (5.33). The self-scattering term is readily picked out as the averaged scattering operator, and M is dominated by $\langle C \rangle = \langle W \rangle - \Gamma_0$, which is the averaged inscattering.

5.5. EXTRACTION OF MOMENT EQUATIONS

We want now to develop moment (or balance) equations for the relevant dynamic variables. These equations will turn out to be of generalized retarded form due to the nature of the generalized transport equations. In order to achieve this, we will have to deal with the memory of the initial state in (5.17),

$$D(t) = iPL \, e^{itQL} Q f(0). \qquad (5.63)$$

This term is sometimes described as a correction to the otherwise instantaneous accelerative effect of the field, due to the need to break up the initial state correlations induced by the scattering process.[35] It also can be viewed as a propagation (by the exponential propagator) of $Qf(0)$, the portion of the initial state which is not projected onto our variables of interest, to states where when acted on by the complete Liouville operator L, we obtain components which are projected onto the variables of interest by the projection operator P. This is similar to the earlier description of the scattering of particles onto the appropriate free path in our path integral formulation. Indeed, we can interpret this term as a renormalization of the driving force term. To see this, we utilize the stationarity of the equilibrium state to write the memory term as[33]

$$D(t) = PLQ \, e^{itL_0'Q} QFQ f(0), \qquad (5.64)$$

where $L_0' = L_0 - F$ neglects the field effect. We now introduce $\mathbf{Y}(t)$ as the solution to the operator equation

$$F[f(0) - Pf(0)] = \mathbf{F} \cdot (L'\mathbf{Y}) \tag{5.65}$$

and

$$D(t) = PL'Q \, e^{itL_0'Q} QL'\mathbf{F} \cdot \mathbf{Y}. \tag{5.66}$$

The operator prefactor is just the low-field (equilibrium) scattering operator of (5.21). In the equilibrium state, we can take $Pf(0) = \langle f(0) \rangle$, and the expected operation of the field is to have $\mathbf{Y} = \partial f(0)/\partial \mathbf{p}$, as in low-field transport. Then we can multiply (5.22) by either p or E and integrate over the momenta just as in the BTE case.[24,26,27] The driving terms have to be integrated by parts, and because of (5.49) the memory function vanishes for the momentum equation. The results are ($mv_d = \langle p \rangle$)

$$\frac{dv_d}{dt} = eF - \int_0^t dt' \, X_v(t') v_d(t - t') \tag{5.67}$$

and

$$\frac{dE}{dt} = eFv_d(t)[1 - X_v'(t)] - \int_0^t dt' \, X_E(t')[E(t - t') - E_0], \tag{5.68}$$

where the decay functions are defined as relaxation functions and are related to the autocorrelation functions for velocity and energy.[20] Here, $X_v'(t) = X_v(t)/X_v(0)$ is the normalized form.

Although delay of the mean energy rise is expected from a pure ballistic analysis (the carriers must actually move in real space through the field if they are to gain energy from it, and therefore the energy cannot rise until a nonzero average velocity has been established), that found here is different and elongated. For example, if $\dot{v} = qF/m - v_d/\tau_m$, then one would expect the first bracketed term in (5.68) to be $1 - \exp(t/\tau_m)$. However, as is clear from (5.68), the full form of the velocity autocorrelation function must be used, which differs from the simple exponential in the hot-electron problem.[36] These correlation functions will be further discussed in Chapter 10.

At first glance, it is tempting to conclude that this analysis shows that ensemble techniques involve an averaging process that introduces correlations among the carriers. This, however, is not the case. Averaging is introduced into actual ensemble Monte Carlo methods only when one explicitly takes an average. In the absence of Coulombic carrier-carrier interactions, every carrier evolves independently of the other carriers. The single particle two-time correlations are not introduced through the ensemble technique but instead were never thrown away, even in a simple, single-particle, Monte Carlo calculation. It is this retention of information concerning the microscopic, single-particle, correlations that makes the Monte Carlo techniques more general than, for example, the application of the method of moments to the Boltzmann equation where, as we integrate over a single time distribution function, such correlations are ignored.

As noted earlier, in the limit of long times (the stationary case), (5.62) reduces to the BTE, $D(t) \to 0$, and the set of (5.67) and (5.68) reduce to the equivalent set obtained from the BTE.[27] For this, we clearly must have

$$\frac{1}{\tau_m} = \int_0^t M_v(t') \, dt' \tag{5.69}$$

and

$$\frac{1}{\tau_E} = \int_0^t M_E(t') \, dt', \tag{5.70}$$

where M_v and M_E are memory functions of the sort discussed here. A consequence of this is the expected result that the BTE can be used only when changes in v_d and average energy are slow on the scale of the memory functions.

5.6. APPLICATION TO A DEGENERATE SYSTEM

An advantage of the EMC technique is that it allows for the inclusion of the Pauli exclusion principle and degeneracy.[37] Although the EMC is a semiclassical technique that simulates electrons moving as classical particles, in reality they are fermions. They obey the Pauli exclusion principle, and for any given state descriptor R_i above we can have at most two electrons occupying that state, and these two electrons must have opposite spins. Following the simple theory of the Fermi gas,[38] the number of available k-space states for a system enclosed in a volume V is $V/(2\pi)^3$. The region in k-space occupied by the electrons when the temperature is absolute zero is called the Fermi sphere. The radius of this sphere is the Fermi wave vector $\mathbf{k}_F$, and it is related to the electron concentration n by

$$\mathbf{k}_F = (3\pi^2 n)^{1/3}. \tag{5.71}$$

The surface of this sphere is the Fermi energy or level. At higher temperatures, thermal fluctuations move electrons from states inside the Fermi sphere to states outside the sphere. Therefore in metals, it is only these electrons, lying in a region within $k_B T_0$ (where T_0 is the ambient temperature) of the edge of the Fermi energy which take part in collision and conduction processes. Similar effects happen in degenerately doped semiconductors. Since we are dealing with energies for which the Fermi-Dirac function is approximately one-half, there is a high probability that a state lying in this conduction region in k-space is already occupied. The Pauli exclusion principle then limits the probability of other electrons making a transition via some phonon or other scattering process into this state.

The probability of an electronic transition from a state $\mathbf{k}$ to a state $\mathbf{k}'$ is proportional to three quantities. First, it is proportional to $f(\mathbf{k})$, the probability that the state $\mathbf{k}$ was originally occupied. Second, it is proportional as well to $1 - f(\mathbf{k}')$, the probability that the target state $\mathbf{k}'$ is empty. Last, it is proportional to $S(\mathbf{k}, \mathbf{k}')$, the probability that an electron in state $\mathbf{k}$ will make a transition to state $\mathbf{k}'$, which is generally calculated from the Fermi golden rule (see Chapter 1). Thus the transition rate is

$$P(\mathbf{k}, \mathbf{k}') = S(\mathbf{k}, \mathbf{k}')f(\mathbf{k})[1 - f(\mathbf{k}')]. \tag{5.72}$$

The standard Monte Carlo procedure works within the approximation $f(\mathbf{k}') = 0$. That is, it considers all final states as being unoccupied and therefore is only strictly applicable in nondegenerate conditions. The problem faced in extending this procedure to cover the degenerate case is that we must include $f(\mathbf{k}')$ in our computation of the scattering rates. Prior to running the simulation, however, we do not know what $f(\mathbf{k}')$ is. There are two basic approaches one can follow. In the first, one recomputes $P(\mathbf{k}, \mathbf{k}')$ on the fly. The second is more efficient. It extends early work in which an attempt was made to include the exclusion principle in single-particle Monte Carlo.[39]

This procedure[37] uses a self-consistent algorithm in which the rejection technique, discussed in Chapter 4, is used to eliminate transitions to occupied final states. (Since we are tracking particles, we are automatically including only occupied initial states.) It is important in this procedure to carefully ensure that the ensemble is used in a fashion that allows it to be interpreted as a probability function—that is, as a statistical representation of a properly normalized distribution function. There are N electrons present in the ensemble and we must relate this number to n, an electron density which is an input parameter to the simulation. Therefore we have an effective real-space volume V which equals N/n. This volume is assumed to be a cube, and the associated $\mathbf{k}$-space density of allowed wave vectors with a single spin is $V/(2\pi)^3$. In the simulation a grid of blocks is established in the three-dimensional $\mathbf{k}$-space, with the spacings Δk_x, Δk_y, and Δk_z. The maximum number of electrons that can occupy a block is $N_b = 2(\Delta k_x \Delta k_y \Delta k_z)V/(2\pi)^3$. The factor of 2 accounts for both spins. Therefore, we can transform the ensemble into a representation of a probability function by dividing the actual number of electrons found inside a block by N_b.

When a scattering event occurs in the EMC simulation, the block b which contains the final state for the selected transition is identified. The above normalization is used then in the rejection technique. A random number r, uniformly distributed between 0 and 1, is selected. If $r >$ (block occupancy)$/N_b$, the transition is accepted and the electronic state updated to the one selected. In the other case, the scattering event is treated as an additional self-scattering event, and the electronic wave vector is left unchanged. Note that here we are assuming that the transition rates and scattering rates were generated in a process in which the exclusion principle was ignored.

Figure 5.3 shows the effect of the exclusion principle on the electron dynamics. The light curves refer to the classical case; that is, they are obtained

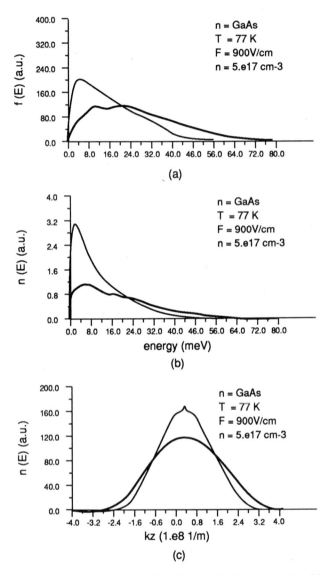

FIGURE 5.3. Ensemble Monte Carlo calculation of energy distribution function (a), energy occupation number (b), and momentum distribution function (c) with (bold curves) and without (light curves) the Pauli exclusion principle. After Lugli and Ferry.[37]

with the standard EMC simulation. The bold curves indicate that the results of the inclusion of degeneracy in the manner described in the previous section. An electric field of 900 V/cm is applied along the z-axis. The lattice temperature is 77 K, and the electron and impurity concentrations are taken to be equal to 5×10^{17} cm^{-3}.

5.7. DRIVE AND FREQUENCY DEPENDENCE OF THE COMPLEX MOBILITY

Many semiconductor diodes intended for use at millimeter- and submillimeter-wave frequencies contain a region of undepleted epitaxial material. The boundary position and total length of this undepleted epitaxial material will vary in a way determined by the total current flowing in the device. It is important to characterize the finite conductivity of this material because its effect on the diode terminal properties can be significant, especially at high (e.g., millimeter-wave) frequencies. In this section some earlier approaches to this problem are reviewed.[40,41] Monte Carlo simulations are then used to characterize uniform, undepleted epitaxial GaAs material as a function of frequency and signal level.[42]

5.7.1. Analyses of Undepleted Epitaxial Material

The simplest analysis assumes no transient effects and a constant carrier mobility. The resulting equivalent circuit of undepleted material of length W, cross-sectional area A, and dielectric constant ε is shown in Figure 5.4(a). This circuit consists of a simple electronic conductance connected in parallel with the

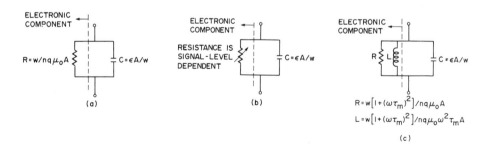

FIGURE 5.4.   Several different equivalent circuits for undepleted semiconductor regions. (a) assumes low fields and low frequencies. (b) assumes arbitrary fields and low frequencies. (c) assumes a constant momentum relaxation time.

cold geometric capacitance. An extension of this analysis uses an arbitrary velocity-field characteristic,[40] in which case the conductance becomes signal-level-dependent, as shown in Figure 5.4(b).

The problem can also be approached by using a constant momentum relaxation time τ_m in a momentum balance equation. This yields a complex-valued, frequency-dependent mobility[42]

$$\mu(\omega) = \mu_r(\omega) + i\mu_q(\omega) = \frac{1}{1 + (\omega\tau_m)^2} - \frac{i\mu_o\omega\tau_m}{1 + (\omega\tau_m)^2}, \qquad (5.73)$$

where μ_o is the low-frequency, low-field mobility. The equivalent circuit implied by this expression is shown in Figure 5.4(c). Here the particle current is represented by a conductance and by an inductive susceptance as well. This inductive behavior represents the lag between time variations in field (voltage) and carrier drift velocity (current). The frequency dependence of the real (in-phase) and imaginary (quadrature phase) mobilities is shown in Figure 5.5. Note that the real mobility deteriorates monotonically with frequency while the imaginary mobility peaks at $\omega\tau_m = 1$. These analyses lead to the expectation that the conductance of a semiconductor should decrease with increases in either frequency or drive level and that there is a significant inductive component to the particle current flow at high frequencies. Monte Carlo analysis has validated these general conclusions, but additionally introduced some surprises at high drive levels.[42]

5.7.2. Monte Carlo Studies of the Admittance of a Semiconductor

This problem is approached in a Monte Carlo simulation by allowing the electric field to become time-dependent. The carrier motion is simulated over many rf cycles of a periodic field variation and then averaged to produce a mean velocity versus rf phase angle. This velocity is Fourier-analyzed to produce real and imaginary components at the fundamental frequency and at a variety of

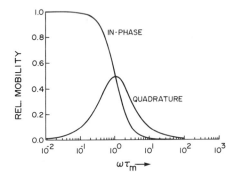

FIGURE 5.5. Frequency dependence of in-phase and quadrature-phase mobilities calculated using a constant momentum relaxation time.

harmonics as well. Mobilities are produced by dividing the fundamental velocity components by the rf field amplitude. The case considered has a zero dc field level and uses fundamental frequencies of 50, 250, and 1000 GHz.

At low drive levels the results essentially agree with the scenario presented in Section 5.7.1. However, for high drive levels at frequencies of 50 and 250 GHz, the imaginary mobility becomes capacitive rather than inductive. This can be understood by examining the mean velocity versus time, shown in Figure 5.6. There is a periodic velocity overshoot seen which arises from the same effects as the step function overshoot of Figure 5.2. The spikes in the velocity lead rather than lag the field in phase, and this leads to the capacitive component.

Strong harmonics are also produced by these spikes. This could serve as the basis for a millimeter-to-submillimeter-wave harmonic generator. Mazzone and Rees[43] have carried out a detailed study of this possibility. Their results indicate a bound of 150 GHz on a GaAs multiplier output frequency. This bound is very sensitive to details of the intervalley scattering process. A similar analysis was performed earlier for Si by Zimmermann et al.[44] Their investigation, however, used two different formats than the one used above. Zimmermann et al. first used fixed bias and rf field amplitudes with a varying frequency to generate the result shown in Figures 5.7 and 5.8. In a second study they fixed both the frequency and the ratio of the rf field amplitude to dc field level. They then varied the dc field level to produce the data which are compared with experimental data in Figure 5.8. As can be seen, excellent agreement was obtained. An earlier analysis of this sort for GaAs was performed by Lebwohl.[45]

In Figure 5.2, we presented the very commonly produced velocity overshoot. In Figure 5.9 we show a different transient phenomenon, the velocity undershoot. Actually such undershoots were predicted by Rees[46] several years before Ruch's pioneering velocity overshoot study.[3] Slow electrons, however, seem to excite less interest than fast ones, and therefore this sort of transient has received little investigation, while everyone has done a velocity overshoot study. This transient is associated with a stepdown in field from a level of 6 kV/cm to a variety of lower-field levels. Whether the final steady-state velocity is greater or less than

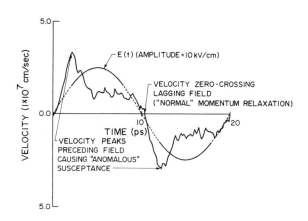

FIGURE 5.6. Mean carrier velocity as a function of phase angle for a field amplitude of 10 kV/cm and a frequency of 50 GHz.

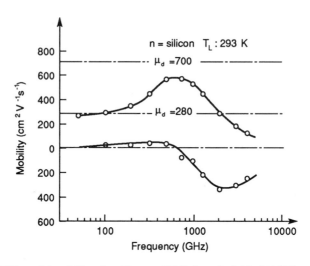

FIGURE 5.7.   Differential mobility of *n*-silicon at 293 K for a dc field of 10 kV/cm and an ac field of 1 kV/cm applied along the ⟨111⟩ direction as calculated by Zimmermann *et al.*[44]

the initial velocity, in all cases the transient starts with a sharp drop in velocity to a level below the eventual steady-state velocity. As we note later, this effect arises from the importance of negative velocity carriers in the Γ-valley near the intervalley scattering threshold energy when a field is applied. Rees used this step response as a tool for the estimation of a frequency-dependent mobility. He viewed a step from 6 to 5 kV/cm as an estimate of the ideal step response at 5.5 kV/cm. Fourier analysis yielded the frequency-dependent small-signal mobility shown in Figure 5.10.

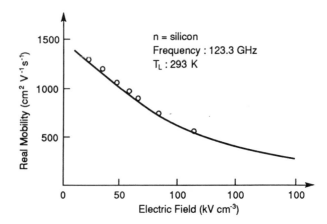

FIGURE 5.8.   Real part of the differential mobility in Si at 123.3 GHz as a function of electric field. (The ac field is a tenth of the dc field.) The curve is a Monte Carlo calculation, and the circles are experimental results, with an uncertainty of about 5%. After Zimmermann *et al.*[44]

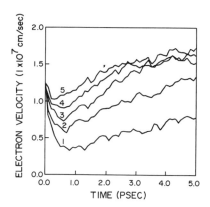

FIGURE 5.9. Transient electron velocity in response to a step-down in field. In all cases the initial state was a steady state at 6 kV/cm.

5.8. ITERATIVE SOLUTION TO THE BOLTZMANN EQUATION

In his calculation, Rees used an iterative solution to the time-dependent but spatially uniform Boltzmann equation. This approach is very closely related to the path integration and iterative procedures already discussed in this chapter. Here we use the notation encountered in other discussions[47,48] of the Rees iterative approach, but we regularly establish the connection with the notation used earlier in this chapter.

The Boltzmann equation is

$$\frac{\partial f(\mathbf{k}, t)}{\partial t} + \frac{e\mathbf{F}}{\hbar} \cdot \nabla_k f(\mathbf{k}, t) + \lambda(\mathbf{k}) f(\mathbf{k}, t) = \int d\mathbf{k}' \, f(\mathbf{k}', t) S(\mathbf{k}', \mathbf{k}). \qquad (5.74)$$

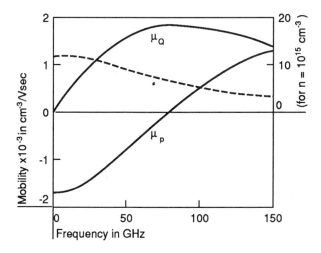

FIGURE 5.10. Frequency dependence of the differential mobility of electrons in GaAs as calculated by Fourier analysis of step-response data (such as that of Figure 5.9) calculated by an iterative solution to the Boltzmann equation. After Rees.[46]

This equation is obtained as a limiting case of the earlier (5.22). Rees then introduced the new variables

$$\mathbf{y} = \mathbf{k} - \frac{e\mathbf{F}}{\hbar} t, \qquad t = \tau. \tag{5.75}$$

These new variables represent the collisionless trajectory of the electron, the path variables of (5.28). The Boltzmann equation then becomes

$$\frac{\partial}{\partial \tau} f\left(\mathbf{y} + \frac{e\mathbf{F}}{\hbar}\tau, \tau\right) + \lambda\left(\mathbf{y} + \frac{e\mathbf{F}}{\hbar}\tau\right) f\left(\mathbf{y} + \frac{e\mathbf{F}}{\hbar}\tau, \tau\right)$$

$$= \int d\mathbf{k}' f(\mathbf{k}', \tau) S\left(\mathbf{y} + \frac{e\mathbf{F}}{\hbar}\tau, \mathbf{k}\right). \tag{5.76}$$

This differential equation then is solved, just as our earlier path integral forms were, by use of the integrating factor $\exp\{\int^\tau \lambda(\mathbf{y} + (e\mathbf{F}/\hbar)x)\, dx\}$. Multiplying through by this factor and integrating, we obtain

$$\left[f\left(\mathbf{y} + \frac{e\mathbf{F}}{\hbar}\tau, \tau\right) \exp\left\{\int^\tau \lambda\left(\mathbf{y} + \frac{e\mathbf{F}}{\hbar}x\right) dx\right\} \right]_{\tau=\tau_1}^{\tau=\tau_2}$$

$$= \int_{\tau_1}^{\tau_2} d\tau' \exp\left\{\int^{\tau'} \lambda\left(\mathbf{y} + \frac{e\mathbf{F}}{\hbar}x\right) dx\right\} \int d\mathbf{k}'\, f(\mathbf{k}', \tau') S\left(\mathbf{y} + \frac{e\mathbf{F}}{\hbar}\tau', \mathbf{k}\right), \tag{5.77}$$

which in the original variables can be written as

$$f(\mathbf{k}, t) = f\left(\mathbf{k} - \frac{e\mathbf{F}}{\hbar}(t - t'), t'\right) \exp\left\{\int_{t'}^{t} \lambda\left(\mathbf{k} - \frac{e\mathbf{F}}{\hbar}(t - x)\lambda\right) dx\right\}$$

$$+ \int_{t'}^{t} d\tau'' \exp\left\{\int_{t''}^{t} \lambda\left(\mathbf{k} - \frac{e\mathbf{F}}{\hbar}(t - x)\right) dx\right\}$$

$$\times \int d\mathbf{k}'\, f(\mathbf{k}', \tau'') S\left(\mathbf{k} - \frac{e\mathbf{F}}{\hbar}(\tau' - t''), \mathbf{k}'\right). \tag{5.78}$$

The value of t' can be taken to be $-\infty$, and this equation then becomes

$$f(\mathbf{k}, t) = \int_0^\infty d\tau'' \exp\left\{\int_0^t \lambda\left(\mathbf{k} - \frac{e\mathbf{F}}{\hbar}x\right) dx\right\} \int d\mathbf{k}'\, f(\mathbf{k}', t - \tau'') S\left(\mathbf{k} - \frac{e\mathbf{F}}{\hbar}t'', \mathbf{k}'\right). \tag{5.79}$$

Rees developed a two-stage iteration procedure for finding the steady-state distribution function equivalent to (5.79). This distribution function is

$$f(\mathbf{k}) = \int_0^\infty d\tau'' \exp\left\{\int_0^t \lambda\left(\mathbf{k} - \frac{e\mathbf{F}}{\hbar}x\right) dx\right\} \int d\mathbf{k}' f(\mathbf{k}') S\left(\mathbf{k} - \frac{e\mathbf{F}}{\hbar}t'', \mathbf{k}'\right). \quad (5.80)$$

The two-stage procedure is very similar to the Monte Carlo process and to our earlier iterative procedures for developing general transport equations. Given an initial guess of the distribution function, $f'(k)$, one substitutes this into (5.80) and first evaluates the scattering integral to obtain an intermediate function

$$g(\mathbf{k}) = \int d\mathbf{k}'\, S(\mathbf{k}, \mathbf{k}') f(\mathbf{k}'), \quad (5.81)$$

while the second stage is specified by

$$f(\mathbf{k}) = \int_0^\infty dt\, g\left(\mathbf{k} - \frac{e\mathbf{F}}{\hbar}t\right) \int_0^\infty d\tau'' \exp\left\{-\int_0^t \lambda\left(\mathbf{k} - \frac{e\mathbf{F}}{\hbar}x\right) dx\right\}. \quad (5.82)$$

Obviously, the second stage propagates the state along the collisionless trajectory or path in which the first stage terminated. If f_n is the nth iterative, then the iterative process can be written schematically in the form

$$f_{n+1} = f_n S P_0, \quad (5.83)$$

where S denotes the transition matrix $S(k, k')$, and P_0 is a matrix determined by the integral

$$P_0(\mathbf{k}, \mathbf{k}') = \int_0^\infty P_0(\mathbf{k}', \mathbf{k}, t)\, dt, \quad (5.84)$$

where the integrand $P_0(\mathbf{k}', \mathbf{k}, t)$ is the probability that an electron initially at k' will be found at k at a time t later without being scattered. This is essentially the function $S_1(p_0, p, t)$ of Section 5.4. Thus the entire process can be written as

$$f_n = f_0 (S P_0)^n, \quad (5.85)$$

and the steady-state distribution function satisfies

$$f = \lim_{n \to \infty} f_0 (S P_0)^n. \quad (5.86)$$

The paper in which Rees introduced the iterative technique[47] was also the paper in which he introduced self-scattering. It is not a surprise therefore to learn that self-scattering eases this iterative technique, just as it did in our earlier discussions. Rees discovered that if the total scattering rate (real plus self-scattering) Γ is set to be a constant that is at least twice as large as the largest physical scattering rate, then

$$f_{n+1}(\mathbf{k}) = f_n(\mathbf{k}) + \frac{1}{\Gamma} \frac{\partial f_n(\mathbf{k})}{\partial t}, \tag{5.87}$$

thus allowing iteration to be viewed as a time step in a time-varying problem. This is how the mobilities of Figure 5.10 were generated. When self-scattering is included, the basic iteration equation, (5.80), becomes

$$f_{n+1}(\mathbf{k}) = \int_0^\infty dt\, e^{-\Gamma t}$$

$$\times \int d\mathbf{k}'\, f_n(\mathbf{k}') \left(S\left(\mathbf{k} - \frac{e\mathbf{F}}{\hbar} t, \mathbf{k}'\right) + [\Gamma - \lambda(\mathbf{k}')]\delta\left(\mathbf{k}' - \mathbf{k} + \frac{e\mathbf{F}}{\hbar} t\right) \right). \tag{5.88}$$

5.9. TRANSIENT DYNAMIC RESPONSE EXPERIMENTS: FREQUENCY- AND TIME-DOMAIN STUDIES

5.9.1. Introduction

In the rest of this chapter we discuss various attempts at experimentally observing a transient dynamic response. There are two main classes of experiments to be discussed. We first discuss experiments in which rapid time variations in an electric field or carrier generation are used to excite the TDR. Then we discuss the second major variant, experiments in which small sample size is used to excite a TDR. In this second experimental class, dc measurements are commonly used, while obviously either time-domain or frequency-domain measurements are used in the first class of experiments. For reasons which will become clear, generally the first methods are assumed to be exploring a velocity overshoot, while the second set of experiments are commonly described as examining ballistic transport.

The material system which we will most commonly consider is gallium arsenide. Electron transport transients in GaAs generally are dominated by either a net transfer of electrons from the Γ-conduction-band valley to the L- and X-valleys or, alternatively, by a net transfer from the higher-energy valleys back to the Γ-valley. The first possibility plays an important role in the most commonly discussed transport transient; the velocity overshoot first predicted by Ruch[3] starts with electrons in the Γ-valley in an equilibrium state. On the sudden

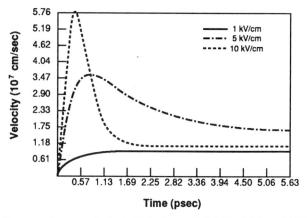

FIGURE 5.11.   Transient electron velocity in GaAs for applied fields of 1, 5, and 10 kV/cm, assuming an initial Maxwellian distribution.

application of an electric field, these electrons are accelerated to high-velocity states in the Γ-valley and then scatter to lower-velocity states in the L- or X-valleys. A typical Monte Carlo calculation of this type of transient is shown in Figure 5.11. The conduction-band parameters of Table 5.1 were used, and the details of the scattering mechanisms included have been discussed by Osman and Ferry.[49] The second possibility, the one in which electrons must transfer from the L-valley back to the Γ-valley, plays an important role in the velocity undershoot of Rees,[46]

TABLE 5.1.   Parameters for GaAs Monte Carlo Program

Parameter		Γ	L	X
Density (g/cm³)	5.37			
Energy band gap at 300 K (eV)	1.43			
High-frequency dielectric constant	10.92			
Static dielectric constant	12.9			
Velocity of sound (cm/s)	5.22×10^5			
Number of valleys		1	4	3
Effective mass ratio		0.063	0.222	0.58
Nonparabolicity factor (eV⁻¹)		0.61	0.46	0.20
Valley separation from Γ-valley (eV)			0.29	0.49
Polar optic-phonon energy (eV)		0.0364	0.0364	0.0364
Acoustic deformation potential (eV)		7.0	7.0	7.0
Coupling constant	Γ to		7	10
(10^8 eV/cm)	L to	7	10	5
	X to	10	5	10
Intervalley phonon energy (eV)	0.0311			
Effective mass ratio for holes				
Heavy-hole band	0.45			
Light-hole band	0.082			
Split-off band	0.17			

discussed in Section 5.7.2. In some instances, this depressed velocity could have a detrimental effect on device performance (e.g., Ref. 42). Its physical origin, however, is well known and when seen in Gunn device simulations is sometimes referred to as the Jones–Rees effect.[50] This is illustrated in Figure 5.12. When carriers enter the Γ-valley from the L-valley they can take on either a positive or a negative velocity. It is the presence of the field, however, that establishes a difference between positive and negative velocity carriers. Electrons entering with a positive velocity gain energy from the field and therefore are ballistically placed into states above the intervalley scattering threshold. Electrons entering the Γ-valley with a negative velocity, however, lose energy to the field and are ballistically placed into states below the intervalley scattering threshold. The result is that the transfer from L- to Γ-valley predominantly gives the negative velocity states, thus producing the transient dip seen in Figure 5.9. As noted by Jones and Rees,[51] such a transfer during the accumulation transit mode of Gunn devices gives rise to a carrier cooling effect by aiding the transfer out of the higher-energy valley and may augment Gunn device performance.

The strategy which would seem to be indicated by the theoretical studies, that of stepping the voltage applied across a semiconductor sample and then measuring the terminal current, has several obvious disadvantages. First, it is extremely difficult to make the voltage applied to a semiconductor sample perform one of these steps. Second, even if it were possible, there would be a large capacitive contribution (associated with this time-varying voltage) to the measured terminal current. This capacitive contribution would complicate the analytical process by which a time-varying conduction current is extracted from the measured terminal current. In view of these difficulties, a better alternative is to perform experiments which rely on using very short optical pulses to suddenly generate electron-hole pairs in an already extant electric field. This in fact is a commonly used strategy for experimental time-domain studies of TDR.

The optical experiments remove the first of the difficulties mentioned above by allowing us to rapidly place the system into an initial state. The challenge in an optical experiment is the measurement of the transient photoresponse in which

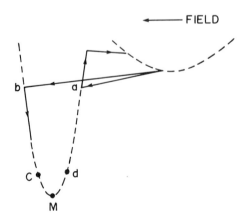

FIGURE 5.12. The Jones–Rees effect. Carriers entering the Γ-valley with negative velocity lose energy and fall below the intervalley scattering threshold.

the system moves from this initial state into a final state. Many commonly used spectral techniques do not provide information on the average momentum of the carriers. Yet it is the momentum that needs to be measured in any experimental study of a velocity overshoot-like behavior. Several "probes" have been used in efforts at determining conduction currents and momentum behaviors. Time-resolved terminal current measurements,[52,53] time-resolved reflectivity measurements,[54] and time-resolved absorption measurements[55] have all been used. With the exception of the all-optical reflectivity measurements of Nuss *et al.*,[54] a bias is applied in all cases. This in turn requires that the sample be embedded in an electric circuit. The response then is strongly affected by the mechanisms in which current continuity is maintained during the transient. Therefore we start our discussion by reviewing transient current continuity.

A bias-dependent field component is needed in a velocity overshoot study. Therefore, terminals are needed even if the terminal current is not observed. At these terminals there will be a "surface" charge that steps the field down from the value it has inside the semiconductor to the nearly zero value that it has inside the external circuit. When electron-hole pairs are generated in this applied field, they separate and move in opposite directions. As they do so, a time varying contribution to the field arises from the space-charge movement. Locally, inside the sample, there is a displacement current associated with this field variation. This local displacement current is essential for current continuity. While there is a particle current only in regions where there are moving particles, the field, which evolves everywhere inside the sample, creates a spatially dependent, time-varying displacement current. The time-varying field at the terminals induces the terminal surface charge to vary in time. Therefore, a current must flow in the bias circuit. This is the induced current of Ramo and Shockley[56,57] upon which microwave time-of-flight measurements of carrier drift velocities are based. As this current flows in the external circuit, there will be a change in the voltage applied across the sample. This time-varying voltage also creates a displacement current contribution at all points inside the sample and a capacitive contribution to the current flowing at the terminals. Therefore, there are two sources of displacement current flow inside the sample. The space-charge contribution is needed, but the capacitive contribution is an undesirable parasitic.

5.9.2. Frequency-Domain Studies

The central problem with a velocity overshoot study is that the overshoot lasts for only a few picoseconds, thus making measurements difficult. While we discuss several experiments in which direct time-domain measurements are attempted, we first discuss experiments which work in the frequency domain. Our discussions of theory have emphasized the existence of a frequency dependence to a carrier mobility and the importance of velocity correlation functions. Since both a mobility [(4.108)] and a velocity overshoot [(4.107)] have been associated with this autocorrelation function, there is a natural hope that frequency-dependent mobilities will be associated with a velocity overshoot.

Teitel and Wilkins[58] have investigated this possibility, and Allen and co-workers[59,60] have attempted the measurement.

The starting point of the Teitel and Wilkins analysis is the familiar energy-momentum balance equations, discussed in Chapter 4. These are

$$\frac{dv}{dt} = \frac{eF}{m^*} - \frac{v}{\tau_m} \tag{5.89}$$

and

$$\frac{dE}{dt} = eFv - \frac{E - E_{th}}{\tau_E}, \tag{5.90}$$

where E_{th} is the thermal energy determined by the lattice temperature. They noted that Das and Ferry[61] had previously found an ac conductivity using (5.89) and (5.90). The Das-Ferry conductivity is

$$\frac{\sigma(\omega, F_0)}{\sigma_{dc}(F_0)} = \frac{G_e - G_m + i\omega\tau_m(E_0)}{[G_e + i\omega\tau_m(E_0)][1 + i\omega\tau_m(E_0)]}, \tag{5.91}$$

where

$$G_e = (E_0 - E_{th}) \frac{\tau_m^2(E_0)}{\tau_E(E_0)} \frac{d}{dE}\left(\frac{1}{\tau_m}\right)_{E_0}, \tag{5.92}$$

$$G_m = \tau_m(E_0) \frac{d}{dE}\left(\frac{E - E_{th}}{\tau_m}\right)_{E_0}, \tag{5.93}$$

$$E_0 = eFv_0\tau_E(E_0) + E_{th}, \tag{5.94}$$

$$v_0 = \frac{eF}{m^*} \tau_m(E_0), \tag{5.95}$$

and

$$\sigma_{dc}(E_0) = \frac{ne^2}{m^*} \tau_m(E_0). \tag{5.96}$$

The subscript 0 denotes a dc quantity, while

$$\sigma(\omega, E_0) = \frac{nev_1}{F_1}, \tag{5.97}$$

where v_1 and F_1 are small sinusoidally varying quantities of frequency ω, denoted as the small-signal velocity and field phasors. The ratio of (5.91) does not become unity under the condition $\omega = 0$ because the numerator refers to a differential relation while the denominator is a chordal relation.

There are four basic solution regions to these equations, and these are shown in Figure 5.13. As can be seen, if G_e is less than G_m or -1, no stable steady-state solutions exist. If $G_m > G_e$, a negative differential conductivity occurs. The separation into regions I and II, however, is the central point of the analysis. In region II

$$\frac{d}{d\omega} \{\text{Re}[\sigma(\omega, F_0)]\} \neq 0, \tag{5.98}$$

and there can be no peak in the ac conductivity, while in region I we are guaranteed that there is such a peak. Teitel and Wilkins then argued that the existence of a peak in conductivity is a sufficient condition for concluding that there is a velocity overshoot. This is done by taking the inverse Fourier transform of the velocity.

This analysis, while intriguing, rests on the validity of the simple balance (5.89) and (5.90). While questions can be raised on the validity of these equations, the essential technique of Teitel and Wilkins, however, is to determine an ac conductivity, relate it to a mobility by assuming that the carrier density is

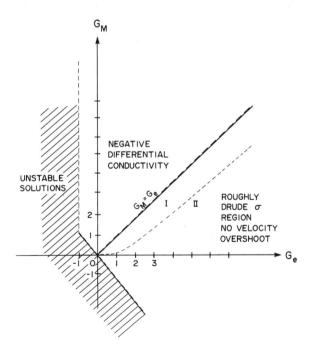

FIGURE 5.13.   The four basic solution regions to equations (5.96) and (5.97).

time-invariant, and then use the inverse Fourier transform of this mobility as an estimate of a time-varying average carrier velocity. This is the crucial portion of the analysis, and it does not rely on (5.89) and (5.90). Instead the critical assumption here is that the carrier density and field are constant.

Allen and co-workers[59,60] have used this analysis to interpret their experimental data. They used a swept frequency measurement of far-infrared transmission to determine a frequency-dependent conductivity. They applied the above analysis to obtain data that indicated the existence of a velocity overshoot in silicon MOSFET at 1.5 K and in a GaAs epilayer at room temperature. The central argument that is presented against accepting these results as unambiguous experimental indications of a velocity overshoot is that the carrier and field profiles inside these samples are not actually constant but will vary in position. This argument is particularly applicable to the GaAs system, in which a grid is imposed on top of an epitaxial layer doped at about 10^{17} cm^{-3}, a situation in which strong space-charge structures are expected.

Glover had, in earlier work,[62] used essentially the same techniques to interpret a 34.4-GHz measurement of the conductivity of GaAs and InP samples in terms of a single relaxation time, and compared his experimental GaAs result with the work of Rees.[46] The trends were comparable with an energy relaxation time of about 2.5 ps for GaAs for fields less than 0.6 times the threshold field F_{th} followed by a linear increase with increasing field. The relaxation time was about 6 ps at the threshold field, whose value for GaAs was about 3.8 kV/cm. In Glover's case, the accuracy of a balance equation with a single relaxation time does affect his measured time constant value.

5.9.3. Time-Domain Studies

Shank et al.[55] have used optical pump and probe measurements to explore velocity overshoot as well. The pump and probe method used to examine overshoot involves photoexcitation by short optical pulses. As illustrated in Figure 5.14, a beam from a passively mode-locked dye laser is split into two beams; one excites the semiconductor layer, while the second is first spectrally broadened and used to probe the absorption spectrum at delayed times.

The experiments were performed on GaAs at 77 K for the structure shown in Figure 5.15, with pump and probe (both at 805 nm) incident through the etched window. A variable optical delay line provided the timing between pumping and probing. A voltage is applied across the sandwich structure and modifies the optical absorption as described by Franz[63] and Keldysh.[64] The optical absorption is further modified by carriers optically injected near the band edge. The optically injected electrons and holes drift to opposite ends of the sample, thereby altering the net field across the device. The small space-charge-induced field perturbs the Franz-Keldysh effect, and the temporal evolution to a "steady state" (which is not truly steady but decays as the carriers recombine) is monitored. Note that in this experiment what actually is observed is the displacement current created by the motion of the charge carriers in an applied field.

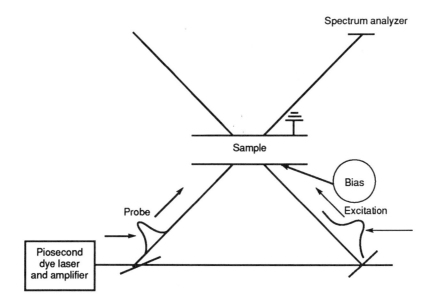

FIGURE 5.14.   Schematic of the basic pump-probe method of transient optical measurements.

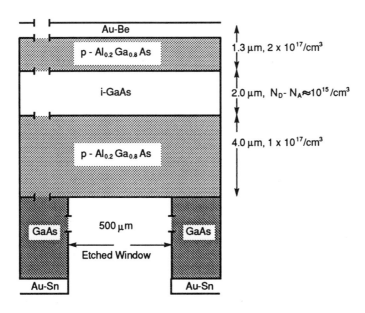

FIGURE 5.15.   Sketch of sample used in an optical measurement of a velocity overshoot.

The experiments measure the optical absorption spectra as a function of relative delay time. The differential optical absorption spectra is then obtained by subtracting the spectra before and after a specified relative delay time. Differential spectra for a 20-ps delay is shown in Figure 5.16 for three applied voltages. Note the dependence of the period of the oscillation on applied field. The amplitude of the absorbance change was obtained by adding the area under the positive and negative portions of the differential absorption spectra, and this is shown as a function of time in Figure 5.17. The dotted lines in Figure 5.16 are data, while the solid line is a fit determined by using the equation

$$\frac{\Delta\alpha(t)}{\Delta\alpha(\infty)} = \frac{\int v_e\, dt - \int v_h\, e^{-\alpha_p d}\, dt}{d(1 - e^{-\alpha_p d})} + \frac{1 - e^{-\alpha_p \int v_h\, dt} + e^{1-\alpha_p d}[1 - e^{\alpha_p \int v_e\, dt}]}{\alpha_p d(1 - e^{-\alpha_p d})}, \quad (5.99)$$

where α_p is the pump absorption constant and d is the width of the region over which the field is applied. The velocities in (5.99) are time-dependent, and while Shank *et al.* did not feel it necessary, within the resolution of their data, to incorporate the full time-dependent velocity overshoot calculation, it was necessary for at least two cases to assume a two-valued electron velocity-time relation. For the 14-kV/cm result the dashed curve represents a fit with a single value of electron velocity chosen to match the data near $t = 0$ where

$$\frac{\partial}{\partial t} \frac{\Delta\alpha(t)}{\Delta\alpha(\infty)} \simeq \frac{v_e + v_h}{d}. \quad (5.100)$$

Here v_h is taken to be 10^7 cm/s. There is an apparent decrease in the slope of the absorbance, and a better fit, represented by the solid curve, is obtained from a two-velocity fit. For the 22-kV/cm result, the dashed curve is for a single value of velocity with $v = 4.4 \times 10^7$ cm/s for $t < 1.1$ ps but 1.2×10^7 cm/s for longer times. For the 55-kV/cm result, a single fit of 1.3×10^7 cm/s appears adequate.

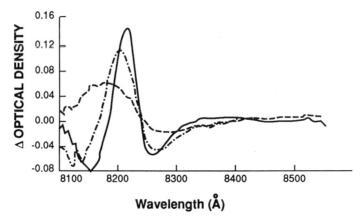

FIGURE 5.16.   Induced optical density change in GaAs 20 ps after excitation by a pulse at 803 nm. Solid line is a sample biased at 14 kV/cm, dot-dash line is a sample bias of 22 kV/cm, and the dashed line is a sample bias of 55 kV/cm. After Shank *et al.*[55]

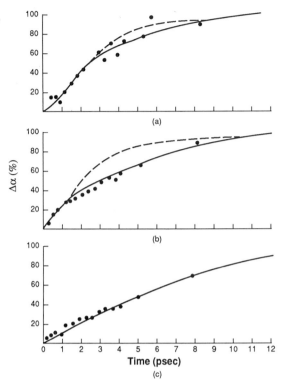

FIGURE 5.17.   Induced absorbance as a function of time: (a) 14 kV/cm; (b) 22 kV/cm; (c) 55 kV/cm. After Shank *et al.*[55]

There are several points worth noting. First, the slopes of the absorbance curves are higher at lower bias levels, suggesting[55] the presence of longer overshoot relaxation. Further, the initial overshoot velocity is higher for the 22-kV/cm measurement than the 14-kV/cm measurement. Both of these factors are consistent with calculated overshoot curves. The interpretation is less direct with the 55-kV/cm measurement, where the initial slope is significantly below the lower field measurement. Shank *et al.* argued that velocity relaxation occurs during a shorter time interval for such high fields and that their time resolution did not allow them to see the overshoot clearly. However, because of the increased value of the high-field overshoot velocity (near 10^8 cm/s), some variation in the slope of absorbance is expected. At 55 kV/cm, the influence of the AlGaAs layers, particularly of the boundaries, and of the finite hole relaxation must be considered.

A second all-optical technique was developed by Nuss *et al.*[54] In their experiment there was no bias voltage applied to the sample. They used the time-varying field associated with an optical pulse itself to accelerate the photo-generated carriers. Their experiment therefore does not closely resemble the normal velocity overshoot conditions. They photogenerated carriers with a pump pulse of 625 nm and then measured with subpicosecond resolution a time-resolved optical reflectivity of the system. From this data they extracted a time-varying mobility which exhibited no overshoot but took several picoseconds to rise to a final value. Since there is no electric field imposed, the lack of an overshoot is not troublesome. The slow rise in mobility was the significant result. Nuss et al. hypothesized that as they were photogenerating carriers high in the Γ-valley, above the Γ-L transfer threshold, the carriers were transferring to the L-valley very rapidly after the photogeneration. The mobility, therefore, started low and then slowly rose as the carriers returned to the Γ-valley. Osman and Grubin[65] have performed a Monte Carlo study in which photogeneration by the same wavelength in a relatively low electric field (500 V/cm) was simulated. They saw exactly the sort of valley transfer behavior hypothesized by Nuss et al.

A similar behavior is present in high excitation photoluminescence experiments. There a hot plasma is generated in an unbiased sample, and time-resolved luminescence is used to watch the system relax. The relaxation occurs relatively slowly. Two competing theories for this have been proposed. In one, the hot carriers are hypothesized to emit optical phonons sufficiently rapidly, creating an increase in the phonon population. Subsequently, the hot carriers reabsorb the phonons that they have emitted, thus increasing the phonon absorption rate and slowing the energy relaxation.[66] This is generally known as the hot-phonon effect, but it is unlikely to explain this situation because it takes several picoseconds of optical phonon emission before the phonon population rises significantly. The second explanation is essentially the same as the intervalley transfer picture of Nuss et al. Shah et al.[67] have performed a comparison between experimental studies of this sort and Monte Carlo calculations. They found that the intervalley scenario did explain the data well, but that care had to be taken in the selection of the Γ-L coupling coefficient, a parameter which is generally used as an adjustable parameter in Monte Carlo studies.

Others have tried to use time-resolved terminal measurements of the conductivity of carriers photogenerated into an already extant field. Early variations on this have been reported by Hammond[52] and Mourou et al.[53] In both experiments a gap was left in a transmission line on top of a semi-insulating GaAs substrate. A pump pulse was focused on the gap, and as the photocurrent in the gap evolved, a time-varying voltage wave was transmitted down the transmission line. The analysis of Auston[68] was used in both experiments to explain why a photoconductive overshoot should produce a transmission line overshoot. The two experiments both used 620-nm-wavelength pumps and differed primarily in the temporal resolution of the measurement of the voltage wave. Hammond used an ion-bombarded second gap as a high-speed photoconductive sampler while Mourou et al. abutted an electro-optic sampling crystal against their sample, extended the transmission line onto this sampling crystal, and used a second pulse to electro-optically sample[69] the voltage wave. The temporal resolution of Hammond was about 6 ps, which complicated his interpretation. However, his results were consistent with the general idea of having no overshoot at very low fields, a temporally extended overshoot (lasting several picoseconds) at medium fields, and a temporally sharp overshoot at high fields. Mourou et al. had temporal resolution of about 0.5 ps and clearly saw an overshoot very similar to that of Figure 5.2.

There are important complications in analyzing both of these experiments. First, in the experiment of Mourou *et al.* there is an impedance mismatch between the photoconductive gap and the sampler. The second is that it is unlikely that good ohmic contacts and uniform fields were attained in either system. A third problem is that the subject of transient carrier transport in semi-insulating GaAs has not been extensively studied. Last, the potential role of trapping in semi-insulating GaAs is present in both experiments. In short, while it is clear that a photoconductive overshoot occurred, it cannot be unambiguously associated with a velocity overshoot.

Meyer *et al.*[70] performed a significantly improved version of such a transient photoconductivity experiment. All measurements were done using nominally undoped $(n = 5 \times 10^{15}\,\mathrm{cm}^{-3})$ high-mobility GaAs. Two microns of undoped material was grown on semi-insulating substrates, followed by a 50-nm layer of highly doped $(n^+ = 2 \times 10^{18}\,\mathrm{cm}^{-3})$ GaAs. Various test structures were patterned by using lift-off photolithography. Furnace-annealed NiAuGe contacts were made to the doped cap layer. A calibrated GaAs etch was used to remove the doped cap layer in the photoconductive gaps and between the transmission lines.

Two laser sources were used to excite and probe the transient photoconductivity. For the first set of measurements a laser with a wavelength of 620 nm, a repetition rate of 100 MHz, a pulse width of 60 fs, and an average power of 5 mW per beam was used. In subsequent experiments a laser which generated 300-fs pulses at 760 nm at a repetition rate of 100 MHz and an average power of 10 mW was used.

Measurement of the transient voltage waveforms was accomplished using reflection-mode electro-optic sampling.[71] In this technique a thin plate of $LiTaO_3$ with a high-reflectivity coating on one surface is placed on top of the GaAs sample with the coating in contact with the GaAs. A small window was etched in the coating to allow for transmission of the excitation beam, which was focused symmetrically on the photoconductive gap. The probe beam was focused between the transmission lines a short distance from the gap. Fringing fields from the substrate extend into the electro-optic superstrate and are detected as a change in polarization of the probe beam. The probe beam is optically biased to ensure that the polarization changes linearly with the electric field. A dc voltage was applied to the transmission line, and the corresponding optical change was recorded to calibrate the measurement. An optical delay line changes the pump/probe delay, and the subsequent time-dependent signal is recorded with an rf mixer, lock-in amplifier, and signal averager.

The temporal resolution of such an experiment is limited by four factors.[72] It can be written as a sum-of-squares convolution of these factors as

$$\tau_{\mathrm{exp}} = [2\tau_1^2 + \tau_0^2 + \tau_e^2 + \tau_i^2]^{1/2}, \qquad (5.101)$$

where τ_1 is the laser pulse width, τ_0 is the transit time of the optical probe pulse across the electric field lines, τ_e is the electrical transit time across the optical probe beam waist, and τ_i is the intrinsic electro-optic response time. The value of τ_0 is determined by the transmission line geometry, and τ_e is determined by the optical beam size. For the coplanar geometry used here, τ_0 is governed by the penetration depth of the field into the substrate. Assuming a τ_i of 50 fs[73] for lithium tantalate, a probe beam spot size of 5 μm, stripline dimensions of 10 μm, and a laser pulse width of 50 fs, a practical resolution limit of 150 fs is obtained.

Several other points of interest should be noted. First, the exact delay between the pump and probe pulses is not known. Therefore, one should be careful not to assume that the time origin shown on the following experimental data actually corresponds to the temporal incidence of the pump pulse onto the gap. Second, the resolution figure quoted above limits the ability to resolve two separate features on a single experimental trace. There is another important resolution limit, however. A very useful experimental procedure is to first collect a trace for one bias, change only the bias setting, and repeat the experiment without altering the laser system in any way. The critical question then is, how accurately can we measure the temporal shift in a single feature as a function of bias? The resolution just described does not limit this. The bias delay resolution instead is limited primarily by our ability to accurately calibrate the extra path delay added in the experiment, and by τ_1. For the short-wavelength case presented here, bias-induced temporal shifts of 60 fs can be resolved. This mode of experimentation therefore allows us to accurately search for bias-dependent shifts in the photoresponse. One last comment should be made concerning this experimental technique. The electro-optic sampling technique is very sensitive and quite capable of measuring submillivolt changes in line voltage. This allows for the possible use of this technique in low-excitation experiments where carrier–carrier scattering and hot-phonon effects are unimportant.

The particular sample geometry consisted of 50-μm-wide coplanar strip lines separated by a 50-μm spacing, and the photoconductive gap length was also 10 μm. The pump and probe beams were each separately focused to approximately 10 μm, and the probe beam was positioned 20 μm downstream from the photoconductive gap. Results obtained with 620-nm excitation are shown in Figure 5.18, plotted as the transient voltage normalized to the gap bias voltage. The measured transient voltage is only 0.01% of the applied bias voltage, and therefore the associated displacement current is apparently much smaller than the particle current. Two features are clearly present in the data: a significant photocurrent overshoot that occurs at high biases but not at moderate or low biases, and a much faster rise time of the photocurrent for high biases.

For the reasons described earlier, it was desirable to repeat the experiment at a longer excitation wavelength. Photocurrent transients obtained with 760-nm excitation are shown in Figure 5.19. Once again, a photocurrent overshoot is observed at high bias and not at low bias. Unfortunately, a larger overshoot was not observed because the 300-fs pulse width of the pump and probe in this case limited the temporal resolution of the measurements. No bias-dependent delay

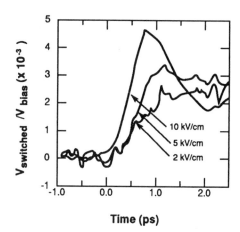

FIGURE 5.18. Transient voltage waveform generated by the photoconductive switch, normalized to the applied dc bias, for an excitation wavelength of 620 nm.

was observed here. While this is consistent with the Monte Carlo studies, the poorer temporal resolution in this case may have obscured the observation of such a delay.

5.9.4. Monte Carlo Studies of Transient Photoconductivity

The simplest approach to understanding the potential for using such experiments for studying transient carrier transport is to perform a Monte Carlo study. A set of valence-band parameters are used, and carriers are photogenerated out of these bands into the conduction bands by photons of a specified wavelength. A spatially uniform field is assumed, and one then studies the transient response of the photogenerated electrons in this field. Several results of this sort have

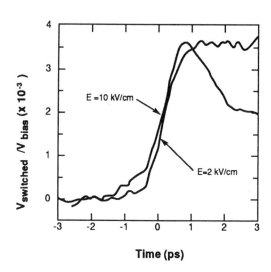

FIGURE 5.19. Transient photoconductivity results for excitation wavelength of 760 nm.

already been reported,[17,53,65,74,75,76] with a parameter sensitivity study reported by Chamoun et al.[77]

Four different sets of parameters were used in the sensitivity study of Chamoun et al. Special attention was given to the Γ-L deformation potential and its effect on the velocity overshoot phenomena. The Monte Carlo program used for the simulation includes carrier photogeneration out of the heavy-hole, light-hole, and split-off bands. The optical transitions are calculated using the small wave-vector approximations for the wave functions described by Kane.[78] All the relevant carrier-phonon scattering processes were taken into account, including the deformation potential and the polar coupling to both the acoustic and optical modes. Attention was restricted to low-excitation experiments. Carrier-carrier scattering was not included, since it is not important at laser excitation intensities below 5×10^{17} cm^{-3}.[79] Furthermore, hot-phonon effects are also not considered because no hot-phonon buildup can occur at the femtosecond time scale and low photon density that is of interest here.[80]

The four parameter sets used were the parameters of Osman and Ferry,[49] Shah et al.,[57] Wysin et al.[75] (who used the deformation potential parameters of Brennan and Hess[81] in conjunction with a pseudopotential calculation of effective masses), and Taylor et al.[82] These sets include the normal mixture of experimentally obtained and theoretically evaluated parameters. Not all the parameters have been measured, and the unmeasured parameters must be inferred indirectly or extrapolated in some manner.[83] The parameters differ mainly in the Γ-L deformation potential and in the valence-band effective masses used. Chamoun et al. found that all four sets produce qualitatively similar pictures, with an overshoot seen for long wavelength (near-band-edge excitation). The peak velocity differed by less than 20% across the parameter sets. As wavelength decreased, the minimum field needed to create an overshoot increased. Potentially measurable differences in this field were obtained as the parameter sets varied.

Examination of this data revealed that parameter sets which included stronger Γ-L coupling predicted that the overshoot would require larger fields at shorter wavelengths than did parameter sets with weaker coupling. Essentially, stronger coupling creates situations in which carriers simply immediately transfer to the L-valley, and the TDR is then dominated by the return to the Γ-valley in a fashion somewhat similar to that discussed in the experiment of Nuss et al.[54] Chamoun et al. particularly investigated the role of the Γ-L deformation potential, $D_{\Gamma L}$, in influencing the velocity overshoot phenomena at $E = 10$ kV/cm for 2-eV photoexcitation. In this investigation they were guided by the experimental data of Shah et al.,[67] who had already performed a subpicosecond luminescence experiment to determine the intervalley deformation potential ($D_{\Gamma L}$) in GaAs. In that experiment, GaAs and InP samples were excited by a subpicosecond pulse and the luminescence intensity was measured. The luminescence intensity for GaAs increased very slowly in contrast with that of InP. Since there is no significant intervalley scattering in InP at the excitation energy used, the slow rise in luminescence in GaAs was attributed to the return of the L-valley electrons to the Γ-valley. Their experimental results were compared with an ensemble Monte

Carlo calculation and the Γ-L deformation potential was determined to be $(6.5 \pm 1.5) \times 10^8$ eV/cm.

Keeping within the error range of Shah *et al.*, Chamoun *et al.* repeated a computation of the transient electron velocity for carriers photogenerated with a wavelength of $\lambda = 620$ nm into a field of 10 kV/cm. Three Γ-L deformation potential values (5×10^8, 6.5×10^8, and 8×10^8 eV/cm) were used. The results are shown in Figure 5.20. The velocity curve for $D_{\Gamma L} = 5 \times 10^8$ eV/cm shows a significant overshoot compared with the other two values of $D_{\Gamma L}$. The velocity curve for $D_{\Gamma L} = 6.5 \times 10^8$ eV/cm shows a slight peak, while that for $D_{\Gamma L} = 8 \times 10^8$ eV/cm just increases to a steady-state velocity. It therefore seems that the existence of an overshoot at $E = 10$ kV/cm for 2.0-eV excitation can only be determined experimentally. Such an experiment would help determine the $D_{\Gamma L}$ value. Since it would look at the departure of the electrons from Γ-valley to L-valley, it would complement the experiment of Shah *et al.*, which observed the L-to-Γ transition.

Even when there is a velocity overshoot following a short-wavelength photoexcitation, it differs in detail from that of the conventional Ruch-like overshoot. When a short-wavelength laser pulse (i.e., $\lambda = 620$ nm) excites a GaAs sample, electrons are photogenerated high in the central valley above the threshold energy for intervalley scattering. These electrons have no net momentum initially. Under the influence of a high electric field (i.e., greater or equal to 20 kV/cm), electrons with initial negative velocity will either be ballistically accelerated to states below the threshold energy for intervalley scattering, E_{int}, or scatter to one of the satellite valleys. The electrons that fall below E_{int} are trapped in the central valley and keep losing energy to the field as long as they have negative velocities.[17]

Examining the drift velocity curves for 2.0-eV excitation (Figure 5.21), we note a shift in the initial rise of the velocity that depends on the value of the electric field. In order to explain this phenomena, the electrons photogenerated

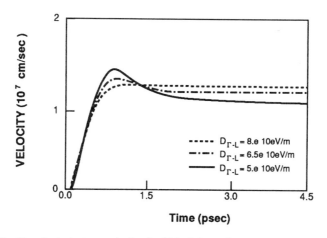

FIGURE 5.20. Transient electron velocity for Γ-L deformation potential values of 5, 6.5, and 8×10^{10} eV/m as given in Shah *et al.*[67] The applied field is 10 kV/cm, and the photon energy is 2.0 eV.

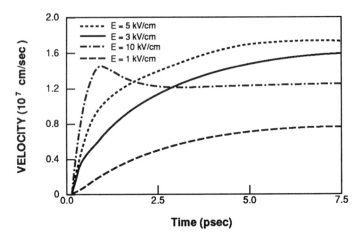

FIGURE 5.21. Transient electron velocity in GaAs obtained using a Monte Carlo model with the parameters set of Table 5.1. The applied fields were 1, 3, 5, and 10 kV/cm. The electrons were photoexcited using a laser pulse with photon energy of 2 eV and a 100-fs FWHM.

were classified into six groups, as shown in Figure 5.22. Groups 1, 2, and 3 are electrons photogenerated from the heavy-hole, light-hole, and split-off bands, respectively, with an initial negative velocity and groups 4, 5, and 6 electrons come from the same valence bands but with an initial positive velocity. The time evolution of the fraction of each of these groups of electrons in the Γ-valley is shown in Figure 5.23 and Figure 5.24 for $E = 20$ kV/cm, excitation energy of 2.0 eV, and pulse width (FWHM) of 20 fs.

Since group 1 electrons are generated high in the band, they will either transfer to the satellite valley or lose energy to the field and fall below E_{int}. Thus,

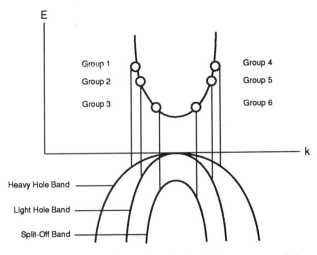

FIGURE 5.22. Energy-band diagram illustrating the six different groups of electrons that can be photogenerated out of heavy-hole, light-hole, and split-off valence bands.

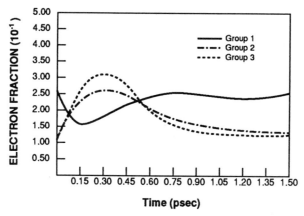

FIGURE 5.23.   Temporal evolution of the fractional electron population in the Γ-valley for electrons photogenerated in groups 1, 2, and 3 of Figure 5.22. A field of 20 kV/cm is applied, and the photon energy is 2 eV.

their fraction will undershoot and then increase to steady-state value. The fraction of group 4 electrons will undergo a similar behavior to that of group 1, but the population undershoot is more pronounced because these electrons possess positive momentum initially and the majority rapidly leave the central valley.

The fractional populations of group 2 and group 3 both overshoot their steady-state value because they are trapped in the Γ-valley below E_{int}. On the other hand, the corresponding positive velocity groups 5 and 6 start low in the Γ-valley and will stay there until they gain enough energy from the field to transfer to the satellite valleys. Therefore, their fractions will increase during the period of time in which they are being accelerated into the high-energy states and then

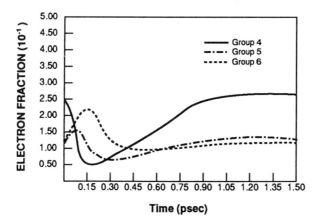

FIGURE 5.24.   Temporal evolution of the fractional electron population in the Γ-valley for electrons photogenerated in groups 4, 5, and 6 of Figure 5.22. A field of 20 kV/cm is applied, and the photon energy is 2 eV.

decrease to a steady-state value as a result of intervalley scattering. The main point to be concluded from Figures 5.23 and 5.24 is that electrons from groups 1, 2, and 3 constitute the majority of the Γ-valley electrons in the first few hundred femtoseconds. This interplay between field acceleration and intervalley scattering is very similar to that of the Jones-Rees effect.

The average velocity in the Γ-valley of groups 1, 2, and 3, shown in Figure 5.25, starts at a large negative value, becomes positive, and then overshoots before reaching a positive steady-state value. Since they constitute the majority of Γ-valley electrons, their negative velocity in the first few hundred femtoseconds will slow the initial rise of the drift velocity. The initial negative average value of the velocity of these electrons is determined by the wavelength and the band structure. It, therefore, is field-independent. However, as the value of the applied field increases, the velocity will rise faster. This causes the bias-dependent shift to an earlier rise of the velocity with higher applied field as shown in Figure 5.21. It is this portion of the electrons that produces an overall velocity overshoot.

It is easy to see why such a delay is not expected when one starts out with a thermal distribution. The initial energy is much lower than the threshold for intervalley transfer. It is, therefore, not possible to create a sizable population of negative velocity electrons in the Γ-valley. The Monte Carlo results of Figure 5.21 confirm this. The velocity for the thermalized distribution rises sharply and shows an overshoot, while the one with the photogenerated carriers does not. Extending the same reasoning, it becomes evident that parameterized models of the transient photoconductive process,[84] will not show this effect. These models are based on an *a priori* assumption of the form of the distribution function which is evaluated by taking the moments of the time-dependent Boltzmann equation. This procedure, in essence, leads to the masking of the true momentum distribution. For more realistic simulations using the analytic approaches, one needs to approximate the electron distribution function more accurately. This is what Jones and Rees did in their work, where they used an iterative solution of

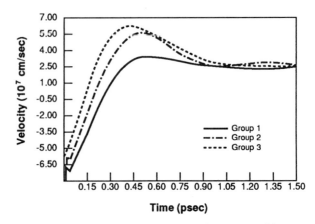

FIGURE 5.25. Transient electron velocity for electrons photogenerated in groups 1, 2, and 3 but remaining in the Γ-valley. A field of 20 kV/cm is applied, and the photon energy is 2 eV.

the Boltzmann equation in which the distribution function was constructed from an appropriately chosen basis set.

The short-wavelength case is an example of a situation in which (4.107) cannot be used to relate the time-evolving ensemble average velocity to an integral of the autocorrelation function. The failure is rather obvious, as in this case the initial velocity evolution may even start with a negative swing. Yet, the autocorrelation is initially positive-valued, and therefore its time integration must initially be nonnegative. The reason for the failure of (4.107) is that in its derivation we assume a memory function in the RLE which does not depend on the present state of the carrier. Yet, as the above discussion clearly showed, a high-energy carrier faces a significantly different scattering environment than does a low-energy carrier. The nonlinearity introduced by the sharp threshold energy for intervalley scattering was ignored in the derivation of (4.107).

In summary, the Monte Carlo studies show that the nature of the transient response depends on both the wavelength of the excitation and the magnitude of the applied electric field. For wavelengths long enough that no electrons are generated into states lying near or above the energy threshold for intervalley scattering, a quite conventional velocity overshoot occurs. For shorter wavelengths, new features appear as a result of a ballistic trapping of negative velocity electrons in the Γ-valley. There will be a bias-dependent delay in the rise of the velocity associated with the time required to accelerate these negative velocity electrons into positive velocity states. If a velocity overshoot occurs, it occurs because of a velocity overshoot of these ballistically selected Γ-valley electrons. The minimum field required for the existence of a velocity overshoot becomes wavelength-dependent and will be larger than is expected from a conventional velocity overshoot study. The exact value of this field cannot be predicted with more than 20% accuracy inside the parameter variation allowed by our present knowledge. Therefore experimental determination of this value is needed, and this type of experimental knowledge can then be used to more accurately constrain the parameter specification used in Monte Carlo studies.

5.9.5. Circuit Effects in Transient Photoconductivity

In order to more fully understand this experiment, the role of the circuit should be explored. Here we will follow Arnold's[85] extension of the basic Auston model[69] of the photoconductive gap for our simulations of the photoconductivity experiment. The original Auston model for the photoconductive gap consisted of a capacitor in parallel with a linear conductance. In that model, the characteristic impedance of the transmission lines had been taken to be frequency-independent. Reflection effects at the transmission line were neglected because the transient switching time of interest was much less than the transit time through the transmission lines. The circuit is shown in Figure 5.26. Here the gap is represented by a capacitance C_g connected in parallel with a conductance G. The capacitance represents the capacitive current flow contribution discussed earlier in the context of current continuity (a parasitic effect here), while the

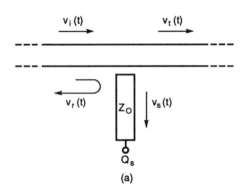

(a)

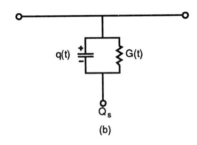

(b)

FIGURE 5.26.   Auston's[68] equivalent circuit for a photoconductive switch. $G(t)$ is the photoconductor, $q(t)$ is the charge on the gap capacitance, and Z_0 is the characteristic impedance of the transmission line. In (a) incident, reflected, and transmitted waves are shown, while (b) illustrates the gap equivalent circuit.

conductance $G(t)$ is the induced terminal current discussed earlier. Our goal is to measure this current. If the charge q on the gap capacitance is used as the state variable, the equation of state for the system then is

$$\frac{dq}{dt} + \frac{1}{2Z_0 C_g}[1 + 2Z_0 G_g(t)]q(t) = \frac{v_i(t)}{Z_0},\qquad(5.102)$$

where Z_0 is the transmission line impedance, C_g is the gap capacitance, and $v_i(t)$ is the incident voltage wave. The reflected and transmitted voltage waves are given by

$$v_r(t) = \frac{1}{2C_g}q(t)\qquad(5.103)$$

and

$$v_t(t) = v_i(t) - \frac{1}{2C_g}q(t).\qquad(5.104)$$

Equation (5.102) can be solved to yield

$$q(t) = \frac{1}{Z_0}\int_{-\infty}^{t} dt'\, v_i(t')\exp\left\{\frac{1 + 2Z_0 G(t')}{2Z_0 C_g}dt'\right\}.\qquad(5.105)$$

Auston discusses various other switching circuit configurations and several special cases of (5.105). The one of greatest interest to us is a small-signal analysis which corresponds to a low-excitation version of our photoconductive experiment. Then the photoconductance $G(t)$ is modeled by

$$G(t) \simeq G_{DK} + g(t), \qquad (5.106)$$

where G_{DK} is the dark conductance and $g(t)$ is a small perturbation on this conductance as a result of our low-level excitation. Then, if the incident voltage wave comes from a simple dc source, V_b, the measured transmitted line voltage is

$$v_t(t) = \frac{Z_0 G_{DK} V_b}{1 + 2Z_0 G_{DK}}$$

$$\times \left\{ 1 + \frac{1}{1 + 2Z_0 G_{DK}} \int_{-\infty}^{t} dt' \, g(t') \exp\left[\left(\frac{1}{2Z_0 C_g} + \frac{G_{DK}}{C_g}\right)(t - t')\right] \right\}.$$

$$(5.107)$$

The first term is the dark current contribution, while the second term arises from the small-signal photoconductive response.

There are several significant assumptions made in this analysis. First, we are still assuming that the instantaneous photocurrent is proportional to the instantaneous gap voltage; that is, that the small photocurrent is

$$i_{ph}(t) \simeq g(t) V_{gap}(t). \qquad (5.108)$$

Nuss et al.[54] make a similar assumption in the analysis of their data. As they note, and as Auston notes in his work, in the TDR regime there is no particular reason for such an assumption to be true. In the following we present several ways of eliminating an assumption such as (5.108). The second significant assumption made here is that the transmission line impedance is constant over frequency. We also discuss the role of that assumption as well.

In the model in Figure 5.27, the conductance is replaced by a photoconductive element, and the characteristic impedance of the transmission lines is taken to be frequency-dependent. The Monte Carlo model for electron transport in GaAs of Table 5.1 is used to simulate the photoconductive element. The inclusion of the Monte Carlo correctly builds in all the nonlinearities associated with transient transport in the photoconductive switch and allows us to obtain the appropriate photocurrent contribution. However, a simple spatially constant field was assumed, and therefore space-charge effects will not appear here. The Monte Carlo routine is embedded in a circuit simulation program in order to obtain a complete model and calculate the output voltage. The circuit program uses the impulse-response method of the time-domain analysis.[85] Prior to photoexcitation, the voltage across the gap is

$$V_G(0) = V_{Bias} - I_{Dark} R_L, \qquad (5.109)$$

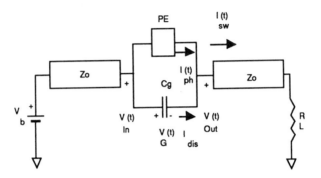

FIGURE 5.27.   Generalization of Auston's equivalent circuit for photoconductive switching studies.

where I_{Dark} is measured experimentally. When the laser source is turned on, the input and output voltages are given as the sum of a dc component and a time-varying component:

$$V_{in}(t) = V_{Bias} - V_{tv}(t);$$ (5.110)

and

$$V_{out}(t) = I_{Dark} \times R_L + V_{tv}(t).$$ (5.111)

As is obvious from (5.111), the time-varying part need not be small compared with the initial steady-state component. The time-varying total current $I_{total}(t)$ is given by

$$I_{total}(t) = C_{gap} \frac{d}{dt} [V_{in}(t) - V_{out}(t)] + I_{ph}(V_{in}(t) - V_{out}(t)).$$ (5.112)

The photocurrent I_{ph} is a function of the voltage across the gap and thus behaves as a voltage-controlled current source. This particle photocurrent corresponding to the voltage drop across the gap can be obtained by using the Monte Carlo program. It is based on the equation

$$I_{ph}(V_{in}(t) - V_{out}(t)) = qnAv_d,$$ (5.113)

where n is the electron concentration, A is the area across which the current flows, and v_d is the ensemble averaged velocity during the appropriate time step.

The left-hand side of (5.112) is determined by the characteristic of the transmission line and is evaluated through the convolution integral

$$I_{total}(t) = \int_0^t V_{tv}(\tau) * Y_0(t - \tau) \, d\tau + I_{Dark}.$$ (5.114)

In the above equation, $Y_0(t)$ is the impulse-response (or inverse transform of the characteristic admittance) of the transmission line. Implicit in the above equation is the assumption that the signal flow is unidirectional, and hence the line impedance is not modified by the load termination. This is valid within the time scales of interest here as the transmission line lengths were in the millimeter range. We use the transmission line model of Whitaker et al.,[86] which has been used in modeling dispersion in similar experiments. This transmission model, though somewhat empirical, satisfies causality, with the function $Z_0(\omega) - 1$ having poles only in the lower half-plane. This satisfies the condition that $G(\tau)$, defined as

$$G(\tau) = \frac{1}{2\pi} \int_{-\infty}^{\infty} [Z_0(\omega) - 1] \exp(-i\omega\tau) \, d\omega, \tag{5.115}$$

is zero for all τ less than zero. Comparing (5.112) and (5.114), we get an equation for the unknown $V_{tv}(t)$:

$$C_{gap} \frac{d}{dt} [V_{in}(t) - V_{out}(t)] + I_{ph}(V_{in}(t) - V_{out}(t))$$

$$= \int_0^t V_{tv}(\tau) * Y_0(t - \tau) \, d\tau + I_{Dark}. \tag{5.116}$$

The above equation was solved numerically by the Runge–Kutta method with forward differencing.

The geometry of the structure simulated is the following: the gap length is $10 \, \mu$m, and the transmission line width is $50 \, \mu$m. The separation between the two transmission lines is $50 \, \mu$m, and the probe beam is $20 \, \mu$m down the line from the gap. The laser power is the $5 \, $mW per beam for an excitation wavelength of $620 \, $nm. The carrier concentration density used is $5 \times 10^{15} \, $cm^{-3}. The capacitance value for this case is $1 \, $fF.

The results obtained from this simulation for the output voltage using an initial bias of 10 and 20 V and excitation wavelength of 620 nm are shown in Figure 5.28. The corresponding electron drift velocity curves are shown in Figure 5.29. For both bias cases, we notice that the shape of the velocity is translated into the shape of the output voltage.

We also considered the problem from a different perspective and extracted the photocurrent from the experimentally measured voltage. This was done by removing the Monte Carlo model from the circuit simulation and calculating the photocurrent by using the output voltage as an input parameter. The magnitude of the extracted photocurrent was less than that predicted by the Monte Carlo model. We believe this discrepancy arises from the assumption of a spatially uniform field across the gap. In the $n^+ i n^+$ structure used, the field is nonuniform and takes on positive and negative values, depending on the position. Electrons in some regions of the gap will possess negative velocity with respect to the

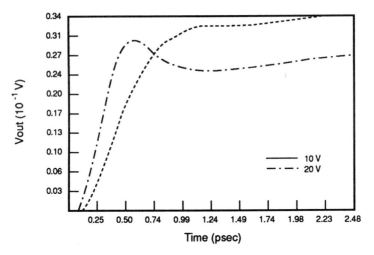

FIGURE 5.28. The transient output voltage waveform, normalized to the dc bias, obtained from the circuit simulation. 620-nm excitation was assumed. The voltages illustrated are applied biases.

direction of the field, and this will reduce the value of the total drift velocity and of the photocurrent. More exact modeling requires the inclusion of a Poisson solver in the Monte Carlo to find the field with respect to position is necessary and will yield results closer to those seen experimentally.

In summary, existing experiments clearly show that electro-optic sampling has both the temporal resolution and sensitivity to measure transients excited by transient transport in the photoconductive switch. Circuit analysis clearly shows that a transient-transport-produced photocurrent transient can create similar transmission line voltage transients. Monte Carlo studies show that a carefully

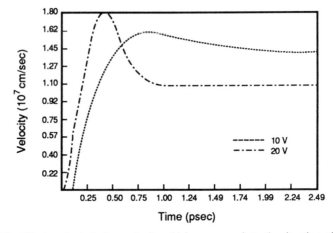

FIGURE 5.29. The transient electron velocity which corresponds to the situation of Figure 5.28.

conducted version of such experiments may be able to more completely restrict the value of the Γ-L coupling coefficient. However, a carefully conducted experiment requires the use of low photoexcitation levels in structures specifically designed to produce quasi-one-dimensional electric fields.

5.10. DC STUDIES IN SUBMICRON STRUCTURES

The velocity overshoot shown in Figure 5.2 can be translated into a velocity-versus-distance plot by integration. Laval et $al.$[87] attempted to demonstrate this length dependence in another experiment by using optical excitation. For these measurements the device structure consisted of a microstripline deposited on a lightly doped GaAs epitaxial layer. The upper strip has a gap, on each side of which are heavily doped contact regions. A voltage is applied across this gap. The gap is then uniformly illuminated, and measurements are made of the photoinduced current. Measurements were made on devices with gap lengths of 385, 1.05, and 0.4 μm. The experimental results are shown in Figures 5.30 and 5.31.

The data are interpreted by using an assumed uniform generation of carriers across the gap and an assumed field $E_x(y, t) = V(t)/L$, where $V(t)$ is the time-varying gap voltage. The photoinduced current is thought to arise from a thin layer of high photoconductivity near the surface and to be directly proportional to an assumed, spatially uniform time-varying average carrier velocity. The assumption is that the long-gap device should be well modeled by the steady-state velocity-field characteristic, but that the shorter-gap devices may show evidence of TDR. The data for 385 μm shows a photocurrent peak for biases near those which produce a field, as described above, which lies near the fields where a peak is seen in the steady-state velocity field curve for GaAs. The peak-to-valley ratio, however, is much larger than that of the velocity-field characteristic. This suggests that the assumptions of spatially uniform fields and carrier densities are

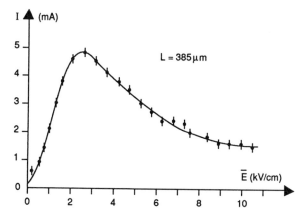

FIGURE 5.30. Photocurrent versus mean electric field for a 385-micron-long diode. This curve is relatively independent of the optical excitation wavelength. After Laval et $al.$[87]

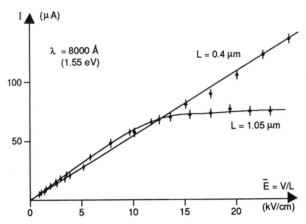

FIGURE 5.31.   Photocurrent versus mean electric field for the 1.05- and 0.4-micron-long diodes for a photon energy of 1.55 eV. After Laval *et al.*[87]

not valid. This could arise for a variety of reasons, including contact and surface effects. The intensity and wavelength of the incident light were varied for the long-gap device without altering the photocurrent bias dependency in a significant fashion. The results for the shorter devices were markedly different from those of the long-gap device, perhaps as a result of TDR and perhaps because the contacts are relatively more important in the shorter structures. The shortest device displayed no saturation and had a nearly linear current-voltage characteristic. The 1.05-μm device, however, did show saturation and seemed to exhibit a higher low-field mobility than did the shortest device. As wavelength decreased, there was a marked reduction in photocurrent for the 1.05-μm device, attributable to increased injection of carriers into heavy-mass satellite valleys. The data are shown in Figure 5.32.

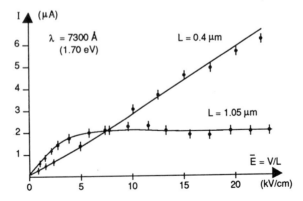

FIGURE 5.32.   Photocurrent versus mean electric field for the 1.05- and 0.4-micron-long diodes for a photon energy of 1.7 eV. After Laval *et al.*[87]

5.11. HIGH-EXCITATION EFFECTS

In the presence of a strong monochromatic source, one, in general, would expect saturation due to state filling also to be a factor. State filling of the conduction and valence bands, as opposed to band filling,[88] arises from monochromatic radiation and results in a δ-function-like spike in the distribution function (as a function of energy).[89] The occurrence of such state filling depends on the generation rate and the length of time that the excitation occurs in a specific energy state. Normally, it is to be expected that such effects will occur under picosecond-laser-pulse excitation, although one would expect the time width of the resulting saturated absorption to be of the same scale as the laser pulse. In Ge, it is found that the state filling that would normally be expected to occur is restricted by the process of energy-gap narrowing, or band renormalization as it is sometimes called, induced by the high electron-hole-pair density.[90]

The shift in the phonon frequencies, induced by the free carriers, also causes a gap narrowing. Brooks[91] relates the energy-gap variation to the change in the lattice vibration frequencies from ω_i to ω_i' when an electron-hole pair is excited across any particular gap. The gap change arises from a change in chemical potential and can be expressed as

$$\Delta E_g = \sum_i [f(\omega_i', n) - f(\omega_i', 0)], \tag{5.117}$$

where $f(\omega, n)$ is the standard formula for the free energy of an oscillator of frequency ω_k, which itself is a function of n (or p). The variation of the energy gap at high densities (high temperatures) was shown by Heine and Van Vechten[92] to be dominated by the anharmonicity in the TA phonons. The TA modes depend critically upon the covalent nature of the bonding in the diamond structure. Without this bonding, the TA modes become unstable. The generation of electron-hole pairs removes the bond charge, thus destabilizing and softening the TA mode. It is significant that estimates of the peak electron density in picosecond-laser experiments at the damage threshold is within a factor of 5–8 of that necessary to completely destabilize the TA mode. Heine and Van Vechten have shown that the density dependence of the anharmonic TA mode may be expressed as

$$\omega_{TA}' = \omega_{TA}\left[1 - \frac{f_{cv}n\varepsilon_0 e_0^*}{4(\varepsilon_0^* - \varepsilon_0)N_A}\right], \tag{5.118}$$

where f_{cv} is the bond-charge shift per electron-hole pair, ε_0^* is the dielectric constant of the competing β-tin phase, and N_A is the atomic density.

As mentioned, the strong energy-gap renormalization leads to a gap narrowing which is sufficiently strong that state filling probably does not occur. Indeed, it has been suggested[93] that under extreme conditions where the electron temperature greatly exceeds the lattice temperature, anomalously large changes in the band gap will occur. This will in turn lead to a gradient in T due to the large rate of phonon emission that is localized and to gradients in the electrochemical potentials of the electrons and holes. As the energy gap is a minimum at the sites of the highest density, these gradients oppose the normal diffusion of the excited carriers, since they are essentially self-trapped in a potential well.

Energy-gap narrowing at high concentrations of electrons and holes arises primarily from two principal mechanisms: self-energy, or exchange energy, contributions to the band gap and the free-carrier-induced shifts of the phonon energies. The narrowing of the energy gap due to the interacting nature of the free carriers has been considered by Inkson,[94] using a dynamically screened potential. The introduction of free carriers alters the quasi-Fermi level so that the exchange energy throughout the band must change. He estimated that at low temperatures the gap closure is given by

$$\Delta E_g = \frac{2e^2}{\pi\varepsilon_0}\left\{k_F + a\left[\frac{\pi}{2} - \tan^{-1}\left(\frac{k_F}{a}\right)\right]\right\},\tag{5.119}$$

where

$$a = \frac{\lambda}{\varepsilon_0^{1/2}}\tag{5.120}$$

and λ is the screening wave vector. Now, (5.119) is a zero-temperature approximation, and its validity at higher temperatures is restricted. However, k_F reflects primarily an estimate of the number of interacting carriers within the conduction band (or valence band). An estimate of this effect can be found by deriving k_F from the Fermi energy, given by

$$E_F - E_C = k_B T_e \ln\left(\frac{n}{N_C}\right),\tag{5.121}$$

where

$$N_C = \frac{1}{4}\left(\frac{2m_e^* k_B T_e}{\pi\hbar^2}\right)^{3/2}\tag{5.122}$$

is the effective density of states in the conduction band, m_e^* is the density-of-states effective mass, and T_e is the electron temperature.

While the above high-excitation effects are indeed possible, they are expected only under extremely high excitation conditions. Thus, they have not been included in any analyses of the transport experiments of interest here. Other "high-excitation" effects are expected at "lower high excitations," and they have been at least partially included in analyses of these experiments. We review these effects, in particular hot phonons and electron-hole interactions, here.

In this analysis we consider only the initial transients in the carrier dynamics. As a result, carrier recombination is not an important consideration for the present. Though recombination rates have to be large in order to turn off the device by carrier loss, typical time constants for Cr-doped GaAs are well beyond 10 ps.[95] The pulse duration and shape also cease to be a factor with the advent of subpicosecond pulses, but we include it for completeness. The charge separation effects will, however, be ignored. Such effects require that the charge carriers are collectively displaced relative to each other. For the subpicosecond times of interest here, such relative separation of the positive and negative charges is negligible. We analyze the situation where the excitations are strong enough to create a carrier density of 10^{17} to 10^{18} carriers per cubic centimeter in bulk GaAs.

We study the transient response of the photogenerated electron-hole plasma in a uniform field through bipolar Monte Carlo simulations. In contrast to other Monte Carlo studies of transient photoconductivity,[17,74] we include the features needed to examine high-excitation experiments. In particular, electron-hole scattering in a bipolar plasma and the hot-phonon effect have been incorporated. Electron–electron interactions have been left out because they only give rise to second-order effects. Since such carrier–carrier collisions conserve the net momentum of the electronic system, the ensemble velocity is left unchanged. Their only effect is on the carrier distribution function which indirectly changes the relative magnitude of the other scattering mechanisms.

Recent experiments on the transient carrier transport have already indicated that electron-hole interactions are important. Degani et al.[96] observed that minority carriers in p-type InGaAs do not exhibit an overshoot. Shah et al.[97] obtained negative absolute electron mobilities in GaAs quantum wells, while Tang et al.[98] noted a sharp reduction in the minority electron mobility in Si at low fields. Theoretical simulations by Osman et al.[99] have shown that the electron-hole interaction provides an important mechanism for energy exchange during the initial thermalization process. All these results underscore the importance of the electron-hole scattering. This interaction, by providing an alternative channel for the hot-electron energy loss, has the following effects. First, it cools the electrons and tends to keep them in the central valley. Second, intervalley transfer rates are affected, leading to changes in the temporal evolution of the valley populations.

In high-excitation experiments, nonequilibrium or hot-phonon effects also become important. They are included here as well and are important in resolving questions about the experimental transient mobility data. For example, the time-resolved reflectivity experiment of Nuss et al.,[54] obtained a carrier density dependence of the mobility rise time which was believed to have been a result of the hot-phonon effect.

A three-valley electron and a three-band hole model has been used for the present Monte Carlo simulations. The material parameters used for the holes were a combination of those given by Hess et al.[81] and Wiley.[100] The electron parameter set is that used earlier in this chapter to study transient photoconductivity. Initial optical generation and distribution of the carriers in k-space takes into account anisotropic effects. Carrier degeneracy has been suitably included, using the rejection technique discussed earlier. The bipolar EMC has all the relevant carrier-phonon and electron-hole interactions. Only single-mode LO and

TO couplings have been considered, and all plasmon–phonon interactions are ignored for the present. A static but time-evolving screening model of Osman and Ferry[49] has been used for all the polar interactions. Only intraband electron-hole processes have been included, leaving out possible multiband scattering, as discussed by D'yakonov *et al.*[101]

Hot-phonon effects are treated by using the EMC algorithm proposed by Lugli *et al.*[102] Both polar optical and intervalley phonon populations have been modified since the photoexcitation levels and electric-field strengths cause large Γ-L transfer. This can lead to significant perturbations in the intervalley phonon populations despite the large wave vectors of the zone boundary phonons and the big volume of phase space associated with it.

The result of an EMC simulation for the transient electron velocity in bulk GaAs at 300 K is shown in Figure 5.33 for a carrier concentration of 5×10^{18} cm^{-3}. The photoexcitation energy was 2 eV, the laser pulse width 30-fs FWHM, and the electric field 2.5 kV/cm. The shape of the optical pulse was taken to be the usual hyperbolic secant. No hot-phonon or electron-hole scattering was included for this EMC simulation.

As can be seen from Figure 5.33, there is an initial delay in the rise time of the ensemble velocity. This is the Jones–Rees effect discussed earlier. Furthermore, as expected, there is no velocity overshoot at this low-field level. The carriers generated from the heavy-hole band are placed high in the Γ-valley and are quickly scattered to the satellite valley before being able to pick up significant energy and momentum from the field. Finally, the time required to reach steady state for such high-energy excitations is well beyond a few picoseconds, and the velocity values at about 2 ps are in excess of 10^7 cm/s.

Shown in Figure 5.34 are electron velocity curves for the same field of 2.5 kV/cm and carrier densities of 5×10^{17} cm^{-3} and 5×10^{18} cm^{-3}. This figure demonstrates the effect of adding the hot phonons. Two features are immediately

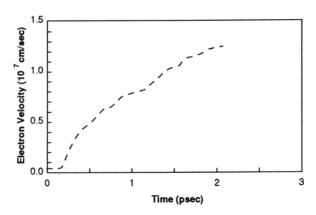

FIGURE 5.33. Electron velocity for a carrier concentration of 5×10^{18} cm^{-3} in GaAs at 300 K. The electric field was 2.5 kV/cm, the optical pulse width was 20 fs FWHM, and the photon energy was 2 eV. Neither hot phonons nor electron-hole scattering were included.

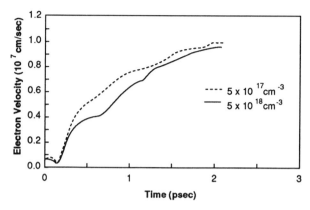

FIGURE 5.34. Electron velocities for carrier concentrations of 5×10^{17} cm^{-3} and 5×10^{18} cm^{-3} with hot-phonon effects included. All other parameters are specified in Figure 5.33.

evident from the curves. First, the velocity values for the concentration of 5×10^{17} cm^{-3} are higher than those obtained for a carrier density of 5×10^{18} cm^{-3}. This clearly shows that the hot-phonon effect works to reduce the velocity through an enhancement of the optical scattering rates. At the higher carrier concentration, the number of net phonon emissions by the hot electrons is greater, leading to a stronger amplification of the phonon modes. The larger phonon population thus created increases the momentum randomization through anisotropic scattering processes. The result is a decrease in the carrier velocity. A comparison of the curves of Figures 5.33 and 5.34 for the carrier density of 5×10^{18} cm^{-3} similarly confirms this effect of the hot phonons. The comparison reveals that the velocity obtained by including the hot-phonon effect is lower than that predicted by the simpler calculation.

The second point of interest regarding the curves of Figure 5.34 is that the difference between the two velocities is highest at intermediate times. The initial behavior is easily understood. Since some time is required for the buildup of the phonon population, there cannot be any significant difference between the two curves initially. Beyond about 0.5 ps, the distinction begins to get noticeable. The decrease in the difference for times greater than 1.5 ps arises as a result of an interaction between several physical factors. The ongoing thermalization process forces the electrons to occupy lower-energy states at later times. At these lower energies, the phonon wave vector associated with any emission/absorption scattering process is larger. The energy dependence of the q vectors is illustrated through the diagram in Figure 5.35. The processes AA' and BB' shown in the figure involve identical energy transfer through optical phonon emission/absorption. Of the two, process AA' occurs for electrons occupying higher-energy states. The phonon wave vector q associated with such a transfer, assuming parabolic bands, is given by

$$q = \left\{ \frac{2m^*}{\hbar^2} [E + E' - 2(EE')^{1/2} \cos \Theta] \right\}^{1/2}, \tag{5.123}$$

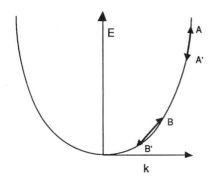

FIGURE 5.35. Schematic diagram of polar optical transitions AA' and BB' for carriers in a parabolic band. The transition AA' involves electrons at a higher energy and requires lower phonon wave vectors than the process BB'.

where E and E' are the initial and final energy states, and Θ is the angle between the initial and final wave vectors. Since the matrix element for the polar optical scattering is inversely proportional to the wave vector, the low-q transitions make significant contributions. This implies that the small-angle scattering events are more important, and the wave vectors for such favored processes decrease with increasing energy in accordance with (5.123). Though the above equation assumes parabolicity, the argument remains valid even for nonparabolic bands.

In order to substantially influence the electron–phonon scattering at later times, the large-q modes would have to be appreciably perturbed. This, however, is not the case. Most of the amplified phonons are created by the energetic hot electrons and have wave vectors which are correspondingly low. As a result, in spite of the low-wave-vector modes remaining sufficiently amplified, they are unable to affect the low-energy electrons at later times. This effectively reduces the difference between the 5×10^{17} and 5×10^{18} cm^{-3} cases beyond 2 ps, as the hot phonons are rendered increasingly ineffective. Extending this reasoning, it is clear that the hot phonons cannot affect the steady-state velocity even at lower temperatures despite their greater amplification. This has been demonstrated in a Monte Carlo analysis.[103]

The wave-vector and temporal dependence of the nonequilibrium phonon population is shown in Figures 5.36 and 5.37. As just described, the small-wave-vector modes are very strongly amplified because of the high-energy electron distribution. Furthermore, since the polar electron-LO phonon scattering strength is inversely proportional to the phonon wave vector, the low-q population is preferentially increased. A small second peak appears in Figure 5.36 and can be attributed to two separate mechanisms. One contribution comes from those electrons which lose an integral multiple of the phonon energy and cascade down to occupy lower-energy states. The other part comes from those low-energy electrons which were photogenerated out of the light-hole band and hence possessed lower kinetic energy. The three curves of Figure 5.37 show the temporal evolution. The short-wave-vector modes peak at about 0.8 ps and decay thereafter. The large-wave-vector mode does not exhibit a decreasing trend even up to 2 ps, since the phonon decay keeps getting replenished by amplification through the low-energy electrons.

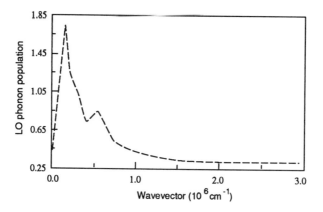

FIGURE 5.36. Wave vector dependence at 2.1 ps of the nonequilibrium LO phonon modes at a carrier concentration of 5×10^{18} cm^{-3} and a temperature of 300 K.

The effect of including the electron-hole scattering together with the hot-phonon effect is shown in Figure 5.38. The same external field of 2.5 kV/cm was used. Comparison of Figures 5.34 and 5.38 shows a significant reduction in the velocity as a result of electron-hole scattering. The velocities at the carrier concentrations of 5×10^{17} and 5×10^{18} cm^{-3} are reduced by factors of about 0.5 and 0.25, respectively with the inclusion of the electron-hole interaction. As can be seen, the steady-state values are largely determined by the electron-hole interaction in a high-density plasma. Furthermore, the rise time for the 5×10^{17} cm^{-3} case appears to be slightly faster than the 5×10^{18} cm^{-3} case. Based upon the time required for reaching 90% of the velocity value at 1.5 ps, the rise-time measure for the two cases turns out to be about 0.70 and 0.45 ps. This trend is in keeping with the experiment of Nuss et al.,[54] which yielded unusually low values of the steady-state mobility and a density dependence of the rise times.

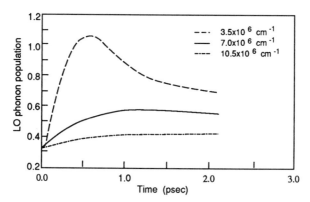

FIGURE 5.37. Temporal evolution for three phonon modes at a carrier concentration of 5×10^{18} cm^{-3} and a temperature of 300 K.

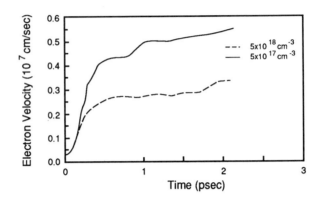

FIGURE 5.38.   Electron velocities for carrier concentrations of 5×10^{17} cm^{-3} and 5×10^{18} cm^{-3} with both hot-phonon and electron-hole scattering effects included. All other parameters are specified in Figure 5.33.

The mobility values in that experiment were extracted from reflectivity measurements, although the applied electric fields were believed to have been lower than 1 kV/cm. It is difficult to make a quantitative comparison directly since the spatial and temporal dependences of the electric fields in the experiment were not known. However, the present results clearly demonstrate that the carrier–carrier effects are more important than the hot phonons at the higher densities. This is contrary to the earlier suggestion advancing the hot phonons as the primary cause of the experimentally observed mobility.[54]

At this point it is perhaps fruitful to look at the role of the intervalley transfer mechanism. If at higher densities, the Γ-valley occupancy were enhanced, then the mobility reduction seen in Figure 5.38 could only be the result of a lowering of the Γ-valley drift velocity from strong scattering. A lower L-valley population, on the other hand, would mean that both electron-hole scattering and intervalley transfer combine to affect the mobility. This point has been studied by Joshi and Grondin.[103]

Figures 5.39 and 5.40 show Joshi's Monte Carlo data for the temporal evolution of the fractional electron occupancy in the Γ- and L-valleys, respectively, for carrier concentrations of 5×10^{17} and 5×10^{18} cm^{-3}. The same electric field of 2.5 kV/cm, a temperature of 300 K, and a laser pulse energy of 2 eV were used, as in Figure 5.38. The main feature seen in Figures 5.39 and 5.40 is the initial rapid transfer of carriers to the L-valley. Since the photoexcitation energy is well above the intervalley transfer threshold, the higher density of final states forces strong intervalley transfer. With increasing time, however, the Γ-to-L transfer begins to decrease. This decrease can be understood as arising from a combination of the following mechanisms. First, the energy relaxation pushes a fraction of the Γ-valley electrons below the intervalley transfer threshold. This is particularly strong at higher carrier densities, where the electron-hole interaction provides an energy-loss channel in addition to the inelastic polar optical scattering.

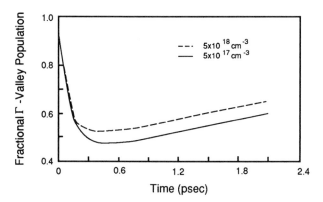

FIGURE 5.39. Fractional Γ-valley populations at electron densities of 5×10^{17} cm^{-3} and 5×10^{18} cm^{-3}. All other parameters are described in Figure 5.38.

The net effect is a decrease in the Γ-to-L-valley transfer. Second, the electrons in the L valley undergo greater electron-hole scattering because of the higher density of states and smaller mass mismatch. This scattering reduces the fraction of the L-valley electrons having energy above the X-valley transfer threshold, and lowers the percentage of the optical deformation potential L-X and L-L scattering events. The net outcome is an increase in the relative magnitude of the L-to-Γ-valley transfer. Finally, energy relaxation also plays an indirect role in promoting intervalley transfer. As the carriers thermalize, the inverse screening length begins to increase with time and leads to enhanced screening of the polar optical scattering. With this screening induced reduction in the polar interaction comes a relative enhancement in the nonpolar intervalley transfer.

    All the above-mentioned effects are strongly density-dependent, and hence it is not surprising to find a higher Γ-valley occupancy at the higher carrier density beyond 0.25 ps. This result eliminates enhanced L-valley transfer as a possible

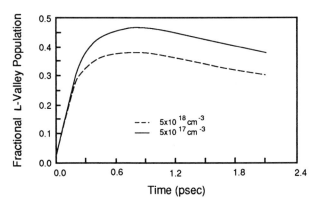

FIGURE 5.40. Fractional L-valley populations at electron densities of 5×10^{17} cm^{-3} and 5×10^{18} cm^{-3}. All other parameters are described in Figure 5.38.

cause for the density-dependent mobility reductions at these high-energy excitations, leaving the momentum randomization through the electron-hole scattering as the primary candidate. Though electron–electron interaction might cause some alteration in the above picture, the changes are not expected to outweigh the electron-hole effects. This is primarily because electron–electron collisions do not affect the net energy and momentum of the system, but have second-order effects through changes in the distribution function.

An additional test is provided by examining the case of long-wavelength excitation where most of the photogenerated carriers are expected to reside in the Γ-valley. Long-wavelength excitations are more interesting from the standpoint of achieving higher velocities and shorter turn-on times. In such a situation, effects arising from intervalley transfer would be strongly reflected in the velocity behavior. Figure 5.41 shows his three velocity curves for a laser excitation of 1.5 eV and a field of 2.5 kV/cm. Two of the curves include electron-hole interaction and were obtained for carrier densities of 5×10^{18} and 5×10^{17} cm^{-3}, while the third at a carrier concentration of 5×10^{18} was without electron-hole scattering. As expected, the addition of electron-hole scattering decreases the velocity. The enhancement in the energy loss from the electron system tends to keep the carriers in the Γ-valley, an effect which washes out the initial velocity overshoot transient. Clearly, the field of 2.5 kV/cm is not strong enough to cause a big transferred-electron effect quickly enough in the presence of electron-hole scattering. By the same reasoning, the velocities obtained at times beyond 0.5 ps, for the density of 5×10^{18} cm^{-3}, are marginally higher than those for a concentration of 5×10^{17} cm^{-3}. Here, the intervalley transfer sets in earlier for the lower concentration.

An apparent point of difference between curves of Figures 5.38 and 5.41 is the velocity value shown at the longer times. Since all the simulation parameters, excepting the initial excitation energy were identical, one expects the final steady states to be the same. This, however, is not seen in these two figures as a consequence of a prolonged transient in one case. While the curves of Figure 5.41 indicate a steady state beyond 0.9 ps, those of Figure 5.38 exhibit a persistent

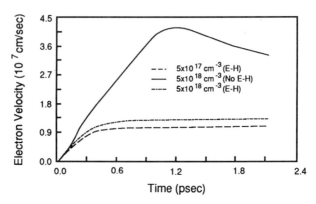

FIGURE 5.41. Velocity transients following excitation by 1.5-eV photons.

rise even at the longer times. For this case the final steady state was reached at times just beyond 9.5 ps, yielding velocity values matching those of Figure 5.41.

High-field results at an excitation of 2 eV are shown in Figure 5.42. Electron-hole scattering reduces the velocity. The initial Jones–Rees trapping effect is lower at the higher concentration of 5×10^{18} cm^{-3} since a lower-satellite-valley transfer takes place. This density dependence of the Jones–Rees effect, discussed elsewhere,[76] is seen more clearly at the high fields. As the difference between the electron and hole momentum increases with time, so does the momentum exchange associated with the electron-hole scattering. The net outcome is a lower peak occurring at an earlier time.

In order to get better quantitative agreement with experimental data, it is necessary to improve both the experimental approach and the theoretical modeling. For example, the fields attained in the transient mobility experiment[54] were not known accurately, and the mobility values were extracted from measurements using a simplified Drude model. The theoretical modeling can be improved by including a treatment of the coupled phonon–plasmon modes and using dynamically screened interactions. The present analysis uses a time-dependent, static screening model. This can be partially justified for the electron-hole scattering since the carriers can respond much faster than lattice ions. However, as shown by Meyer et al.,[104] the electron-hole interaction is modified by the relative motion of the electrons and holes. Whether this modification at the initial times can be neglected due to the small relative carrier velocities remains unresolved. The mixing of the phonon–plasmon modes could also modify the polar optical energy loss rates. As demonstrated by Ridley[105] and Kim et al.,[106] the coupled modes can lead to either a screening or an antiscreening effect which is density-dependent and strongly related to the damping within the system.

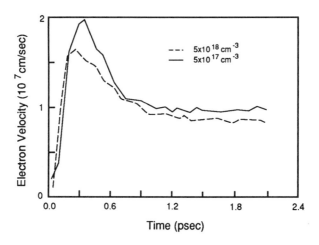

FIGURE 5.42.   Electron velocities for carrier concentrations of 5×10^{17} cm^{-3} and 5×10^{18} cm^{-3} with both hot-phonon and electron-hole scattering effects included. The field is 20 kV/cm, and the photon energy is 2 eV.

5.12. BALLISTIC TRANSPORT IN SMALL STRUCTURES

One of the simplest and oldest models of semiconductor transport is quite simple. One assumes that electrons (or holes) are little charged billiard balls that, while acted on by electromagnetic fields, never scatter. This ballistic transport model is the model that underlies the thermionic emission theory developed by Bethe in 1942[107] for Schottky barrier diodes. It also has a rich history in the theory of impact ionization where it is known as the lucky electron hypothesis, first used in 1961 by Shockley.[108] It was also used by van der Ziel in the mid-1970s in an analysis of infrared detection by thin Schottky barrier diodes.[109] During the 1980s, however, there has been a continued search for device structures in which ballistic or collisionless transport can be clearly seen and hopefully used as a basis for faster switching devices. The main spark of this effort was a 1979 paper in which Eastman and Shur[110] suggested that fast MESFETs could be based on ballistic transport. They additionally attempted to use dc current-voltage characteristics of short diode structures in an effort to experimentally observe ballistic transport.[111]

There was a flurry of papers critical of this concept and these experiments. While obviously the transport is ballistic in between the collisions in any semi-classical model, questions were raised concerning the length and time scales over which one could safely neglect collisions.[1,11,112-114] As in the case with any such arguments, to some degree one saw greater disagreement on the question of how safe is safe enough to allow a simplifying approximation to be made than one did on physics. Several important points, however, were raised. Ballistic transport actually is better viewed as a way of preserving fast electrons than as a way of accelerating them. The original FET structure was abandoned, and greater attention paid to systems, such as heterojunction bipolar transistors (HBTs), in which injection over the top of an energy barrier (sometimes called a ballistic launcher) was used to filter out slow, low-energy electrons. The fast electrons which entered over the barrier then traverse a short region in a hopefully ballistic fashion.

The next point raised was that the experimental signature of ballistic transport sought in the short-diode measurements, a current which is proportional to the voltage raised to the $\frac{3}{2}$ power, unfortunately can be created by a variety of other effects, so many in fact that it would be extremely difficult to do an experiment in which all competing possibilities could be eliminated. This caused a shift in both sample structure and data analysis. This new type of experiment, essentially a spectroscopic study performed in heterojunction systems, was suggested by at least one theoretical group,[115] and nearly simultaneously demonstrated by two competing experimental groups.[116,117]

The last point is more subtle. It was noted that the condition being described for ballistic transport, the absence of inelastic scattering processes, also is the condition needed for the appearance of purely quantum mechanical wave effects since it is the condition needed to maintain phase coherency in a quantum transport model. Evidence of such effects has been seen in several experiments.[118,119]

The type of structure used in the new experimental techniques is illustrated in Figure 5.43. Electrons are injected over the first potential barrier, the emitter, traverse a short base region, and then strike the second potential barrier at the collecting contact. If the electrons are sufficiently energetic, they are collected and contribute to the dc current flow through the system. The collector current I_c is[120]

$$I_c = -\frac{q}{m^*} \int_{p_{per,0}}^{\infty} p_{per} f(p_{per}) \, dp_{per},$$ (5.124)

where m^* is the effective mass, p_{per} is the momentum perpendicular to the collecting barrier, $f(p_{per})$ is the electron distribution incident on the barrier, and

$$p_{per,0} = (2m^* \phi_{bc})^{1/2},$$ (5.125)

where ϕ_{bc} is the collector barrier height. The critical concept in this spectroscopic study is that ϕ_{bc} can be lowered by use of the applied base-collector bias voltage V_{bc}. In fact, one can show that

$$\frac{dI_c}{dV_{bc}} \simeq f(p_{per}).$$ (5.126)

Therefore, simple differentiation of the dc characteristics yields a quantity proportional to the distribution function. The energy of these electrons at the emitter end of the base is controlled by the emitter barrier height. For short bases, a peak is found that corresponds to this injected energy. This peak indicates that a significant fraction of the injected carriers have traversed the base region without experiencing any inelastic scattering processes.

The key point in these spectroscopic experiments is an unambiguous demonstration of ballistic transport across short regions of GaAs. As useful transistors, they still fall significantly behind far more conventional transistors as amplifiers. They, however, are one of the best vehicles for experimental studies of nonstationary hot-carrier transport. The results obtained in these experiments show ballistic behaviors for lengths of 50 to 100 nm, a value much lower than those which originally were expected.

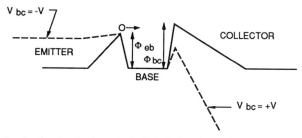

FIGURE 5.43.   Conduction-band schematic for hot-electron transistor experiments. The hot electrons are injected over the emitter-base junction barrier and collected at the base-collector junction. Solid lines are equilibrium, while dotted lines are under bias. After Hayes et al.[120]

Short, but otherwise conventional, transistors also have been useful vehicles for the study of velocity overshoot behavior. This is done by extracting an equivalent velocity from the dc characteristics of the device. Velocity overshoots in devices with gate lengths of under 100 nm have now been seen in Si MOSFETs[121] and GaAs MESFETs.[122] Additionally, a falloff in velocity seems to appear in GaAs MESFETs for gate lengths of 30 nm, as the device becomes too small to allow low-velocity electrons coming out of the source to be accelerated.[122]

REFERENCES

1. See, e.g., the Special Issue on Hot Carrier Effects in Short-Channel Devices, *IEEE Trans. Electron Dev.* **ED-28**, August (1981).
2. See, e.g., H. Grubin, in: *Physics of Submicron Devices* (H. L. Grubin, D. K. Ferry, and C. Jacoboni, eds.), Plenum, New York (1988).
3. J. G. Ruch, *IEEE Trans. Electron Dev.* **ED-19**, 652 (1972).
4. M. Brauer, *Phys. Stat. Sol. (b)* **81**, 147 (1977).
5. T. J. Maloney and J. Frey, *J. Appl. Phys.* **48**, 781 (1977).
6. S. Kratzner and J. Frey, *J. Appl. Phys.* **49**, 4064 (1978).
7. G. Hill, P. N. Robson, A. Majerfeld, and W. Fawcett, *Electron. Lett.* **13**, 235 (1977).
8. D. K. Ferry and J. R. Barker, *Phys. Stat. Sol. (b)* **100**, 683 (1980).
9. K. Hess, in: *Advances in Electronics and Electron Physics*, Vol. 59, Academic Press, New York (1982).
10. C. Jacoboni, in: *Physics of Submicron Devices* (H. L. Grubin, D. K. Ferry, and C. Jacoboni, eds.), Plenum, New York (1988).
11. G. J. Iafrate and K. Hess, in: *VLSI Electronics: Microstructure Science*, Vol. 9 (N. G. Einspruch, ed.), Academic Press, Orlando (1985).
12. D. K. Ferry and R. O. Grondin, in: *VLSI Electronics: Microstructure Science*, Vol. 9 (N. G. Einspruch, ed), Academic Press, Orlando (1985).
13. T. H. Glisson, C. K. Williams, J. R. Hauser, and M. A. Littlejohn, in: *VLSI Electronics: Microstructure Science*, Vol. 4 (N. G. Einspruch, ed.), Academic Press, Orlando (1985).
14. W. Fawcett, A. D. Boardman, and S. Swain, *J. Phys. Chem. Sol.* **31**, 1963 (1970).
15. J. Zimmermann, P. Lugli, and D. K. Ferry, *Solid-State Electron.* **26**, 233 (1983).
16. P. Lugli, J. Zimmermann, and D. K. Ferry, *J. Phys.* **42** (Suppl. 10), C7-103 (1981).
17. R. O. Grondin and M. J. Kann, *Solid-State Electron.* **31**, 567 (1988).
18. See, e.g., R. W. Zwanzig, in: *Lectures in Theoretical Physics* (W. E. Brittin, B. W. Downs, and J. Downs, eds.), Interscience, New York (1961).
19. Remember that the general N-body problem in phase space can be described by a hierarchy of coupled equations for the 1 through N particle distribution functions. To decouple the hierarchy, some additional approximation, such as molecular chaos, must be assumed. See N. N. Bogoliubov, *Lectures on Quantum Statistics*, Gordon and Breach, New York (1967).
20. R. Kubo, in: *Lecture Note in Physics, Vol. 31, Transport Phenomena* (J. Ehlers *et al.*, ed.), Springer-Verlag, Berlin (1974); I. Prigogine and P. Resibois, *Physica* **27**, 629 (1961); H. J. Kreuzer, *Nonequilibrium Thermodynamics and Its Statistical Foundations*, Oxford University Press, London (1981). In each of these, however, (5.16) is reduced to a kernel of a relaxation integral whose form must relate to (5.17) due to the need to reduce to a master equation. Barker[21] shows the exact form.
21. J. R. Barker, *J. Phys. C* **6**, 2663 (1973).
22. P. N. Argyres, *Phys. Rev.* **132**, 1527 (1961).
23. K. K. Thornber, *Phys. Rev. B* **3**, 1929 (1971).

24. D. K. Ferry, in: *Physics of Nonlinear Transport in Semiconductors* (D. K. Ferry, J. R. Barker, and C. Jacoboni, eds.), Plenum, New York (1980).

25. D. K. Ferry and J. R. Barker, *J. Phys. Chem. Sol.* **41**, 1083 (1980).

26. P. Price, *Solid-State Electron.* **21**, 9 (1978).

27. D. K. Ferry, in: *Handbook of Semiconductors*, Vol. 1 (W. Paul, ed.), North-Holland, Amsterdam (1980).

28. J. R. Barker and D. K. Ferry, *Phys. Rev. Lett.* **42**, 1779 (1979).

29. H. F. Budd, *Phys. Rev.* **158**, 798 (1967).

30. J. Klafter and R. Silbey, *Phys. Rev. Lett.* **44**, 55 (1980).

31. V. M. Kenkre, E. W. Montroll, and M. F. Schlesinger, *J. Stat. Phys.* **9**, 45 (1973).

32. H. J. Kreuzer, *Nonequilibrium Thermodynamics and Its Statistical Foundations*, Oxford University Press, London (1981).

33. D. N. Zubarev, *Nonequilibrium Statistical Thermodynamics*, Consultants Bureau, New York (1974).

34. R. W. Zwanzig, *Phys. Rev.* **124**, 983 (1961); see also R. Zwanzig, K. S. J. Nordholm, and W. C. Mitchell, *Phys. Rev. B* **6**, 1226 (1972).

35. J. R. Barker, in: *Physics of Nonlinear Transport in Semiconductors* (D. K. Ferry, J. R. Barker, and C. Jacoboni, eds.), Plenum, New York (1980).

36. P. J. Price, in: *Fluctuation Phenomena in Solids* (R. E. Burgess, ed.), p. 355, Academic Press, New York (1965).

37. P. Lugli and D. K. Ferry, *IEEE Trans. Electron Dev.* **ED-32**, 2431 (1985).

38. O. Madelung, *Introduction to Solid State Physics*, Springer, Berlin (1978).

39. S. Bosi and C. Jacoboni, *J. Phys. C* **9**, 315 (1976).

40. B. B. van Iperen and H. J. Tjassens, *Proc. IEEE* **59**, 1032 (1971).

41. K. S. Champlin and G. Eisenstein, *IEEE Trans. Microwave Theory Tech.* **MTT-26**, 31 (1978).

42. R. O. Grondin, P. A. Blakey, and J. R. East, *IEEE Trans. Electron Dev.* **ED-31**, 21 (1984).

43. A. M. Mazzone and H. D. Rees, *IEE Proc. Pt. 1* **127**, 149 (1980).

44. J. Zimmermann, Y. Leroy, and E. Constant, *J. Appl. Phys.* **49**, 3378 (1978).

45. P. A. Lebwohl, *J. Appl. Phys.* **44**, 1744 (1973).

46. H. D. Rees, *IBM J. Res. Dev.* **13**, 537 (1969).

47. H. D. Rees, *J. Phys. Chem. Sol.* **30**, 643 (1969).

48. W. Fawcett, in: *Electrons in Crystalline Solids*, International Atomic Energy Agency, Vienna (1973).

49. M. A. Osman and D. K. Ferry, *J. Appl. Phys.* **61**, 5330 (1987).

50. D. Jones and H. D. Rees, *J. Phys. C* **6**, 1781 (1973).

51. D. Jones and H. D. Rees, *Electron. Lett.* **8**, 363 (1972).

52. R. B. Hammond, *Physica B* **134**, 475 (1985).

53. G. Mourou, K. Meyer, J. Whitaker, M. Pessot, R. Grondin and C. Caruso, in: *Picosecond Electronics and Optoelectronics II*, Springer Series in Electronics and Photonics, Vol. 24, p. 40, Springer-Verlag, New York (1987).

54. M. C. Nuss, D. H. Auston, and F. Capasso, *Phys. Rev. Lett.* **58**, 2355 (1987).

55. C. V. Shank, R. L. Fork, B. I. Greene, F. K. Reinhart, and R. A. Logan, *Appl. Phys. Lett.* **38**, 104 (1981).

56. S. Ramo, *Proc. IRE* **27**, 584 (1939).

57. W. Shockley, *J. Appl. Phys.* **9**, 635 (1981).

58. S. Teitel and J. W. Wilkins, *J. Appl. Phys.* **53**, 5006 (1982).

59. S. J. Allen, C. L. Allyn, H. M. Cox, F. DeRosa, and G. E. Mahoney, *Appl. Phys. Lett.* **42**, 96 (1983).

60. S. J. Allen, in: *Physics of Submicron Devices* (H. L. Grubin, D. K. Ferry, and C. Jacoboni, eds.), Plenum, New York (1988).

61. P. Das and D. K. Ferry, *Solid-State Electron.* **19**, 851 (1976).

62. G. H. Glover, *J. Appl. Phys.* **44**, 1295 (1973).

63. W. Franz, *Z. Naturforsch* **13**, 484 (1958).

64. L. V. Keldysh, *Sov. Phys. JETP* **7**, 778 (1958).

65. M. Osman and H. Grubin, *Solid-State Electron.* **31**, 471 (1988).

66. W. Pötz and P. Kocevar, *Phys. Rev. B* **28**, 7040 (1983).
67. J. Shah, B. Deveaud, T. C. Damen, W. T. Tsang, A. C. Gossard, and P. Lugli, *Phys. Rev. Lett.* **59**, 2222 (1987).
68. D. H. Auston, *IEEE J. Quantum Electron.* **19**, 639 (1983).
69. J. A. Valdmanis, G. A. Mourou, and C. W. Gabel, *IEEE J. Quantum Elect.* **QE-19**, 664 (1983).
70. K. E. Meyer, M. Pessot, G. Mourou, R. O. Grondin, and S. N. Chamoun, *Appl. Phys. Lett.* **53**, 2254 (1988).
71. K. Meyer and G. Mourou, in: *Picosecond Electronics and Optoelectronics*, Springer Series in Electrophysics, Vol. 21 (G. Mourou, D. Bloom, and C. Lee, eds.), Springer-Verlag, New York (1985).
72. J. A. Valdmanis, *Subpicosecond Electro-Optic Sampling*, Ph.D. Dissertation, University of Rochester (1983).
73. D. H. Auston, K. P. Cheung, J. A. Valdmanis, and D. A. Kleinman, *Phys. Rev. Lett.* **53**, 1555 (1984).
74. A. Evan Iverson, G. M. Wysin, D. L. Smith, and A. Redondo, *Appl. Phys. Lett.* **52**, 2148 (1988).
75. G. M. Wysin, D. L. Smith, and A. Redondo (unpublished).
76. R. Joshi, S. Chamoun, and R. O. Grondin, in: *Picosecond Electronics and Optoelectronics IV* (T. C. L. Gerhard Sollner and D. M. Bloom, eds.) (1989).
77. S. N. Chamoun, R. Joshi, E. N. Arnold, R. O. Grondin, K. E. Meyer, M. Pessot, and G. A. Mourou, *J. Appl. Phys.* **66**, 236 (1989).
78. E. O. Kane, *J. Phys. Chem. Sol.* **1**, 249 (1957).
79. M. A. Osman and D. K. Ferry, *Phys. Rev B* **36**, 6018 (1987).
80. M. Rieger, P. Kocevar, P. Bordone, P. Lugli, and L. Reggiani, *Solid St. Electron.* **31**, 687 (1988).
81. K. Brennan and K. Hess, *Phys. Rev. B* **29**, 5581 (1984).
82. A. J. Taylor, D. J. Erskine, and C. L. Tang, *J. Opt. Am. B* **2**, 663 (1985).
83. D. E. Aspnes, *Phys. Rev. B* **14**, 5331 (1976).
84. V. N. Freire, A. R. Vasconcellos, and R. Luzzi, *Solid. St. Commun.* **66**, 683 (1988).
85. E. N. Arnold, *Time-Domain Analysis of Nonlinear Microwave Circuits Using a Frequency-Domain Description of the Linear Subnetworks*, Masters Thesis, Arizona State University (1988).
86. J. F. Whitaker, R. Sobolewski, D. Dykaar, T. Hsiang, and G. Mourou, *IEEE Trans. Microwave Theory Tech.* **36** (1988).
87. S. Laval, C. Bru, C. Arnodo, and R. Castagne, in: *IEEE Int. Electron Devices Meeting Tech. Digest*, p. 626 (1980).
88. R. N. Zitter, *Appl. Phys. Lett.* **14**, 73 (1969).
89. C. J. Hearn, P. T. Landsberg, and A. R. Beattie, in: *Proc. Sixth Int. Conf. Physics of Semiconductors*, p. 857, Inst. Physics, London (1962).
90. D. K. Ferry, *Phys. Rev. B* **18**, 7033 (1978).
91. H. Brooks, in: *Advance in Electronics and Electron Physics*, Vol. 7, p. 85 (1955).
92. V. Heine and J. A. Van Vechten, *Phys. Rev. B* **13**, 1622 (1976).
93. M. Wautelet and J. A. Van Vechten, *Phys. Rev. B* **23**, 5551 (1981).
94. J. C. Inkson, *J. Phys. C*, **9**, 117 (1976).
95. C. H. Lee, A. Antonetti, and G. Mourou, *Opt. Commun.* **21**, 158 (1977).
96. J. Degani, R. F. Leheny, R. Nahory, and J. P. Heritage, *Appl. Phys. Lett.* **39**, 569 (1981).
97. R. A. Höpfel, J. Shah, P. A. Wolf, and A. C. Gossard, *Phys. Rev. Lett.* **56**, 2736 (1986).
98. D. D. Tang, F. F. Fang, M. Scheuermann, and T. C. Chen, *Appl. Phys. Lett.* **49**, 1540 (1986).
99. M. A. Osman and H. L. Grubin, *Proc. SPIE* **942**, 18 (1988).
100. J. D. Wiley, in: *Semiconductors and Semimetals* (R. K. Willardson and A. C. Beer, eds.), Vol. 10, p. 91, Academic Press, New York (1975).
101. M. D'yakonov, V. I. Perel, and I. N. Yassievich, *Sov. Phys. Semicond.* **11**, 801 (1977).
102. P. Lugli, C. Jacoboni, L. Reggiani, and P. Kocevar, *Appl. Phys. Lett.* **50**, 1521 (1987).
103. R. Joshi and R. O. Grondin, *J. Appl. Phys.* **66**, 4288 (1989).
104. J. R. Meyer and F. J. Bartoli, *Phys. Rev. B* **28**, 915 (1983).
105. R. K. Ridley, *Superlattices and Microstructures* **2**, 159 (1986).
106. M. E. Kim, A. Das, and S. D. Senturia, *Phys. Rev. B* **15**, 6890 (1978).
107. H. A. Bethe, *MIT Radiation Lab Rep.* **43-12** (1942).

108. W. Shockley, *Solid St. Electronics* **2**, 35 (1961).
109. A. van der Ziel, *J. Appl. Phys.* **47**, 2059 (1976).
110. M. S. Shur and L. F. Eastman, *IEEE Trans. Electron Dev.* **ED-26**, 1677 (1979).
111. L. F. Eastman, R. Stall, D. Woodard, N. Dandekar, C. E. C. Wood, M. S. Shur, and K. Board, *Electron. Lett.* **16**, 524 (1980).
112. T. J. Maloney, *IEEE Electron Dev. Lett.* **EDL-1**, 54 (1980).
113. J. R. Barker, D. K. Ferry, and H. L. Grubin, *IEEE Electron Dev. Lett.* **EDL-1**, 209 (1980).
114. H. U. Baranger and J. W. Wilkins, *Phys. Rev. B* **36**, 1487 (1987).
115. P. Hesto, J-F. Pone, and R. Castagne, *Appl. Phys. Lett.* **40**, 405 (1982).
116. A. J. F. Levi, J. R. Hayes, P. M. Platzman, and W. Wiegmann, *Phys. Rev. Lett.* **55**, 2071 (1985).
117. M. Heiblum, M. I. Nathan, D. C. Thomas, and C. M. Knoedler, *Phys. Rev. Lett.* **55**, 2200 (1985).
118. M. V. Fischetti, D. J. DiMaria, L. Dori, J. Batey, E. Tierney, and J. Stasiak, *Phys. Rev. B.* **35**, 4404 (1987).
119. M. Heiblum, in: *High Speed Electronics* (B. Kallback and H. Beneking, eds.), Springer-Verlag, Berlin (1986).
120. J. R. Hayes, A. J. F. Levi, A. C. Gossard, and J. H. English, in: *High Speed Electronics* (B. Kallback and H. Beneking, eds.), Springer-Verlag, Berlin (1986).
121. G. A. Sai-Halasz, M. R. Wordeman, D. P. Kern, S. Rishton, and E. Ganin, *IEEE Electron Dev. Lett.* **EDL-9**, 464 (1988).
122. J. M. Ryan, J. Han, A. M. Kriman, and D. K. Ferry, *Solid St. Electron.* **32**, 1609 (1989).

6

Alloys and Superlattices

In recent years, the concepts of band-gap engineering have become widespread especially for their impact on performance modifications of electron devices and for their suggestion of new methods of device operation. This has primarily arisen from the ability to modify material characteristics through alloying and through the use of superlattices and quantum wells to tune energy levels. If there is a single reason for the advent of alloys, it is the advantages that are offered by the tuning of the fundamental energy gaps (between conduction and valence bands) of the material system to some desired value. This single parameter governs the electrical and optical properties through the intrinsic carrier densities and the fundamental absorption edge. As the investigations of semiconductor materials has progressed, studies of the simple elements Si and Ge moved to the more interesting compounds of GaAs, AlAs, etc., and then to alloys of these compounds. Interestingly enough, the use of molecular-beam epitaxy has once again allowed interesting alloys of Si and Ge to be investigated.

The first semiconductor alloys were composed of a simple binary mixture of two individual elemental semiconductors, such as Ge and Si. Early studies focused on the growth of these alloys in bulk form, but the crystal properties were not good. Subsequent work addressed the so-called pseudobinaries, such as GaAs-InAs or $Ga_{1-x}In_xAs$. Here, the common anion As resides on one of the sublattices of the zinc blende structure, while the cation residing on the second sublattice is a random alloy of Ga and In with an average composition of x parts In and $1 - x$ parts Ga. In these alloys, the band gap of the resulting alloy lies somewhere between those of the two end-point compounds. Unfortunately, the lattice constant also changes, so that the alloy is not lattice matched to either of the two end-point compounds, and growth is complicated by the problem of finding a suitable "seed" crystal—one with a lattice constant nearly equal to that of the desired alloy composition. In Figure 6.1, we illustrate the primary band gap (at the Γ-point of the Brillouin zone) and the lattice constant for GaInAs alloys. A more general plot is shown in Figure 6.2, where we illustrate various compounds and alloys on a plane whose coordinates are the principal optical energy gap and the lattice constant.

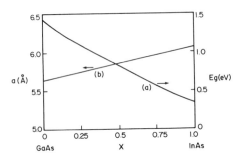

FIGURE 6.1. The energy gap (a) and lattice constant (b) for the alloy InGaAs as a function of composition.

It is obvious from Figure 6.2 that we can extend the concept of a pseudobinary compound one step further. For the ternary compounds, such as GaInAs, alloying was carried out on only one of the two sublattices (e.g., just for the cation sublattice). One can carry out alloying on both of the sublattices. In the quaternary compound $Ga_{1-x}In_xAs_{1-y}P_y$, the cation sublattice is occupied by a random mixture of Ga and In atoms, with average concentration of In given by x, and the anion sublattice is occupied by a random mixture of As and P atoms, with average P concentration given by y. By proper choice of the two composition variables x and y, any site enclosed by the outer boundary in Figure 6.2 can be addressed with a suitable alloy. With these two parameters, we can now adjust both the primary energy gap and the lattice constant to fit a suitable substrate material. One example, indicated by one of the horizontal lines in Figure 6.2, is a quaternary compound of GaInAsP which is lattice-matched to an InP substrate.

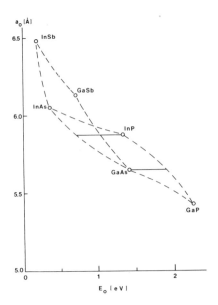

FIGURE 6.2. The values of band gap and lattice constant for the various binary compounds. The solid curves represent the quaternary lattices matched to GaAs and InP.

Thus, the alloy of this example can be varied from pure InP across to $Ga_{0.47}In_{0.53}As$, all the while maintaining a lattice match to the substrate InP so that good quality in the grown layer can be achieved.

The material properties of intrinsic and extrinsic alloys developed by the above procedures are, in general, derived from the properties of the end-point compounds in terms of simple interpolation (and correction) procedures. This model (the virtual crystal model) of the behavior of the alloy assumes that the material is a nearly perfect random alloy with no compositional ordering (or segregation) and relies on the fact that the properties of most semiconductors are dominated by the similarities of these materials rather than by their differences.

The earliest drive to tune the band gap of semiconductor materials was the need to provide compounds for long-wavelength infrared detectors, although this was quickly followed by the desire to provide semiconductor lasers in the visible region of the spectrum. By referring to Figure 6.2 once again, we see that band gaps over the range of 0.18 to 2.4 eV can be obtained by alloying various combinations of the III-V compounds. If we also consider the II-VI compounds, this range can be extended. In particular, a band gap of 0 to 1.6 eV can be achieved in the alloy HgCdTe. Thus, these two groups of pseudo–group IV compounds, all of which have the zinc blende (derived from diamond) type of lattice, allow one to cover the range from the far infrared through to the blue region.

Although the major efforts historically have been directed to the true alloys of these compounds, the development of molecular-beam epitaxy and metal-organic chemical vapor deposition has allowed heterostructures to be epitaxially grown with extremely sharp interfaces. This ability has meant that quantum wells and superlattices, whose individual layers are quite thin, can be tuned to provide further modification of the principal energy gaps. Properties of the quantum wells of narrow-gap materials, interspersed between layers of wider-band-gap material, allow tuning of the actual energy gap through the quantization energy of the thin layers. Later developments of purposely *non-lattice-matched* growth of layers, in order to incorporate further energy level modifications by the incorporation of strain in the layer, has broadened the scope of methods of adjusting the properties of the grown layers. Indeed, we are now approaching an era where it is quite feasible to engineer a desired energy band gap by pursuing not one but several methods of materials preparation so as to optimize a great many different properties of the material structure.

In spite of these varieties of techniques available for modification of the energy structure, the general properties of the resultant structure are still governed by the fact that these materials are similar to one another. Thus, quite general rules can be laid down for the manner in which their individual properties will be governed by the finished structure and the properties of the end-point compounds. As a consequence, it is a result that all of the optical properties and most of the electrical properties are governed by the same principles that are used for the elemental and binary compounds and are, in fact, governed in detail by the resultant electronic band structure of the compound.

In this chapter, we want to concentrate on the electronic structure itself. We first discuss the so-called tight-binding approach to band structure, in which the principal minima of the valence and conduction bands are related to the chemical levels of the individual atoms. This provides a rational basis for the nature of the band structure and how it is modified as various compounds are alloyed together. Then, we discuss the concepts of alloying itself, in considerably more depth than done so far. One method of providing interpolative variations among the alloys is the dielectric method, and we provide an introduction to this approach next. This latter approach is a general approach that concentrates on the similarities of the different materials rather than their differences. In this sense, it is a true interpolative theory, in that once the structure of the end-point compounds is known, excellent predictions for the alloys are immediately possible. As with all such theories of electronic structure, it is not completely accurate. However, it is more transparent than most and does not thrust the reader into the esoteric details encompassed in such methods as the nonlocal pseudopotential approach. We then examine the pseudoalloys that can be achieved with superlattices and quantum wells, as well as the strain-layer effects. In this way, we hope to cover the most important consequences of the general concept of band-gap engineering.

6.1. THE SEMIEMPIRICAL TIGHT-BINDING METHOD

The energy bands of covalent semiconductors have long been studied by the method of *linear combination of atomic orbitals*, where it is assumed that the actual band structure is due to the interaction of the atomic valence electrons of the atoms on the lattice sites. In group IV, or pseudo–group IV (III-V and II-VI elements with the zinc blende lattice), the valence electrons hybridize into orbitals composed of an admixture of the three independent p-type wave functions and a single s-type wave function. This hybridization is termed sp^3 hybridization, and accounts for the four outer shell electrons (on average) contributed by each of the atoms in the lattice. The variation from exactly four outer electrons on each atom leads to the ionic contribution to the bonding in the III-V and II-VI compounds. These hybrids, localized on each atom, then interact between the near neighbors to form the energy bands of the periodic lattice structure. The top of the valence band is primarily composed of p-type wave functions, while the bottom of the conduction band is primarily composed of s-type wave functions (at the zone center Γ-point). Thus, if the interaction energies, between wave functions localized on neighboring atoms, are known, we can make very accurate estimates of the principal energy gaps in any given solid from the atomic energies of the atoms themselves. Harrison[1] has applied this approach quite extensively to investigate the properties of covalent solids and the trends of these properties between various semiconductors.

The basic lattice structure is composed of two interpenetrating face-centered cubic lattices, with one corner at $(0, 0, 0)$ for one lattice and the corresponding corner offset to $(1, 1, 1)$ for the second lattice (the normal lattice constant is $2a$ in this notation). The primitive cell is then a tetrahedron with one atom at the center (say Ga in GaAs) at $(0, 0, 0)$ and four neighbors located at the vertices $(1, 1, 1)$, $(1, -1, -1)$, $(-1, -1, 1)$, $(-1, 1, -1)$ (again, we note that the lattice constant in the basic fcc cell is $2a$ in this notation). These latter are the four nearest neighbors (and would be As in GaAs). Thus, the relevant interaction is between s-type wave functions on the central atom and s- and p-type wave functions on the four atoms at the vertices. At the center of the Brillouin zone $(\mathbf{k} = 0)$, only the s-s and p-p interactions are relevant, and we can write the new energies as

$$E = E_s \pm 4V_{ss}, \tag{6.1}$$

$$E' = E_p \pm 4V_{pp}. \tag{6.2}$$

In these equations, the atomic energies are measured from the vacuum level and are consequently negative values. The interaction potentials V, however, are positive quantities, usually, for nearest-neighbor interactions. Thus, in (6.1), the negative sign gives the bottom of the valence band, while the positive sign gives the bottom of the conduction band. In (6.2), the negative sign gives the top of the valence band and the positive sign gives the top of the conduction band in the simplest theory. In each case, we note that the interaction produces a bonding state of lower energy and an antibonding state of higher energy. The bonding energies lie in the valence band, while the antibonding energies compose the conduction band.

In compound semiconductors, the atoms at the vertices are different from those at the center, and the interaction potentials involve wave functions from two different atoms. Moreover, the atomic energies are different for the two types of atoms. Thus, (6.1) and (6.2) must be replaced by[1]

$$E = \left[\frac{E_s^A + E_s^B}{2}\right] \pm \left\{\left[\frac{E_s^A - E_s^B}{2}\right]^2 + (4V_{ss})^2\right\}^{1/2}, \tag{6.3}$$

$$E' = \left[\frac{E_p^A + E_p^B}{2}\right] \pm \left\{\left[\frac{E_p^A - E_p^B}{2}\right]^2 + (4V_{pp})^2\right\}^{1/2}. \tag{6.4}$$

Here, we have taken the compound as A_{III}-B_V for example, so that the A atom is the cation and the B atom is the anion. Harrison[1] has pointed out that the various interaction potentials can be written as

$$V = \frac{\eta \hbar^2}{md^2}, \tag{6.5}$$

where η is a numerical constant, m is the electron mass, and d is the interatomic distance. For V_{ss}, $\eta_{ss} = -9\pi^2/64 = -1.39$, while for V_{pp}, η_{pp} is a factor of 3 smaller. Slater and Koster,[2] on the other hand, have suggested that the interaction matrix elements be treated as adjustable parameters, with the band extremum points fit to more accurate band calculations by the adjustment of these parameters. This latter method has become known as the semiempirical tight-binding method. One very important point that arises from Harrison's approach, however, is the variation of the interaction potentials with the interatomic distance d. Thus, as the lattice constant changes, either by alloying or by introducing strain, the bands will shift as a result of the shifts in the interaction potentials. This point will reappear in the dielectric theory discussed below.

A second important point arises from consideration of (6.3)–(6.4). Since alloying in the compound semiconductors proceeds by a random mixture on one or the other of the two sublattices, we can replace the appropriate quantity in (6.3) and (6.4) by the mixed quantity. For example, in $Ga_{1-x}In_xAs$,

$$E_j^A = xE_j^{In} + (1 - x)E_j^{Ga}, \qquad j = s, p. \tag{6.6}$$

A similar mixture is used for the interaction potentials. This method of introducing the alloy is called the pseudobinary alloy approach, and we will return to it later.

For points in the Brillouin zone away from the center, discussed above, we must incorporate the properties of the Bloch functions and their plane-wave properties. This leads to a k-dependent angular function that multiplies the interaction potentials. The angular function is a lattice sum over the appropriate neighbors, in this case the four nearest neighbors considered before. Thus, the factor 4 in equations (6.1)–(6.4) represents this sum for the $k = 0$ case. In the general case, we have, for example, in the case of the s-type wave functions:[2]

$$
\begin{aligned}
g_s(k) &= \exp\left[\frac{ik_xa}{2} + \frac{ik_ya}{2} + \frac{ik_za}{2}\right] + \exp\left[\frac{ik_xa}{2} - \frac{ik_ya}{2} - \frac{ik_za}{2}\right] \\
&\quad + \exp\left[-\frac{ik_xa}{2} - \frac{ik_ya}{2} + \frac{ik_za}{2}\right] + \exp\left[-\frac{ik_xa}{2} + \frac{ik_ya}{2} - \frac{ik_za}{2}\right] \\
&= 4\left[\cos\left(\frac{k_xa}{2}\right)\cos\left(\frac{k_ya}{2}\right)\cos\left(\frac{k_za}{2}\right) \right. \\
&\quad \left. - i\sin\left(\frac{k_xa}{2}\right)\sin\left(\frac{k_ya}{2}\right)\sin\left(\frac{k_za}{2}\right)\right].
\end{aligned}
\tag{6.7}
$$

In the case of the overlap of the s-wave function with the p-wave function, we have to worry about the direction of the p-function, and this will change appropriate signs in the sums present in the first line of (6.7). The result becomes

$$
g_{sx}(k) = 4\left[i\sin\left(\frac{k_xa}{2}\right)\cos\left(\frac{k_ya}{2}\right)\cos\left(\frac{k_za}{2}\right) - \cos\left(\frac{k_xa}{2}\right)\sin\left(\frac{k_ya}{2}\right)\sin\left(\frac{k_za}{2}\right)\right],
\tag{6.8}
$$

and the factors for the other p-functions are obtained by a cyclical permutation of the directions $x \to y \to z \to x$. The interaction potential is different when the two atoms are different, since the s- and p-functions are derived from different atomic levels. Equation (6.7) also results for the overlap between two p-functions if they are directed in the same direction; that is, $g_{xx} = g_s$. On the other hand, we must account for the directions once again for the term g_{xy}. This becomes

$$g_{xy}(k) = 4\left[i \cos\left(\frac{k_x a}{2}\right) \sin\left(\frac{k_y a}{2}\right) \sin\left(\frac{k_z a}{2}\right) - \sin\left(\frac{k_x a}{2}\right) \cos\left(\frac{k_y a}{2}\right) \cos\left(\frac{k_z a}{2}\right) \right],$$

(6.9)

and the other terms are generated by a cyclical permutation as above.

All of these potentials and angular factors determine the matrix elements for a large 8×8 matrix (four wave functions from each of two atoms, since the atoms at the vertices are shared with other unit cells). The matrix is of block form, in that it is composed of four 4×4 blocks. The two diagonal blocks contain just the atomic levels of the two atoms (the three p-levels are degenerate with the value E_p), since we are considering only the nearest neighbors. The two corner blocks contain the interaction potentials and angular factors that describe the couplings of the wave functions on one atom with those on the other atom. This 8×8 matrix can then be diagonalized to yield the four valence-band and four conduction-band energy levels for each set of values for (k_x, k_y, k_z). In most cases, however, good fit to more accurate calculations is not obtained with just the nearest neighbors, and second-neighbor interactions must be introduced. It is beyond the present scope to detail all of these angular functions, but the details are found in Refs. 1 and 2. These second-neighbor interactions are those between adjacent Ga atoms, for example. In Figure 6.3, the band structure of GaAs is

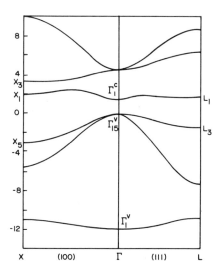

FIGURE 6.3. The band structure of GaAs as determined from the semiempirical tight-binding method. The interactions to the first and second neighbors were included, and the coupling constants are shown in Table 6.1.

TABLE 6.1.   Parameters for SETBM
for GaAs Band Structure

E_s^A	$= -14.54\,\text{eV}$	E_s^B $= -7.14$
E_p^A	$= -5.9$	E_p^B $= -0.72$
E_{ss}	$= -1.41$	E_{xx} $= -0.13$
E_{sp}^{AB}	$= 1.13$	E_{sp}^{BA} $= 1.09$
$E_{xy,2}^A$	$= 0.125$	$E_{xy,2}^B$ $= -0.05$
E_{110}^A	$= 0.04$	E_{110}^B $= 0.275$
E_{011}^A	$= 0.03$	E_{011}^B $= -0.625$

shown for a calculation including second neighbors. The values of the interaction potentials have been adjusted to fit more exact pseudopotential calculations, for which the fit is quite good. In Table 6.1, the values used in the calculation are listed.

The major conclusions we draw from looking at this approach to the band structure is that the major features of the structure arise from the geometrical structure of the crystal, while the variations in the interaction potentials are important but of a detail nature rather than a gross physical difference. The band structure arises from the properties of the atomic levels and the structure of the solid. Indeed, from (6.5) we see that the interaction potentials depend only on the structure through the interatomic distance d. The interaction potentials scale with the lattice parameters, and one can develop quite good insight into variations among tetrahedrally coordinated semiconductors by using just these scaling factors. This is the basis of the dielectric theory we shall investigate.

6.2. THE DIELECTRIC THEORY OF BAND STRUCTURE

If we are to properly make a material choice that will result in improved performance of electronic or optical devices, then it is necessary to understand the requisite properties of the materials and their dependence on the band structure of the material. Some of these properties are the band gap, effective mass, satellite valley separations (in the conduction band), and the behavior of these properties upon alloying if the material is to be a truly "random" alloy, such as a ternary or quaternary solid solution. As we saw in the last section, the properties of Si and the relevant III-V and II-VI compounds are basically very similar, in that they arise from the chemical s- and p-type wave functions of the four (average) electrons contributed by each atom in the unit cell. All of these compounds are tetrahedrally coordinated with a common lattice structure. Silicon possesses the diamond structure, which is an fcc structure with two atoms per basis site. For each of the sites, atoms are located at the lattice site and displaced along the body diagonal a distance of one quarter of the total distance across the cubic cell. The III-V and II-VI materials possess the zinc blende structure, which differs from the former only in that the two atoms per basis point are dissimilar atoms (i.e., one Ga atom and one As atom). In this way the zinc blende lattice is composed of two interpenetrating fcc lattices with one displaced from the first $(a/2, a/2, a/2)$, where $2a$ is the cubic edge.

As a consequence of the above similarities in structure plus the additional chemical constraint that the two basis atoms contribute a total of eight valence electrons (and therefore a full shell) between them, these materials are characterized more by their similarities than by their differences. Phillips[3,4] and Van Vechten[5,6] have utilized fully these similarities among tetrahedrally coordinated covalent materials to develop a general quantum dielectric theory of electronegativity in such systems, which describes not only the relative ionicity and dielectric constants but also gives an excellent description for the ionization potential and interband transition energies. Notably, it is just these quantities that we need to describe the optical properties and effective masses of the alloys. While others, notably Harrison,[1] have also developed general theories for a description of the variations in these materials, the dielectric method remains the most extensively developed (essentially to completion). From the last section, we note that the basic chemical levels lead to the basic properties of the materials and that the interaction matrix elements are scaled by interatomic distances. The dielectric method adopts the basic measurements of certain optical transitions, used to determine the interaction elements in the semiempirical tight-binding approach of the last section, and elaborates on the scaling rules to produce a generally complete interpolation theory.

The interpolative nature of the dielectric theory allows an excellent approach without having to resort to extensive computer calculations. Thus, we will utilize this approach to describe the general band structure variations for Si and the various compound semiconductors. This will then be extended to random alloys in the following section.

The dielectric method is based upon a universal model—a dielectric two-band model, where the two bands are generic representations of the bonding and antibonding orbitals. In this model, it is assumed that there exists an *average* energy gap E_G between these two bands and that the value of this gap may be deduced from the real, static (electronic contribution to the) dielectric constant $\varepsilon(0)$ (we discuss the origin of this quantity in a later chapter). The quantity E_G is not relatable directly to any of the measured optical properties, but is a true average between the bonding (valence) and antibonding (conduction) hybridized sp^3 orbitals. Moreover, the nature of this gap is such that it can be decomposed into homopolar (covalent) and heteropolar (ionic) contributions E_h and C, respectively. These two contributions represent, in turn, the symmetric and the antisymmetric parts of the atomic potentials from the two atoms within the basic unit cell of the materials considered here. The three energies are related by

$$E_G^2 = E_h^2 + C^2. \tag{6.10}$$

In a homopolar material such as Ge or Si, the two atoms within the unit cell are identical (the diamond lattice structure), so that we have $C = 0$ and $E_G = E_h$. This latter value is then taken as a reference value for the appropriate compound semiconductors, and its value may be obtained from the dielectric constant through[5,7]

$$\varepsilon(0) = 1 + \frac{(\hbar\omega_P)^2 DA}{E_G^2}, \tag{6.11}$$

where

$$A = 1 - B + \frac{B^3}{3}, \tag{6.12}$$

$$B = \frac{E_G}{4E_F}, \tag{6.13}$$

E_F is the Fermi energy of the valence electrons, ω_P is the valence plasma frequency, and D is a correction factor to account for some slight interaction of the valence sp^3 hybrids with core d-electrons, where they exist.[5] The heteropolar potential C is related to the valence difference and is given by

$$C = s\left[\frac{Z_a}{r_a} - \frac{Z_b}{r_b}\right] \exp\left\{-\frac{k_s(r_a + r_b)}{2}\right\}, \tag{6.14}$$

where k_s is the linearized Fermi–Thomas screening wave vector, and r_a and r_b are the covalent radii of elements a, b. The factor s also contains a correction factor of approximately[3] 1.5 (in addition to $e^2/4\pi\varepsilon_0$ in MKS units) to account for the fact that the Fermi–Thomas screening is generally stronger than actually found. In Table 6.2, the values of $\varepsilon(0)$, E_F, $\hbar\omega_P$, E_h, C, and D are listed for the major compounds of interest.

TABLE 6.2.   Optical Properties of Some Selected Materials

Material	$\varepsilon(0)$	E_F (eV)	$\hbar\omega_P$ (eV)	E_h (eV)	C (eV)	D
Si	12.0	12.5	16.6	4.8	0	1.0
Ge	15.9	11.5	15.6	4.31	0	1.26
GaAs	10.9	11.5	15.6	4.3	2.9	1.24
GaP	9.1	12.4	16.5	4.7	3.3	1.15
GaSb	14.4	9.85	13.9	3.5	2.1	1.31
InAs	12.3	10.1	14.2	3.7	2.74	1.35
InP	9.6	10.7	14.8	3.9	3.34	1.27
InSb	15.7	8.8	12.7	3.1	2.1	1.42
AlAs	10.3	11.5	15.5	4.38	2.67	1.1
HgTe	6.9	10.1	12.8	2.42	4.0	1.3
CdTe	7.2	10.0	12.7	3.08	4.9	1.3

The strength of the dielectric method lies in its ability to give good fits to the variations among elements of these compounds. The results of the dielectric method can be expressed in seven postulates;[3-6]

1. Any direct energy gap E_i, in the absence of d-level perturbations, is given in analogy to (6.10) as

$$E_i = [E_{ih}^2 + C^2] = E_{ih}\left[1 + \left(\frac{C}{E_{ih}}\right)^2\right]^{1/2},\qquad (6.15)$$

where E_{ih} is related to the value of the gap for the corresponding homopolar material (usually Si). Throughout this section, the subscript h refers to the value for the homopolar compound.

2. The E_{ih} values and all other homopolar variables are assumed to be simple power-law functions of the nearest neighbor distance d, relative to the Si value, as

$$E_{ih} = E_{ih,\text{Si}}\left(\frac{d}{d_{\text{Si}}}\right)^{r_i},\qquad (6.16)$$

where the various r_i are parameters, and differ from the pure value of 2 that arose in the last section.

3. The ionization potential (i.e., the energy difference between the top of the valence band at Γ and the vacuum level) is

$$I = I_h\left[1 + \left(\frac{C}{I_h}\right)^2\right]^{1/2},\qquad (6.17)$$

where I_h is the homopolar value and is scaled as in (6.16).

4. The energy at the top of the valence band at the symmetry point X (the zone edge along any of the [100] directions), the X_4 state in Si, and the X_5 state in the compounds, relative to the vacuum level is independent of C and scales with d as in (6.16).

5. The energy at the top of the valence band at the symmetry point L (the zone edge along any of the [111] directions), the L_3 state, is midway between the values of the energies at Γ and X (for the valence band), as

$$E_{L_3} = \frac{I + E_X}{2}.\qquad (6.18)$$

6. The splittings of the conduction band X levels, X_1 and X_3, in the compounds is generally given by a value that is nearly $0.14C$.

7. Finally, the perturbative effect of the d-levels on the s-like levels of greatest interest, the Γ_1^c and L_1^c of the conduction band, is expressed by decreasing

the E_0 (direct optical gap at the zone center) and E_1 (direct optical gap at the L-point) transitions from the values given by (6.15) to

$$E_i = [E_{ih} - (D - 1)\Delta E_i]\left[1 + \left(\frac{C}{E_{ih}}\right)^2\right]^{1/2}, \tag{6.19}$$

where $i = 0, 1$ here, and ΔE_i is a parameter which is a function of d alone, as in (6.16). We tabulate and discuss the various parameters below.

6.2.1. Silicon

The valence- and conduction-band spectra for silicon are shown in Figure 6.4, which is a result of a tight-binding calculation such as that of Pandey and Phillips[8] (and of the last section). The lowest energy gap is the indirect gap $\Delta_1^c - \Gamma_{25'}^v$. The minima in the conduction band are located neither in the zone center nor at the zone edges, but at a point approximately 85% of the distance to the zone edge at X along the (100) axes. The lowest direct transition at the zone center is not the one usually quoted. Rather, the slightly wider gap $\Gamma_{2'}^c - \Gamma_{25'}^v$ is usually quoted, as it is this gap which incorporates the conduction-band state that converts to the Γ_1^c state that is the conduction-band minimum in the compounds. The E_0, E_1, and E_2 (direct optical transition at X) transitions play an important role in estimating the positions of the upper conduction-band minima in the ternary and quaternary compounds.

The six equivalent minima of the conduction band are ellipsoids of revolution, characterized by a longitudinal mass $m_L = 0.91 m_0$ and a transverse mass $m_T = 0.19 m_0$. These give rise to a density-of-states effective mass

$$m_d = 6^{2/3}(m_T^2 m_L)^{1/3} = 1.06 m_0 \tag{6.20}$$

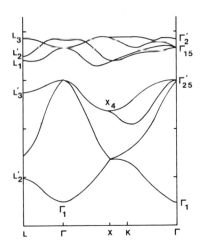

FIGURE 6.4.   The band structure for Si.

and a conductivity mass

$$m_c = \frac{3m_L m_T}{m_T + 2m_L} = 0.26 m_0. \tag{6.21}$$

In Table 6.3, the various band-gap and symmetry point energies are shown for Si along with the index r_i. One difference from the original values appears in this table, in that the value of r_i for E_X has been reduced substantially from its earlier value, in order to give a much better representation of the conduction-band minima for the binaries in the next subsection.

Germanium differs from Si only in the fact that the lowest conduction-band energy is that at L rather than at Δ. The values of the various parameters differ only somewhat from those of Si, although there is a d-level correction for this material. The conduction band is characterized by four equivalent ellipsoids, which have a transverse mass $m_T = 0.082 m_0$ and a longitudinal mass $m_L = 1.64 m_0$. This gives a density-of-states effective mass

$$m_d = 4^{2/3}(m_L m_T^2)^{1/3} = 0.56 m_0 \tag{6.22a}$$

and a conductivity mass

$$m_c = 0.12 m_0. \tag{6.22b}$$

6.2.2. The III-V Compounds

The basic band structure of the III-V elements differs little from Si other than in the positions and ordering of the various minima of the conduction band. In Figure 6.3 we illustrated the band structure of GaAs, but this could also have been the structure for InAs, InP, InSb, or GaSb, or any other of the direct-gap materials, with a scaling of the energy axis. Each of these named materials has Γ, L, X ordering of the conduction-band minima. Only GaP differs, and it is essentially Si-like with X, L, Γ ordering of the gaps. In Table 6.4, the principal energy levels and transitions are listed for these compounds. Also shown are the effective masses for the spherically symmetric minimum of the conduction band (where it is the lowest minimum for the various bands), and the Γ-X and Γ-L

TABLE 6.3.   Properties of the Homopolar Gaps

Parameter	Si value	Ge value	r_i
I_h	5.17	4.9	−1.31
X_4	8.63	8.14	−1.4
E_{0h}	4.1	3.64	−2.75
E_{1h}	3.6	3.27	−2.22
E_{2h}	4.5	4.06	−2.38
ΔE_1	0	3.88	−2.0
ΔE_0	0	10.6	−2.0

TABLE 6.4. The Principal Gaps in the III-V Alloys

Gap	AlAs	GaP	GaAs	GaSb	InP	InAs	InSb
E_{0h}	3.67	4.06	3.67	2.69	3.31	3.07	2.52
ΔE_0	10.9	12.67	10.9	7.81	8.75	7.92	0.27
E_0	3.96	2.75	1.42	0.7	1.35	0.35	0.2
I_h	4.91	5.15	4.91	4.42	4.67	4.5	4.1
I	6.02	6.12	5.7	4.9	5.74	5.27	4.7
X_5	8.51	8.62	8.51	8.0	8.42	8.34	7.94
E_{2h}	4.09	4.46	4.09	3.39	3.74	3.5	2.96
E_2	5.2	5.55	5.01	3.99	5.01	4.44	3.74
L_3	7.25	7.37	7.11	6.45	7.08	6.81	6.32
ΔE_1	4.07	4.69	4.07	2.75	3.33	2.74	2.07
E_{1h}	3.29	3.57	3.29	2.76	3.03	2.85	2.43
E_1	4.36	3.89	3.13	2.41	3.17	2.61	2.16
$\Delta E_{\Gamma X}$	−1.36	−0.16	0.4	0.19	0.98	1.02	0.32
$\Delta E_{\Gamma L}$	−0.85	−0.11	0.3	0.16	0.5	0.72	0.36
m_c/m_0	0.145	—	0.063	0.045	0.072	0.022	0.013
$P^2/2m_0$	18.5	—	21.1	14.9	17.4	15.6	15.2
C	3.62	3.3	2.9	2.1	2.74	2.74	2.1
D	1.07	1.11	1.24	1.24	1.45	1.45	1.43

separations of the conduction band. Many of these parameters depend on the absolute X_5 level in the material. Using the r_i value given in Table 6.3 gives good results except for the Sb compounds, and these have been adjusted from the Ge-based, rather than Si-based, extrapolation. These values then give good $\Delta E_{\Gamma L}$ and $\Delta_{\Gamma X}$ separations that are in good agreement with experimental results, except for GaP, and even here it is not far from experiment.

One line in Table 6.4, for $P^2/2m_0$, arises from $\mathbf{k} \cdot \mathbf{p}$ theory, as it is used in band structure calculations. In this approach, for a spherical band, one can scale the momentum matrix element P and the effective masses through

$$\frac{m_0}{m} = 1 + \frac{P^2}{m_0 E_0}. \tag{6.23}$$

As one can see from the various values of $P^2/2m_0$ in the different materials, P^2 is a relatively constant parameter for these tetrahedrally coordinated compounds. P itself is the momentum matrix element and should show only very weak chemical trends.[4] The slight variation observed in this quantity can be attributed to a slight variation in wave functions. We have used this quantity in estimating the conduction-band effective mass in the alloys discussed below.

6.2.3. Some II-VI Compounds

The basic band structure of the zinc blende II-VI compounds, such as CdTe, differs little from the structure of GaAs. Only HgTe has a significant difference. In this latter material, the lowest conduction band actually lies below the highest

TABLE 6.5.   Principal Gaps for II-VI Alloys

Gap	HgTe	CdTe	Gap	HgTe	CdTe
E_{0h}	0.13	0.13	L_3	6.0	6.35
ΔE_0	0.29	0.29	ΔE_1	3.16	3.16
E_0	−0.3	1.6	E_{1h}	2.4	2.4
I_h	3.5	3.5	E_1	1.72	3.3
I	5.32	6.01	$\Delta E_{\Gamma X}$	3.88	3.42
X_5	6.7	6.7	$\Delta E_{\Gamma L}$	1.34	1.36
E_{2h}	2.94	2.94	m_c/m_0	−0.31	0.096
E_2	4.96	5.71	$P^2/2m_0$	1.27	15.1
C	4.0	4.9	D	1.48	1.30

valence band, so that the character of these two energy levels is reversed. For this reason, this compound is said to possess a negative energy gap. When alloys of HgTe and CdTe are prepared, the resultant band gap can be brought very close to zero (it passes through zero in the simplest approximation, but second-order effects usually prevent a value exactly equal to zero from being obtained in practice). Thus, materials with very small band gaps can be prepared. It is for this reason that these alloys have found extensive usage in far infrared detectors. In Table 6.5, the principal gaps and transitions are listed for these two compounds. Also shown are the effective masses for the spherically symmetric conduction bands and the momentum matrix elements. We note that these compounds are composed of materials that lie low in the periodic table, and so there is a significant admixture of d-level wave functions in the valence energy levels. This has been reflected in the table itself. Again, we have extrapolated from the values of r_i appropriate to Ge, rather than those of Si. We also note that the homopolar energies are extrapolated from those of gray Sn, the cubic form that is closest to Si and Ge.

6.3. PSEUDOBINARY ALLOYS

As discussed in the preceding sections, the zinc blende structure is composed of two interpenetrating fcc lattices. Ternary alloys, such as GaInAs, have been formulated by a smooth mixing between the two constituents. In such $A_xB_{1-x}C$ alloys, all of the points of one fcc sublattice are occupied by type C atoms, but the points of the second sublattice are shared by the atoms of type A and type B, such that[9]

$$N_A + N_B = N_C = N,$$

$$x = \frac{N_A}{N_C} = C_A, \tag{6.24}$$

$$1 - x = \frac{N_B}{N_C} = C_B.$$

In this arrangement, a type C atom may have all type A neighbors or all type B neighbors, but on the average will have x type A neighbors and $1 - x$ type B neighbors. In effect, the structure is fcc structure of mixed A-C and A-B molecules, complete with interpenetrating molecular bonding. This structure composes a "pseudobinary" alloy with the properties determined by the relative concentrations of A and B atoms.

In recent years, quaternary alloys have also appeared as $A_x B_{1-x} C_y D_{1-y}$. Here, C and D atoms now share the first sublattice. This new compound is still considered as a pseudobinary compound composed of $A_x B_{1-x} C$ and $A_x B_{1-x} D$ molecules, which are somewhat more complex than the simple ones discussed above. Still, we assume that a true random mixing occurs so that the properties can be easily extrapolated from those of the constituent compounds. Then a general theory of pseudo-binary alloys can be applied equally well to quaternary alloys as well as to ternary alloys. If these compounds are truly smooth mixtures, then the alloy theory will hold, but if ordering occurs we can expect changes. For example, $In_{1-x} Ga_x As$ may be a smooth alloy composed of a random mixture of InAs and GaAs. However, if perfect ordering were to occur, particularly near $x = 0.5$, the crystal structure would become a chalcopyrite—a superlattice on the zinc blende structure with significant c/a distortions of the basic unit cell. In this latter case, we would expect changes to occur in the band structure due to Brillouin-zone folding about the point $(0, 0, \pi/2c)$. We return to this point later.

Consider now a pseudobinary disordered alloy (in the above sense, not in the perfect crystal sense) of A-C and B-C molecules on a Bravais lattice. We treat here ternaries, rather than quaternaries, but the treatment is general. We sum the crystal potentials over the A and B atoms as[9]

$$U(\mathbf{r}) = \sum_A U_A(\mathbf{r} - \boldsymbol{\tau}) + \sum_B U_B(\mathbf{r} - \boldsymbol{\tau}), \tag{6.25}$$

where $\boldsymbol{\tau}$ defines the lattice sites of the A-B sublattice. We can now decompose the total crystal potential into a symmetric and an antisymmetric part. The former is termed the "virtual crystal" potential and is given by

$$U_1(\mathbf{r}) = \sum [C_A U_A(\mathbf{r} - \boldsymbol{\tau}) + C_B U_B(\mathbf{r} - \boldsymbol{\tau})]. \tag{6.26}$$

The antisymmetric part is a random potential, which is presumed to be sufficiently small, and is given by

$$U_2(\mathbf{r}) = \sum [U_A(\mathbf{r} - \boldsymbol{\tau}) - U_B(\mathbf{r} - \boldsymbol{\tau})]C_\tau, \tag{6.27}$$

where $C_\tau = 1 - C_A$ for an A atom and $-C_A$ for a B atom at the site τ.[10] The virtual crystal potential (6.26) is the basis for smoothly varying the properties from the A-C crystal to the A-B crystal. The random part can contribute either to scattering of the carriers (alloy scattering and an imaginary part of the total self-energy of the carriers) or to bowing of the energy levels in the mixed crystal (a real part of the total energy).

The function C_τ was introduced by Flinn[10] primarily to discuss local ordering in binary alloys. It has the sum-rule properties

$$\sum_\tau C_\tau = 0, \qquad \sum_{\tau'} C_{\tau'}C_{\tau-\tau'} = NC_A(1 - C_A)\alpha_\tau, \tag{6.28}$$

where α_τ is the so-called Cowley–Warren order parameter. Equation (6.28) is the source of the $x(1 - x)$ terms that appear in the bowing of the energy gaps in alloys, as C_A is related to the quantity x. The order parameter is a major part of the bowing energy term that multiplies the $x(1 - x)$ contribution. We can see this by further noting that the matrix element of U_2 is just

$$|\langle \mathbf{k}'|U_2|\mathbf{k}\rangle|^2 = \sum_\tau \alpha_\tau I(\mathbf{k}', \mathbf{k}, \tau) \simeq C_A(1 - C_A)\alpha_0 I(\mathbf{k}', \mathbf{k}, 0), \tag{6.29}$$

where I is an overlap integral between the Bloch functions of the initial and final states. In the last expression, we have assumed that the solutions are indeed truly random so that there is no correlation from one site to the next (a rather strong assumption that is questionable). Therefore, only the term for $\tau = 0$ contributes, and the last expression represents a weak perturbation on the virtual crystal. This perturbation leads to a small self-energy, whose real part contributes the bowing effect, while the imaginary part is the "alloy scattering" that arises in the carrier transport.[11]

Hall[12] has shown that even short-range order can invalidate the approximation leading to the last part of (6.29). If the exact terms for the matrix elements are used in an energy calculation, they yield only pair interactions with α_τ giving the number of correlated pairs separated by τ. The individual A-A, A-B, and B-B pairs have the weighting specified by the factors C_τ. In this case, α_0 still corresponds to the self-energy discussed above and mainly leads to the bowing effects. This bowing can, in fact, work to stabilize an ordering trend in that bowing usually lowers the energy of the alloy. If this lowering of the energy is greater than that achieved by ordering, then the alloy is the more stable configuration. On the other hand, correlation between differing lattice sites will more strongly affect the imaginary parts of the self-energy, reducing the scattering strength of the alloy on the carriers. With these thoughts, we now return to consideration of the virtual crystal approximation for the alloy.

Van Vechten and Bergstresser[13] have shown that the dielectric method discussed above can be applied to the alloy problem quite generally. In the spirit of the virtual crystal approximation, the various parameters (C, D, E_{0h}, E_{ih}, etc.) are assumed to vary linearly with composition just as the crystal potential in (6.26) varies linearly with composition. In particular, the variation of C as the alloy is formed automatically incorporates an averaged antisymmetric potential contribution, so that the bowing of the band gap is automatically incorporated in the extrapolation. In Figure 6.5, we illustrate this for $In_{1-x}Ga_xAs$ by plotting the direct and indirect gaps for this material. We also use (6.23) to estimate the

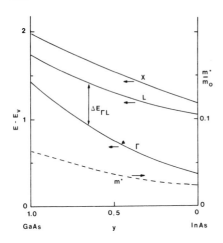

FIGURE 6.5. Variation of the principal gaps and the conduction-band effective mass for the alloy InGaAs.

effective mass for the conduction-band electrons in the alloy. The accepted gap variation in this compound is[14]

$$E = 0.35 + 1.07x - 0.28x(1 - x), \tag{6.30}$$

although some give a bowing parameter of 0.46 eV[15] rather than the 0.28 used. The difference is small (5%) and (6.30) agrees with Figure 6.5 to within a few percent over the entire alloy range. We note that the L-valley is separated from the Γ-valley in InAs by only 0.72 eV. This value is in general agreement with experimental evidence obtained by photoemission and Raman scattering.

The form (6.30) is the general form that has been found for most alloys, and, while there are some minor deviations from this form, it is a very good first approximation to the alloy band gap in any random alloy (with different coefficients, of course). The coefficient of the $x(1 - x)$ term is the bowing parameter and measures the deviation of the materials from the simplest virtual crystal approximation. In Table 6.6, we list the bowing parameter found on the direct gap at Γ in the alloys of the III-V and II-VI elements. The experimental values are also given for comparison. One should note that the Sb compounds, which lie low in the periodic table and have large d-level interactions, all show large bowing parameters. In fact, the alloy between InAs and InSb has a smaller band gap at the $x = 0.5$ composition than either of the two end-point compounds.

The alloy of GaAs and AlAs has become one of the most important compounds for both optical and electronic applications. In Figure 6.6, the principal energy gaps are shown for this alloy. In this case, the lowest conduction band at the AlAs end is at the X-point, which differs considerably from GaAs. This leads to different behavior in the alloys, but is a typical result when an indirect-gap material is alloyed with a direct-gap material.

The II-VI alloy of HgTe and CdTe is illustrated in Figure 6.7. In this case, as mentioned above, the principal gap is negative at the HgTe end, so that alloying

TABLE 6.6.   Bowing Parameters in Alloys

	c (DM)	c (exp)	c (EPM)	c (SETBM)
GaP–GaAs	0.06	0.17–0.21	0.014	0.2
GaP–InP	0.49	0.5–0.76	0.05	0.7
GaP–GaSb	1.34		0.13	
GaAs–InAs	0.68	0.28–0.46	0.03	0.4
GaAs–AlAs	0.16	0.2–0.6		
GaAs–GaSb	1.4	0.8	0.05	
GaSb–InSb	1.24	0.36–0.66	0.01	
InP–InAs	0.06	0.1	0.01	0.1
InP–InSb	1.61		0.06	
InAs–InSb	0.62	0.6	0.02	
InAs–AlAs	0.42			
HgTe–CdTe	0.40	0.4		0.36

of these compounds allows one to pass through a point of zero band gap. For this reason, this material has been used quite extensively for far-infrared detectors.

In the quaternary alloys, it is necessary to extrapolate the band gap and lattice constant data from that of the ternaries. There are many possible quaternary materials. However, to grow these properly, it is necessary to lattice match the alloy to a suitable substrate, which for InGaAsP, for example, is typically either InP or GaAs. We earlier showed the variation of energy gaps and lattice constants for the various III-V materials in Figure 6.2. While there are many quaternaries that are possible, we will concentrate on the single quaternary InGaAsP, as it has received the most attention to date. This is because of its possible use as a high-velocity semiconductor in electronic devices and for its direct optical band gap lying in the near infrared for fiber-optic applications. We treat the composition that is lattice-matched to InP.

Vegard's law provides for a linear extrapolation of lattice constant in the alloy. This implies, for example, that we find the value of x in InGaAs that provides a lattice match to InP, and then select the range of x and y that maintain

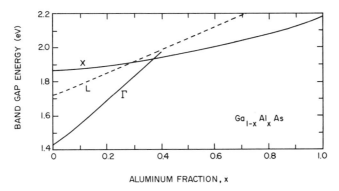

FIGURE 6.6.   Variation of the principal gaps and the central valley mass for the alloy GaAlAs.

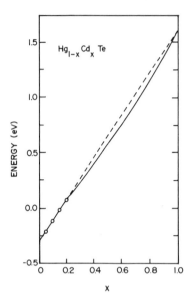

FIGURE 6.7. Variation of the principal gaps and the central valley mass for the alloy HgCdTe. After Haas *et al.*,[41] by permission.

this match across the quaternary. Moon *et al.*[16] have applied Vegard's law to this quaternary with the result

$$a = 0.587 + 0.018y - 0.042x + 0.002xy \quad \text{nm.} \qquad (6.31)$$

For lattice matching to InP ($a = 5.87$), we require

$$1.8y - 4.2x + 0.2xy = 0$$

and

$$x = \frac{0.43y}{1 - 0.048y}, \qquad (6.32)$$

with $0 < y < 1$. The dielectric theory can equally be applied to the quaternary, and direct and indirect gaps are shown in Figure 6.8 along with the effective mass of the carriers in the lowest conduction band. The experimentally determined direct band gap is[16]

$$E = 1.35 + (0.101y - 1.101)y + (0.758 - 0.28y)x(1 - x)$$

$$- (0.101 + 0.109x)y(1 - y) \quad \text{eV.} \qquad (6.33)$$

This result agrees with the band plotted in Figure 6.8 to within 3% over the entire range illustrated.

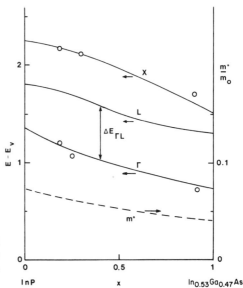

FIGURE 6.8. Variation of the principal energy gaps and the conduction mass effective mass for the quaternary InGaAsP, lattice-matched to InP.

For further comparison, the energy-gap data of Grinyaev et al.[17] are also shown in Figure 6.8. These latter data were calculated using a nonlocal pseudopotential method, and the agreement is quite good, not only for the direct gap but also for the indirect gap to the X-point. Many experimental measurements have been made of the direct gap, and the spread in these measurements is more than the error listed above for the theory. We note from the figure that the separation of the Γ minimum from the subsidiary valleys at L is only 0.55 ± 0.05 eV over the entire range, so we expect that this quaternary compound will not provide great improvements over the binary compounds for high speeds, at least not from just the transfer characteristics of these subsidiary valleys.

6.4. ALLOY ORDERING

We now want to turn to the fact that these compounds might not be perfectly random alloys, but might possess some ordering in their structure. The basis of ordering in otherwise random alloys lies in the fact that the ordered lattice, whether it has short-range or long-range order, may be in a lower energy state than the perfectly random alloy. In a random alloy, $A_x B_{1-x} C$, the average cohesive energy will change by

$$E_{\text{coh}} = E^B_{\text{coh}} + x(E^A_{\text{coh}} - E^B_{\text{coh}}). \tag{6.34}$$

While the A-C compound is losing energy, the B-C compound is gaining energy, and the cohesive energy of the random alloy is a simple average of its constituents.

For example, in $In_xGa_{1-x}As$, the cohesive energy is the average of those of InAs and GaAs, but the gain of energy by InAs in the alloy is exactly offset by the loss by GaAs, at least within the virtual crystal approximation which leads to the linear extrapolation of (6.34).

If any short-range order persists, however, this argument no longer holds. Rather, the ordered GaAs regions undergo a loss of energy as their bonds are stretched in the alloy, while the ordered InAs regions gain energy as the bonds are compressed (here we mean gain in energy in the sense that the crystal is compressed and the equilibrium state has a lower energy). Since the cohesive energy varies as $1/d^2$ (just as other energies in the tight-binding models), a net increase in cohesive energy in semiconductor crystals is a very simplified calculation. We return to this point shortly.

We can as well calculate an "average" energy of the bonding valence electrons with the dielectric method. This is given by[18]

$$E_{av} - E_b = E_F\left[0.68 + 3B^2\left(1 + \ln\left\{\frac{B}{2}\right\}\right) - 4B^3\right], \quad (6.35)$$

where B was given in (6.13) and E_b is the absolute level of the bottom of the valence band. Thus, we can use (6.35) to calculate the average energy of the bonding electrons if we can determine the width of the valence band.

The range of energy levels incorporated in the valence band of tetrahedrally coordinated semiconductors is just sufficient to account for the $4n$ electrons, where n is the atomic density, that make up the valence band. Therefore, this width must be related to the valence Fermi energy used above in the dielectric method. In practice, the width E_b (we take the top of the valence band as $E = 0$) is exactly E_F in Si, but is increased in Ge due to the admixture of d-level states. Thus, we can use a form similar to (6.19) if we identify E_{bh} with E_F, so that

$$E_b = \frac{E_F + (D-1)\,\Delta E_b}{[1 + (C/E_F)^2]^{1/2}}. \quad (6.36)$$

We note here that the role of the heteropolar energy C is to *narrow* the band just as it widens the gaps in the previous sections. Using this equation, we may calculate the values of the average energy for the alloys from the dielectric method. These values are shown in Table 6.7 for several of the diamond and zinc blende semiconductors of interest.

With these numbers in hand, we can now compute the equivalent values for the ordered alloy relative to the random alloy for the composition $x = 0.5$ (equal concentrations). Recall that a decrease of the value of E_{av}, or an increase of E_{coh}, indicates that ordering of the alloy is energetically favored. These results are shown in Table 6.8. The data on AlGaAs are mixed, but the energy change is so small that ordering would occur only at very low temperatures if at all. On the other hand, if ordering were incipient in this alloy, the self-energy correction mentioned above could work to stabilize the ordered structure at much higher

TABLE 6.7. Average Energies of Valence Band

Material	$\Gamma_{25}^v - \Gamma_1^v$	$E_{av} - \Gamma_1^v$	E_{coh}
Si	12.5	6.75	2.32
Ge	12.7	5.59	1.94
AlAs	11.9	6.09	1.89
GaP	12.4	6.37	1.78
GaAs	12.3	6.0	1.63
GaSb	11.2	5.18	1.48
InP	11.4	5.66	1.74
InAs	11.3	5.2	1.55
InSb	10.6	4.6	1.4
HgTe	9.52	5.16	0.53
CdTe	9.06	4.86	1.06

temperatures (say room temperature). In this case, only realistic total energy calculations can shed much light on the question. The experimental situation has not been effectively investigated as yet, particularly with the care devoted to InGaAs, as discussed below. However, there is some indication that ordering can occur during molecular-beam epitaxy of this compound for some temperatures of the substrate.

There is still another method of estimating the tendency to develop ordering in the alloys. This latter method relies upon extensions of Pauling's heats of formation to the dielectric method.[4] The spectroscopic formula for the heat of formation has the form

$$-\Delta H = d^{-3}(\text{ionicity})(\text{metallization}) + \text{constant},\tag{6.37}$$

where the ionicity factor is given by the ionic nature of the compound

$$f_i = \frac{C^2}{C^2 + E_h^2}.\tag{6.38}$$

TABLE 6.8. Energy Changes in the Alloys

Alloy ($x = 0.5$)	ΔE_{av} (meV)	ΔE_{coh} (meV)	T_c (K)
GaAlAs	−1.0	−2.1	−48
InGaAs	−89	23.4	−900
InAsP	−48	8.2	6300
GaInSb	−137	29.7	1950
InGaP	652	−110	70
GaAsP	101	−18.5	1880
InAsSb	122	−17.8	1740
GeSi	4.0	−32.9	
HgCdTe	685	0	−6000

The metallization factor arises from the tendency of the bond orbitals to hybridize in such a way that the average energy gaps are close to true energy gaps measured in optical transitions. In the dielectric theory, we can define a metallization factor as[4]

$$f_m = 1 - b \left(\frac{E_G}{\langle E \rangle} \right)^2, \tag{6.39}$$

where

$$\langle E \rangle = \frac{E_0 + E_1}{2} \tag{6.40}$$

and E_0 and E_1 are the optical energy gaps discussed previously. The factor b is adjusted to match the observed metallization in gray Sn and is $b = 0.05$. These can now be combined to give the heat of formation as

$$\Delta H = \Delta H_0 f_i \left(\frac{d_{Ge}}{d} \right)^3 \left[1 - b \left(\frac{E_G}{\langle E \rangle} \right)^2 \right] + \text{constant}. \tag{6.41}$$

The value $\Delta H_0 = -68.6$ kcal/mole $= 2.86$ eV per atom pair. In a regular mixture of the two constituents of the pseudobinary compound, we would expect a linear variation of the heat of formation from one end of the alloy to the other. However, there is often a bowing of this parameter as well, so that

$$\Delta H_{\text{mix}} = \alpha x (1 - x), \tag{6.42}$$

where α is the bowing parameter for the heat of formation. The presence of the term in (6.42) means that it costs enthalpy to dissolve one semiconductor in another to form the alloy. However, mixing of the two semiconductors is favored by the ideal entropy term at high temperatures, so that there is a critical temperature below which the alloy will tend to order. This critical temperature above which the component semiconductors are miscible in all proportions is given by

$$T_c = \frac{\alpha}{2R}, \tag{6.43}$$

where $R = 0.026$ eV per atom pair is the universal gas constant. Therefore, by calculating the variation in E_0 and E_1, we can evaluate (6.41)–(6.43) throughout the alloy and determine this critical temperature. The limits obtained in this way are generally in good agreement with experiment, except for InGaAs as discussed below, and these transition temperatures are also shown in Table 6.8.

In the case of InGaAs, InGaSb, and InAsP, all indications suggest that the alloy will favor phase separation and ordering at room temperature. Indeed, this tendency to order in the InGaAs and InAsP compounds may be the reason that a suspected miscibility gap exists in the InGaAsP quaternary alloy in the range $0.7 < y < 0.9$, as suggested by both crystal growth[19] and Raman scattering studies of the phonon spectrum.[20] The actual ordering can be quite subtle, however. For example, recent experimental results on the x-ray absorption fine structure (EXAFS) of InGaAs by Mikkelsen and Boyce[21] indicate that the Ga-As and In-As nearest-neighbor distances remain nearly constant at the binary values for all alloy concentrations. The average anion–cation distance follows Vegard's law and increases by only 0.0174 nm. In addition, the cation sublattice strongly resembles a virtual crystal (alloying actually occurs on this sublattice), but the anion sublattice is very distorted and there are two very distinct As-As distances, which differ by 0.024 nm. The distribution of these second-neighbor distances about the two values has a nearly Gaussian profile. Such a structure can be accommodated in a model crystal that resembles closely a local chalcopyrite distortion, and can explain the bowing of the band structure itself.[22] These latter authors suggest that in alloys in which the atoms of one sublattice are greatly different in size, the system will always prefer to adopt an arrangement in which the local bonds are preserved as essentially those in the binary constituents.

Zunger and his co-workers[23-25] have carried the above ideas much further in studying the ordering of III-V alloys. In most of the above arguments, we have treated really only the driving forces introduced by stretching one compound and compressing the other to arrive at the alloy. We have not taken into account the above tendency for bond lengths to remain very nearly at their binary value. For this to occur, there must be a relaxation of the common constituent atom within the unit cell, as well as a possible charge transfer among the atoms. In fact, Srivastava et al.[23] find that the latter factors are the major ones in alloy ordering. For a $A_x B_{1-x} C$ alloy, the four cations of type A and B can assume five different nearest-neighbor arrangements around the C atom: A_4C, A_3BC, A_2B_2C, AB_3C, and B_4C. These authors denote these as the $n = 0, 1, 2, 3, 4$ arrangement, where n denotes the number of B atoms. These arrangements form perfectly ordered crystal structures of the zinc blende ($n = 0, 4$), luzonite or famatinite ($n = 1, 3$), and CuAu-I or chalcopyrite ($n = 2$) type. The choice of the particular structure of a given n-value depends on which is the lowest-energy configuration. In any case, a disordered (random) alloy is a statistical mixture of these structures. In Figure 6.9, we show the results of total energy calculations[25] for the GaInP alloy as well as the appropriate crystal structure. This work suggests that highly ordered alloys can generate new types of superlattices with very short periods. Indeed, recent experiments have observed the highly ordered ($x = 0.5$) structure in both GaAsSb[26] and GaAlAs.[27] In the latter case, the structure seems to be the CuAu-I lattice, while both the CuAu-I and chalcopyrite structures are seen in the former case. The famatinite structure has been observed in the InGaAs alloy for the cases of $n = 1, 3$.[28] It is clear from this that the random alloys are anything but that, and are a quite complicated crystal, far from the simple virtual

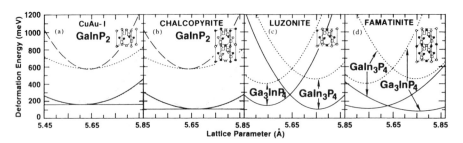

FIGURE 6.9. Deformation energies (per eight atoms) of ternary $Ga_n In_{4-n} P_4$ structures. For (a) and (b) (dash-dot-dash line) u and η unrelaxed; (dashed line) η relaxed; (dotted line) u relaxed; (solid line) full (u and η) relaxed. For (c) and (d) (dashed line) unrelaxed; (solid line) full relaxation. Here, η is the c/a ratio and u is the cell-internal (C atom) displacement parameter ($u = 1/4$ is unrelaxed). After Mbaye et al.,[25] by permission.

crystal structure first envisioned. Mbaye et al.[29] have calculated the phase diagrams, from first-principles band structure methods utilizing the total energy, for the alloys. They find that the strain, discussed above, actually stabilizes the ordered stoichiometric compounds. Moreover, they find that disordered alloys, ordered alloys, and miscibility gaps can all occur, and that metastable ordered phases can actually occur within the miscibility gaps, which has been observed in $GaAs_{0.5}Sb_{0.5}$.[30]

If ordering does occur, e.g., one axis in the Brillouin zone will have a new zone-edge boundary (only for long-range ordering, though) and this is expected to occur at $\pi/2c$, as opposed to π/c, for the $x = 0.5$ alloy ($a = b = c$ in the zinc blende structure if there is no tetragonal distortion in the ordered structure). This zone folding is the equivalent of the creation of the chalcopyrite structure, but the consequence of this on the transport properties of the carriers is expected to be rather negligible in direct-gap materials since the zone folding is unlikely to affect the valley separations. However, the lattice dynamics will be modified and can be probed by Raman scattering. These ordered structures can be expected to show two LO modes, one extending from each of the binary constituents, and likely to be coexisting near $x = 0.5$. This is especially prevalent in InAsP, and the two modes are indeed thought to arise from the two ordered parts of the lattice. This multimode behavior carries over strongly into the quaternaries. Near the miscibility gap in InGaAsP, distinct phonon spectra are not observed.[20] Rather a broad background of scattering is found, presumably from weak multiphonon modes.

One final problem of ordering in the alloys can be best illustrated in terms of the metastable alloys $(III-V)_{1-x}(IV)_{2x}$, which have been grown by chemical vapor deposition.[31] These materials show an extremely large bowing in the band-gap variation, and even exhibit a "kink" or V-shaped variation in the gap. This kink occurs near $x = 0.3$ in almost all of the alloys studied. The cause of this kink is felt to be the extreme of the ordering tendency, an order–disorder phase transition.[32] For small x, the crystal structure is essentially zinc blende with a small amount of group IV (usually Ge) sitting on each of the two sublattices, but randomly distributed. In this case, the zinc blende lattice is fully ordered. On the other hand, when x is near 1, the crystal structure is predominantly the diamond lattice with small amounts of the group III and V elements (Ga and As, for example). Since there is no GaAs (for example) substrate for reference, the individual Ga and As atoms have no reference for determining which of the two sublattices is preferred for group III and which is preferred for group V. Therefore, each will randomly pick either of the two sublattices, and each sublattice will have both atom types sitting on it. Thus, the zinc blende part of the alloy is totally disordered, and large numbers of Ga—Ga and As—As bonds will form as the percentage of GaAs in the alloy increases. Finally, as x approaches a critical value, experimentally determined to be near 0.3, the phase transition occurs for the disorder to the ordered structure. In Figure 6.10, the principal energy gap E_0 for the alloy $(GaAs)_{1-x}Ge_{2x}$ is shown as a function of the compositional parameter x. The presence of the phase transition is clearly indicated by the deep V-notch in the behavior of the principal gap. The dashed curve is the virtual crystal approximation that is expected for a smooth, random alloy.

6.5. ALLOY SCATTERING

In a semiconductor alloy, the scattering of free carriers due to the deviations from the virtual crystal model has been termed alloy scattering. This concept was introduced in the preceding sections. The general treatment of alloy scattering has always, at least in previous works, followed an unpublished, but well-known result due to Brooks, and extended by Makowski and Glicksman.[33] While this scattering mechanism generally supplements the normal phonon and impurity scattering, it has on occasion been conjectured to be sufficiently strong as to be the dominant scattering mechanism. The work of Makowski and Glicksman, however, showed that the scattering was quite weak and was probably only important in the InAsP system, although even here it was likely to be much weaker than experimental data would suggest. These authors utilized a scattering

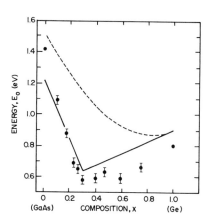

FIGURE 6.10. The variation of the direct gap in the metastable alloy $(GaAs)_{1-x}Ge_{2x}$. From Newman and Dow.[32]

potential given by the difference in the band gaps of the constituent semiconductors. Harrison and Hauser[9] later suggested that the scattering potential was related to the differences in electron affinity. However, as pointed out by Kroemer,[34] the electron affinity is a true surface property, but no qualitatively useful theory exists for it and it is a notoriously bad indicator of bulk potentials, even in heterojunctions. Its use in scattering theories for carrier transport in bulk materials should therefore be treated with a degree of scepticism. A subsequent effort has suggested that the proper estimator for the disorder potential can be calculated from the previously specified ionic potential in (6.14), as this scattering arises from the same source as the bowing in the band gap itself.[35]

The electron scattering rate in ternary semiconductors due to the random potential of the alloy, for nonparabolic bands, is given by[9]

$$\Gamma_A = \frac{3\pi}{4}\left[\frac{(m/2)^{3/2}}{\hbar^4}\right]U^2 S(\alpha)\gamma(E)\gamma'(E), \qquad (6.44)$$

where

$$\gamma = E\left(1 + \frac{E}{E_g}\right), \qquad (6.45)$$

E_g is the appropriate energy gap causing the nonparabolicity, and $S(\alpha)$ is an energy-dependent parameter which describes the effect of ordering on the scattering rate. In this theory, $S = 0$ for a perfectly ordered ternary and $S = 1$ for a perfectly random alloy. Most theories assume that $S = 1$.

Besides the effect of ordering, the most significant parameter in (6.44) is the scattering potential U. The derivation of this equation is based on the Mott inner-potential arguments, which arise from a short-range potential, but the general form is equivalent to treating U as a pseudodeformation potential and following an argument similar to acoustic phonon scattering. It is U which is determined by the various approaches discussed above.

The scattering potential that leads to disorder scattering is just the aperiodic contribution to the crystal potential that arises from the disorder introduced into the lattice by the random siting of constituent atoms. In the virtual crystal approximation, the perfect zinc blende lattice is retained in the solid solution. Thus, the bond lengths are equal, and the homopolar energy gap does not make a contribution to the random potential U. The random potential then arises solely from the fluctuations in the heteropolar gap C. For a binary compound, the heteropolar energy is given by (6.14). In a ternary solid solution $A_xB_{1-x}C$, we can extend this to[35]

$$C_{AB} = bZ_A(r_A^{-1} - r_B^{-1})\exp\left[-\frac{k_s(r_A + r_B)}{2}\right], \qquad (6.46)$$

where we assume the valence of the A and B atoms is the same and the other parameters have been previously defined. This can now be used to calculate the aperiodic potentials used for alloy scattering. The results for a number of alloys

TABLE 6.9. Alloy Scattering Parameters

Alloy	U (eV)	ΔE_{aff} (eV)	ΔE_{BG} (eV)
GaInAs	0.5	0.83	1.07
InAsP	0.36	0.5	1.0
GaAsP	0.43	0.07	0.83
InAsSb	0.82	0.31	0.18
InAlAs	0.47	1.32	1.79
AlAsP	0.64	0.08	0.27
InGaP	0.56	0.4	1.08
InAlP	0.54	0.9	1.08
InGaSb	0.44	0.53	0.52
InPSb	1.32	0.19	1.17
GaPSb	1.52	0.06	1.57
InGaAsP	0.29	0.26	0.54
InGaPSb	0.54	0.17	0.56
InAlAsP	0.28	0.42	0.58

are presented in Table 6.9. Also shown, for comparison, are the equivalent values estimated by the discontinuity in band gap and by the difference in electron affinity. There is a weak dependence of U on the composition as well, but this is small when compared with the $x(1-x)$ term.

As mentioned above, it is generally found that the role of alloy scattering is very weak, although there are often strong assertions from experimentalists that reduced mobility found in alloys must be due to "alloy scattering." In fact, only the work of Makowski and Glicksman was sufficiently careful to exclude other mechanisms. In general, little effort is taken to accurately include the proper strength of optical phonon scattering, due to the complicated multimode behavior of the dielectric function in the alloy, or to include dislocation or cluster scattering that can arise in impure crystals.

6.6. QUANTUM WELL SUPERLATTICES

One of the principal reasons for developing alloy semiconductors is the desire to tailor the principal band gaps to desired values appropriate for the application at hand. A disadvantage of the alloy approach is that several of the alloys are subject to ordering and clustering of the constituent members. This is complicated by the fact that the growth of the alloy is often difficult, and good alloys are primarily obtained only by epitaxial growth with one of the new approaches—molecular-beam epitaxy or organometallic chemical vapor deposition. While quite good alloy compound layers can be grown by these techniques, the range of alloys is mainly limited to those that can be matched to the substrate in this heteroepitaxial growth process.

As long as one is investing in the heteroepitaxial growth process, it is worth considering other additional controls on the material parameters that can be achieved. One major new approach is the use of quantum well structures in multiple-layer heterostructures. These structures use the quantum size effect, which arises when the layer thickness becomes comparable to the de Broglie wavelength of the carriers of interest, to restrict the motion in the direction perpendicular to the growth planes. Here, layers of a narrow-band-gap material are placed alternately between layers of a wider-band gap material. Then, carriers in the narrow-band-gap material are affected by the quantum size effect. They respond much as a particle in a finite-barrier potential well, and their lowest eigenenergy (in the direction perpendicular to the growth planes) is raised over the bottom of the conduction band (or lowered below the top of the valence band). This quantum size effect raises the transition energy of optical transitions across the principal energy gaps, thus achieving some of the goals of, for example, alloying GaAs with AlAs. The increase of transition energy is obtained for this latter system when thin layers of GaAs are sandwiched between thicker layers of GaAlAs. Since the relevant carriers are in the GaAs, the properties of the GaAlAs barrier layers can be adjusted to suit the desired barrier without concern as to the transport quality of the layer. However, if we are to make this system work in the mode of band-gap engineering, another new parameter has now to be determined. This new parameter is the fraction Q_e of the energy gap discontinuity that lies in the conduction band (and therefore $1 - Q_e$ lies in the valence band). In principle, we can ascertain this value from the absolute levels of the top of the valence band; i.e., from the ionization energies of the compounds and alloys given in Table 6.4. While these are determined from the dielectric theory, which is an interpolative theory and not a first-principles theory, the results are about as good as any current theory can yield. In fact, the cause of the discontinuity distribution is just not well understood at present. We will return to this point.

6.6.1. The Schrödinger Equation

The solution of the Schrödinger equation for a particle in a one-dimensional box is one of the most elementary problems in introductory quantum mechanics. The problem is complicated somewhat in this case, however, because the conduction bands of these materials are quite nonparabolic in nature for states above the band extrema. This complicates the energy eigenvalues somewhat, but not the approach. The Schrödinger equation and its solutions for an infinitely deep potential well are given by (Figure 6.11)

$$\frac{d^2\Psi}{dz^2} = -\left(\frac{2m}{\hbar^2}\right)E\Psi,$$

$$\Psi_n = A_n \sin\left(\frac{n\pi z}{L}\right), \tag{6.47}$$

$$E_n = \frac{E_g}{2}\left[\left(1 + \frac{4E_0 n^2}{E_g}\right)^{1/2} - 1\right],$$

with $n = 1, 2, 3, \ldots$ and $E_0 = \pi^2 \hbar^2 / 2mL^2$, and L is the thickness of the quantum well. This is written for electrons in a positively directed potential well and explicitly includes the hyperbolic band approximation for nonparabolic bands. For the infinite well, the wave functions must vanish at the boundaries.

If the well is not infinite, but has a height V_0, then the wave functions penetrate into the barrier regions some distance, and a transcendental equation relating the matching of the wave functions at the interface must be solved to yield the energy eigenvalues. In the well region, the above equation still holds, but the energy is usually written in terms of the wave vector (in the z-direction)

$$k = \frac{[2mE(1 + E/E_g)]^{1/2}}{\hbar}. \tag{6.48}$$

Within the barrier regions, the wave function decays nearly exponentially $(E < V_0)$, so that the Schrödinger equation becomes

$$\frac{d^2\Psi}{dz^2} - w^2\Psi = 0, \qquad w = \frac{[2m(V_0 - E)(1 + E/E_g')]^{1/2}}{\hbar}, \tag{6.49}$$

where E_g' is the energy gap in the wider-band-gap material. As usual, one requires continuity of the wave function and its derivative at the interfaces at each side of the quantum well (the situation is more complicated if the barrier region is sufficiently thin that the wave functions from each well can tunnel through to the adjacent well). In Figure 6.12, the dependence of the bound state energies for a normalized potential are shown as a function of its strength for the GaAs-GaAlAs system. We note that the normalization on energy is given by the material-independent, and dimensionless, quantity

$$E_v\left(\frac{2mL^2}{\hbar^2\pi^2}\right),$$

where E_v is either V_0 or an energy E. Hence, we note that the effect of the potential barrier can be achieved either by adjusting the band offset or by adjusting

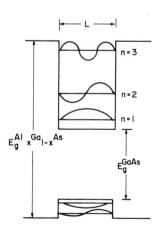

FIGURE 6.11. Depiction of the energy levels formed in the quantum wells in the conduction and valence bands of GaAs/GaAlAs.

the thickness of the quantum well itself. Since, the latter enters quadratically, it is the source for a great deal of the error observed in experimentally measuring the band offsets from quantum well data.

It is clear from the foregoing that the quantum well structure can be used to complement the alloy properties in a manner to adjust the band gaps to any desired values. While most of this has been carried out in the GaAs-GaAlAs system, it is applicable to any of the multiple-semiconductor systems. Indeed, its use in a superlattice whose layers are GaAs and Ge would preclude the strong order–disorder transition which dominates the behavior of the band-gap energy in the alloy between these materials. This has been suggested in the HgCdTe system as well, where lack of long-term stability of the alloy compound is a serious problem for device applications. Use of a superlattice composed of alternating layers of HgTe and CdTe to achieve a material with the same narrow band gap (0.1 eV) has been suggested as the prime way to achieve stable device performance.[36] The quantum well energy transitions are shown in Figure 6.13 for this latter system.

6.6.2. Strained-Layer Superlattices

Strained-layer superlattices are high-quality superlattices grown from purposely lattice-mismatched materials in each of the layers. These materials can be grown with essentially no misfit dislocation generation at the interfaces if the layers are sufficiently thin.[37] In this case, the mismatch between the layers is totally accommodated by uniform lattice strain, and these layers can be grown from a wide variety of material systems. For example, one can grade GaAs by alloying GaP up to an alloy with 2.5% of the latter compound. Then alternate layers of GaAs and $GaAs_{0.95}P_{0.05}$ are no longer lattice matched. The GaAs layers are in compression while the latter compound layers are in tension. The strain modifies the band structure further from the quantum well behavior (due to the lattice-constant dependence of the band parameters discussed in the previous

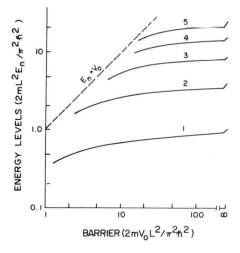

FIGURE 6.12. Dependence of the bound state energies on the well depth in reduced energy units.

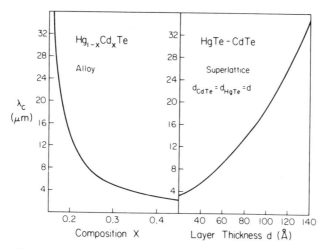

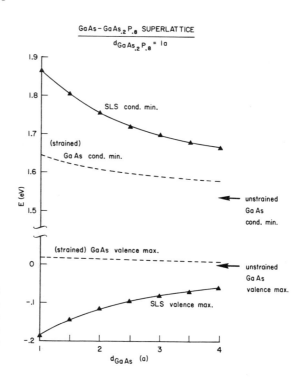

FIGURE 6.13.   The predicted energy levels for a superlattice of HgTe sandwiched between layers of CdTe. After Smith et al.,[36] by permission.

sections). Here, though, the layers must be quite thin in order to not have the strain relax by misfit dislocation generation. In Figure 6.14, we show the lowest-energy levels in the quantum wells in strained GaAs that have been calculated for such a strained-layer superlattice.[37]

FIGURE 6.14.   Variation of the lowest-energy bands in the quantum well of GaAs placed between 20% GaAsP as a strained-layer superlattice. After Osbourn,[37] by permission.

In the strained-layer system, the lattice constants for the two different types of layer that make up the superlattice are equal in the two directions parallel to the heterointerface. However, the lattice constant in the direction perpendicular to the interface differs for the two materials. Thus, there are straight channels through the lattice in some crystalline directions, such as the $\langle 100 \rangle$ directions, while the $\langle 110 \rangle$ directions undergo a tilt at the interface. This difference in lattice constant in the two directions introduces tetragonal distortion into the crystal structure, which further complicates the band calculations. In thicker layers, there is a tendency for this tetragonal distortion to begin to relax,[38] so that there is a critical thickness beyond which the strained layer is not stable. Data on InGaAs-GaAs strained-layer superlattices suggests that the theoretical expression[39,40]

$$\varepsilon = \frac{w(1 - \sigma/4)[\ln(w) + 1]}{2\pi(1 + \sigma)}, \tag{6.50}$$

where $w = a/2^{1/2}h$, h is the critical thickness, ε is the strain, a is the lattice constant, and σ is Poisson's ratio, can well represent the experimental situation with a value of $h = 25$ nm.

The strain also creates modifications of the band structure that is not expected from normal superlattice considerations. The valence band of most III-V materials is twofold degenerate with a light-hole and a heavy-hole band degenerate at the zone center. The superlattice normally splits this degeneracy, but this can be accomplished by the strain as well. However, the quantum well effect lowers the light-hole band more than the heavy-hole band. For expansion of the lattice in the interface plane, this effect is exaggerated, as the strain also contributes to lowering of the light-hole band relative to the heavy-hole band. On the other hand, compression in the interface plane reverses this trend, with a tendency to lower the heavy-hole band relative to the light-hole band. The two bands can therefore cross at some point away from, but near to, the zone center, thus creating a quite complicated energy dispersion relationship.

REFERENCES

1. W. A. Harrison, *Electronic Structure and the Properties of Solids*, W. H. Freeman, San Francisco (1980).
2. J. C. Slater and G. F. Koster, *Phys. Rev.* **94**, 1498 (1954).
3. J. C. Phillips, *Phys. Rev. Lett.* **20**, 550 (1968).
4. J. C. Phillips, *Bonds and Bands in Semiconductors*, Academic Press, New York (1973).
5. J. A. Van Vechten, *Phys. Rev.* **182**, 891 (1969).
6. J. A. Van Vechten, *Phys. Rev.* **187**, 1007 (1969).
7. D. Penn, *Phys. Rev.* **128**, 2093 (1962).
8. K. C. Pandey and J. C. Phillips, *Solid St. Commun.* **14**, 439 (1976).
9. J. W. Harrison and J. R. Hauser, *Phys. Rev. B* **13**, 5347 (1976).
10. P. A. Flinn, *Phys. Rev.* **104**, 350 (1956).
11. D. K. Ferry, *Phys. Rev. B* **14**, 5364 (1976).
12. G. L. Hall, *Phys. Rev.* **116**, 604 (1959).

13. J. A. Van Vechten and T. K. Bergstresser, *Phys. Rev. B* **1**, 3351 (1970).
14. J. C. Woolley, C. M. Gillet, and J. A. Evans, *Proc. Phys. Soc. London*, **77**, 700 (1961).
15. T. Y. Wu and G. L. Pearson, *J. Phys. Chem. Sol.* **33**, 409 (1972).
16. R. L. Moon, G. A. Antypas, and L. W. James, *J. Electron. Mater.* **3**, 635 (1974).
17. S. N. Grinyaev, M. A. Il'in, A. T. Lukomskii, V. A. Chalyshev, and V. M. Chupakhina, *Sov. Phys. Semicond.* **14**, 446 (1980).
18. J. A. Van Vechten, *Phys. Rev.* **170**, 773 (1968).
19. J. H. Marsh, P. A. Houston, and P. N. Robson, *Inst. Phys. Conf. Ser.* **56**, 621 (1981).
20. A. Pinczuk, J. M. Worlock, R. E. Nahory, and M. A. Pollack, *Appl. Phys. Lett.* **33**, 461 (1978).
21. J. C. Mikkelson and J. B. Boyce, *Phys. Rev. Lett.* **49**, 1412 (1983).
22. A. Zunger and J. E. Jaffe, *Phys. Rev. Lett.* **51**, 662 (1984).
23. G. P. Srivastava, J. L. Martins, and A. Zunger, *Phys. Rev. B* **31**, 2561 (1985).
24. J. L. Martins and A. Zunger, *Phys. Rev. Lett.* **56**, 1400 (1986).
25. A. A. Mbaye, A. Zunger, and D. M. Wood, *Appl. Phys. Lett.* **49**, 782 (1986).
26. H. R. Jen, M. J. Cherng, and G. B. Stringfellow, *Appl. Phys. Lett.* **48**, 1603 (1986).
27. T. S. Kuan, T. F. Kuech, W. I. Wang, and E. L. Wilkie, *Phys. Rev. Lett.* **54**, 201 (1985).
28. H. Nakayama and H. Fujita, *Inst. Phys. Conf. Series* **79**, 289 (1986); *Proc. GaAs and Related Compounds Conf.*, Japan (1985).
29. A. A. Mbaye, L. G. Ferreira, and A. Zunger, *Phys. Rev. Lett.* **58**, 49 (1987).
30. M. J. Cherng, Y. T. Cherng, H. R. Jen, P. Harper, R. M. Cohen, and G. B. Stringfellow, *J. Electron. Mater.* **15**, 79 (1986).
31. Z. I. Alferov, R. S. Vantanyan, V. I. Korol'kov, I. I. Mokan, V. P. Ulin, B. S. Yavich, and A. A. Yakovenko, *Sov. Phys. Semicond.* **16**, 567 (1982).
32. K. E. Newman and J. D. Dow, *Phys. Rev. B* **27**, 7495 (1983).
33. L. Makowski and M. Glicksman, *J. Phys. Chem. Solids* **34**, 487 (1976).
34. H. Kroemer, *Crit. Rev. Solid St. Science* **5**, 555 (1975).
35. D. K. Ferry, *Phys. Rev. B* **17**, 912 (1978).
36. D. L. Smith, T. C. McGill, and J. N. Schulman, *Appl. Phys. Lett.* **43**, 180 (1983).
37. G. C. Osbourn, *J. Appl. Phys.* **53**, 1586 (1982).
38. J. M. Gibson, R. Hull, J. C. Bean, and M. M. J. Treacy, *Appl. Phys. Lett.* **46**, 649 (1985).
39. I. J. Fritz, S. T. Picreaux, L. R. Dawson, T. J. Drummond, W. D. Laidig, and N. G. Anderson, *Appl. Phys. Lett.* **46**, 967 (1985).
40. J. W. Matthews and A. E. Blakeslee, *J. Crystal Growth* **27**, 118 (1974).
41. K. C. Haas, H. Ehrenreich, and B. Velicky, *Phys. Rev. B* **27**, 1099 (1983).

7

The Electron–Electron
Interaction

In recent years, as device sizes have become smaller and carrier densities have become larger, the role of electron–electron scattering and screening have become very important in transport within semiconductor devices. The main difficulty in dealing with the electron–electron interaction lies in the nonlinear behavior of the interaction potential and in the long range of the Coulomb potential. In the past, several approaches to the study of electron–electron scattering have been presented. Almost all of these were based on the assumption that the interaction potential is screened at some characteristic length, usually the Debye length $L_D = 1/\lambda = (\varepsilon k_B T/ne^2)^{1/2}$, and that the distribution function is a Maxwellian (inherent in the use of the Debye length). Yet, in the high densities inherent in modern devices, the Debye length is often smaller than the interelectronic distance. The effects are complicated by the fact that in polar materials, the rather high free-carrier plasma frequency that accompanies the high densities couples strongly to the polar optical phonon vibrations to produce coupled modes of electron–plasmon–phonon interactions. To properly treat all of these effects, we must divorce ourselves from the simple screening approach and treat the full frequency- and wavelength-dependent dielectric function.

The momentum- and frequency-dependent dielectric function has been investigated quite extensively for a variety of physical systems. The main results are that screening is a dynamic process, and the full (q, ω) variation of the dielectric function must be included. In this chapter, we review one form of the dielectric function, the Lindhard function, and discuss the role of electron–electron and electron–plasmon scattering. Several calculations of the effect of this scattering will also be presented to illustrate the importance of properly treating the interaction.

7.1. THE RANDOM-PHASE-APPROXIMATION DIELECTRIC FUNCTION

The dielectric function describes how the various charged species move in response to external (or internal) potentials and provide a screening of those potentials. In the compound semiconductors, like GaAs and AlGaAs, these charged species are the electrons and holes and the atoms themselves, since there is an ionic contribution to the bonding of these atoms. Thus, we can write down the dielectric function as

$$\varepsilon(q, \omega) = \varepsilon(\infty) + \delta\varepsilon_L + \delta\varepsilon_e, \tag{7.1}$$

where the second term is the lattice contribution and the last term is the electronic contribution. We show the dielectric function as being dependent on the wave number q, as its variation provides major effects in the process of screening. Equation (7.1) expresses the fact that, at sufficiently high frequency, neither the electrons nor the lattice can follow any time-varying perturbation, and we get only the high-frequency permittivity. This latter quantity differs from the free-space value by the contribution of the valence electrons, which we discuss below, and is usually termed the "optical dielectric constant" (times the free-space permittivity).

In the absence of the electrons, the permittivity must approach its static value at low frequency which includes the entire lattice and electronic contributions, while for high frequencies the lattice contribution vanishes, and at still higher frequencies the electronic contribution also vanishes. In purely covalent materials, such as Si and Ge, there is no polar contribution to the optical phonon vibrations, and therefore no static lattice contribution to the dielectric function. The difference between the optical dielectric constant and the static value is just the contribution of the polar modes of the lattice vibrations. We can write the lattice contribution as[1]

$$\delta\varepsilon_L = \frac{[\varepsilon(0) - \varepsilon(\infty)]\omega_{TO}^2}{\omega_{TO}^2 - \omega^2}, \tag{7.2}$$

where

$$\frac{\omega_{LO}^2}{\omega_{TO}^2} = \frac{\varepsilon(0)}{\varepsilon(\infty)} \tag{7.3}$$

is the Lyddane–Sachs–Teller relation connecting the longitudinal and transverse optical phonon frequencies in polar material. Thus, at low frequencies $\varepsilon(\omega) \rightarrow \varepsilon(0)$, the static value, while for high frequencies the lattice contribution vanishes as desired.

Generally, the electronic contribution is calculated in the absence of the lattice contribution. For this, we set (7.2) to zero, but the dielectric permittivity that enters into (7.1) is just the free-space permittivity ε_0. To incorporate all of the electronic contribution, we must now derive a form to describe the dielectric contribution of these free carriers (and the valence-band electrons, as well).

7.1.1. The Lindhard Potential

We now will calculate the electronic contribution within the random phase approximation, which essentially means that we will only calculate to lowest order the screening effects of the electrons upon an external perturbing potential. To do this, we first calculate the fluctuation in density caused by a perturbing potential δU. We then calculate the resulting self-consistent potential fluctuation Φ caused by the small fluctuation in density. Finally, we relate the total perturbing potential δU to the applied potential V and the self-consistent potential Φ. From this we obtain the dielectric function. We do this in Fourier transform space, so that the response is at a particular frequency and wave vector.

The perturbing potential at frequency ω and momentum $\mathbf{q}$ may be expressed as

$$\delta U(\mathbf{q}, \omega) = U_0 \cos(i\mathbf{q} \cdot \mathbf{r} + i\omega t) \, e^{-\alpha t}, \tag{7.4}$$

where α is the damping constant to account for the dissipative part of the carrier–carrier interaction. The wave function for an electron state may be expressed as

$$\Psi_k = |k\rangle + b_{k+q}|k + q\rangle, \tag{7.5}$$

where the second term represents the deviation from the equilibrium state. By simple, first-order perturbation theory, we can calculate this latter term as (an e is inserted, since the perturbation is caused by an energy, not by a potential)

$$b_{k+q} = \frac{\langle k + q|e\delta U|k\rangle}{E(k + q) - E(k) + \hbar\omega - i\hbar\alpha} = \frac{eU_0 \, e^{i\omega t - \alpha t}}{E(k + q) - E(k) + \hbar\omega - i\hbar\alpha}. \tag{7.6}$$

The change in electron density that is produced by this perturbation is just

$$\delta\rho = -e \sum_k f(k)[|\Psi_k|^2 - 1], \tag{7.7}$$

from which we can obtain (after some simple algebraic manipulation)

$$\delta\rho = -e^2 U_0 \sum_k \frac{f(k) - f(k + q)}{E(k + q) - E(k) + \hbar\omega - i\hbar\alpha} e^{i\mathbf{q} \cdot \mathbf{r} + i\omega t - \alpha t} + \text{c.c.} \tag{7.8}$$

This fluctuation in charge density produces, in turn, a fluctuation in the potential itself. Since,

$$\nabla^2\Phi = -\frac{\delta\rho}{\varepsilon_0},$$

we have

$$\Phi = \frac{\delta\rho}{q^2\varepsilon_0},$$

and since Φ also varies as the cosine function, we obtain

$$\Phi = -\frac{e^2}{\varepsilon_0 q^2}\sum_k \frac{f(k)-f(k+q)}{E(k+q)-E(k)+\hbar\omega-i\hbar\alpha}U_0, \qquad (7.9)$$

where we have dropped the exponential variations, as they will cancel out. The total potential is now just U_0. But this is composed of the applied potential V plus the response term Φ, or

$$U_0 = V + \Phi,$$

and with (7.9) we can identify the dielectric function, using

$$U_0 = \frac{\varepsilon_0 V}{\varepsilon},$$

as

$$\varepsilon(q,\omega) = \varepsilon_0 + \frac{e^2}{q^2}\sum_k \frac{f(k)-f(k+q)}{E(k+q)-E(k)+\hbar\omega-i\hbar\alpha}. \qquad (7.10)$$

This last expression is termed the Lindhard dielectric function, and includes the free-space permittivity ε_0.

7.1.2. The Optical Dielectric Constant

Before proceeding, we want to separate out the portion of the summation representing the valence electrons. These excitations can only arise from a very high frequency source in which electrons are excited across the band gap. To approach this, we note that these excitations are primarily those involving a shift of k by a reciprocal lattice vector. Now, to proceed, we separate the two sums inherent in (7.10), and in the second we replace $k+q$ by k (and k by $k-q$). Then, the denominator terms are essentially just E_G, where this latter quantity is an effective optical gap in the sense of the homopolar gap introduced by Phillips.[2] Equation (7.10) becomes

$$\varepsilon(q,\omega) = \varepsilon_0 + \frac{e^2}{q^2}\sum_k f(k)\frac{2E(k)-E(k+q)-E(k-q)}{[E(k+q)-E(k)-\hbar\omega][E(k)-E(k-q)-\hbar\omega]}$$

$$\simeq \varepsilon_0 + \frac{e^2}{q^2}\frac{\hbar^2 q^2}{m_0}\frac{1}{E_G^2}\sum_k f(k). \qquad (7.11)$$

The sum runs over all valence electrons, so is just the valence electron density N. Then

$$\varepsilon(\infty) = \varepsilon_0 \left[1 + \left(\frac{\hbar \omega_P}{E_G} \right)^2 \right],\tag{7.12}$$

where the valence plasma frequency is just

$$\omega_P^2 = \frac{Ne^2}{m_0 \varepsilon_0}.\tag{7.13}$$

It is the valence plasma frequency ω_P that is listed in Table 6.2 for a variety of semiconductor materials. The form of (7.12) is that due to Penn,[3] and clearly shows that the optical dielectric function is related to the properties of excitations across the major energy gap between valence and conduction bands. However, as we noted, the energy gap is an average over the Brillouin zone and is not just the fundamental band gap.

Using the above considerations, we can now write the dielectric function in its relatively complete form as

$$\varepsilon(q, \omega) = \varepsilon(\infty) + \frac{[\varepsilon(0) - \varepsilon(\infty)]\omega_{TO}^2}{\omega_{TO}^2 - \omega^2}$$
$$+ \frac{e^2}{q^2} \sum_k \frac{f(k) - f(k+q)}{E(k+q) - E(k) + \hbar\omega - i\hbar\alpha}.\tag{7.14}$$

In this last expression, the sum runs only over the free carriers, since the valence contribution is now included in $\varepsilon(\infty)$.

7.2. PLASMONS AND PHONONS

In previous sections, we discussed the fact that transport of carriers was significantly affected by the interaction between the electrons and the plasmons, which are coherent oscillations of the electron (or hole) gas itself. The plasmon frequency is just the plasma frequency of the free carriers, given by

$$\omega_p^2 = \frac{ne^2}{m\varepsilon}.$$

In this expression, we have not indicated just which of the many values of the dielectric permittivity ε we want to use.

To obtain the form of the plasma frequency, we write the sum over free carriers, that is in (7.14), in the form of (7.11). For small q, and for high frequencies, the denominator is just $(\hbar\omega)^2$, while the numerator is given as

$$2E(k) - E(k+q) - E(k-q) \simeq -\frac{\hbar^2 q^2}{m},$$

where m is the density-of-states effective mass for the appropriate band in which the free carriers are situated. The sum over $f(k)$ is just n, the number of free carriers (there would be two sums if both electrons and holes are present). We can then define the free-carrier plasma frequency ω_p (note the use of lowercase in the subscript here, as opposed to the uppercase in the valence plasma frequency) through

$$\omega_p^2 = \frac{ne^2}{m\varepsilon(\infty)}. \tag{7.15}$$

While most people are readily familiar with the Coulombic scattering of electrons from impurities and from other electrons, they are not so familiar with the scattering of individual electrons by the collective oscillations of the electron gas, represented by the plasmons, as this has only recently been treated to any great extent.[4] The plasma oscillations are longitudinal-mode oscillations, by which we mean that there is a longitudinal charge vibration along the direction of propagation (q direction). In this regard, it is almost the same as the polar mode of the longitudinal optical phonon. As a consequence these two longitudinal oscillations can actually couple to each other, providing for hybrid modes, which are combined oscillations of the lattice and the electron, with characteristics of each. We discuss this just below by treating the total dielectric function (in the small q limit, however). Then, we turn to estimating the coupling strength of these two modes for the electron-coupled-mode interaction.

7.2.1. Plasmon–Phonon Coupling

We can now write the total dielectric function, which includes both of the contributions of the electrons and the lattice, as

$$\varepsilon(\omega) = \varepsilon(\infty) + \frac{[\varepsilon(0) - \varepsilon(\infty)]\omega_{TO}^2}{\omega_{TO}^2 - \omega^2} - \frac{\omega_p^2}{\omega^2}\varepsilon(\infty). \tag{7.16}$$

In treating the role of the dielectric function on transport, we must remember that is always the quantity $1/\varepsilon$ that appears in the scattering rates, in potentials, in screening, etc. Thus, the relative singularities that we must examine are those of the inverse dielectric function, which are the zeros of (7.16). To be fully exact, we would have to also properly include the imaginary parts that arise due to decay of the electronic states by scattering as a fully self-consistent calculation. This is a difficult problem and is beyond the treatment we want to use here, but we will return to it. We can get a quite good feeling for the important effects without resorting to the full, and complicated, treatment. One should also account for the q variation, but this only shifts the effective plasmon frequency by a quantity $[1 + (qv)^2]^{1/2}$, where v is an average (thermal or Fermi) velocity. Again, this effect is not terribly large and does not alter the basic approach we want to follow.

The zeros of the dielectric function are simply found by setting the left-hand side of (7.16) to zero and solving for the relevant values of ω. This process gives us an equation for ω:

$$\omega^4 - \omega^2(\omega_{LO}^2 + \omega_p^2) + \omega_p^2\omega_{TO}^2 = 0. \tag{7.17}$$

If $\omega_p \ll \omega_{TO}$, then (7.17) can be easily found to give the limiting cases

$$\omega_u^2 = \omega_p^2 + \omega_{LO}^2, \qquad \omega_1^2 = \frac{\varepsilon(\infty)\omega_p^2}{\varepsilon(0)} = \omega_p'^2, \tag{7.18}$$

where we have introduced the upper and lower "hybrid frequencies" and the modified plasmon frequency ω_p', which includes the static dielectric constant rather than the optical dielectric constant. On the other hand, if the plasma frequency satisfies $\omega_p \gg \omega_{LO}$, then we arrive at

$$\omega_u^2 = \omega_p^2 + \omega_{LO}^2, \qquad \omega_1^2 = \omega_{TO}^2. \tag{7.19}$$

Although the upper hybrid frequency remains the same functional root, this mode changes its character from phononlike at low density (small ω_p) to more plasmonlike at high density. Conversely, the lower hybrid frequency changes its character from plasmonlike at low density to phononlike at high density. In Figure 7.1, we plot the dispersion diagram for these modes of coupled oscillation. The critical frequency is obviously where $\omega_p = \omega_{LO}$, and for GaAs this occurs at a density of approximately 7×10^{17} cm^{-3}, which is just slightly larger than the doping in most current bulk devices, but below densities that are achieved in some superlattice devices.

In superlattices and quantum wells, the approach is modified slightly. For transport normal to the superlattice layer, the electrons may appear more two-dimensional than three-dimensional. In this latter case, ω_p is modified somewhat to

$$\omega_p^2 = \frac{n_s e^2}{m\varepsilon} q, \tag{7.20}$$

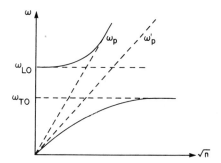

FIGURE 7.1. The dispersion curves for the lower and upper hybrid frequencies as a function of the plasma frequency (square root of the electron density).

which tends to vanish at small values of q. However, there is a minimum value of q corresponding to the thickness of the quantum well layer. Thus, many authors use $q = 1/w$ as a minimum value, for which a two-dimensional layer density of 10^{12} cm^{-2}, in a 100-nm-thick layer, corresponds to a bulk doping of 10^{17} cm^{-3}. For a 10-nm layer, this sheet density corresponds to 10^{18} cm^{-3}, so that the effect cannot be ignored in the quantum well systems.

7.2.2. Scattering Strengths

Both the electron–plasmon and the electron–(polar–)phonon scattering are caused by the perturbing potential that arises from the polarization of the charge oscillation (we treat this in a later section). For scattering of this type, the scattering rate (proportional to the square of the matrix element) contains a variation as

$$\frac{1}{\tau} \sim \omega_0 [\varepsilon_H^{-1} - \varepsilon_L^{-1}], \tag{7.21}$$

where ω_0 is the appropriate hybrid frequency and ε_H and ε_L are the high- and low-frequency dielectric constants appropriate to that mode. The first of these parameters is obviously found from the dispersion curve in Figure 7.1 for the relevant density and plasma frequency. The selection of the latter two quantities creates something of a problem, because we must now determine the relevant "oscillator strength" contained between $\varepsilon(\infty)$ and $\varepsilon(0)$ to assign to each of the modes.

In Figure 7.2, we plot the dielectric function itself. The zeros at ω_1 and ω_u correspond to the excited modes of the two hybrid frequencies. As the density is increased, ω_1 moves up toward ω_{TO} but never passes it, so the basic two-mode behavior is not modified. The distinction as to how much of the total behavior to assign to each of the modes is rather arbitrary. Here, we adopt an approach that has proven successful in SiO_2,[5] which has two dominant polar modes, thus facing the same considerations. We arbitrarily chose a frequency between ω_1 and ω_{TO} and have it represent the cutoff frequency. Then, for the upper hybrid mode,

$$\omega_0 = \omega_u, \qquad \varepsilon_H = \varepsilon(\infty), \qquad \varepsilon_L = \varepsilon(0), \tag{7.22}$$

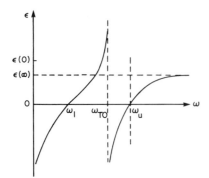

FIGURE 7.2.   The dielectric function of the coupled electron–plasmon–polar phonon system.

which defines the cutoff frequency as that at which the curve has the value of $\varepsilon(0)$. This choice is guided by the fact that these choices are the same values that would occur without the electrons (and plasmons) except for the frequency of the mode. Now, according to (7.22), the upper hybrid mode actually undergoes stronger scattering in the presence of the coupling, and the polar mode can be considered to be *descreened* by the plasmon–phonon coupling.

In a similar manner, we can address the lower hybrid frequency, for which the parameters become

$$\omega_0 = \omega_1, \qquad \varepsilon_H = \varepsilon(0), \qquad \varepsilon_L \to -\infty, \qquad (7.23)$$

and the coupling strength for this tends to saturate as ω_1 approaches ω_{TO}. The rationale for this saturation is that the plasmonlike nature is being transferred to the upper hybrid mode, and as this latter mode increases in scattering strength the lower hybrid mode saturates. At these high carrier densities, the electrons are actually screening out the polar-mode oscillations, so that only the transverse frequency (which is the frequency of the bare lattice vibrations) is seen. The plasmon mode remains, with the lattice vibrations screening these and following them closely. In this case, it is more likely that the optical phonons will scatter through the nonpolar interaction rather than the polar one, but the plasmonlike mode at ω_u will remain. This general observation of the weakening of the lower hybrid mode as a scatterer is reinforced by noting in Figures 7.1 and 7.2 that the zero corresponding to ω_1 approaches very closely to the pole at ω_{TO}. The interaction of this zero and pole tries to produce a general canceling of their individual contributions. To fully understand the behavior in this frequency region, at the high carrier densities, should include the proper wave-vector dependence and requires the full self-consistent treatment.[6]

7.3. SCREENING OF SCATTERING POTENTIALS

When an external, or nonintrinsic, potential is applied to a semiconductor, the electrons and lattice act in a manner to screen this potential. By nonintrinsic potential, we mean a potential arising from a charged impurity atom or even the scattering potential of one electron for another. In general, the applied potential is a Coulomb potential, such as that arising from a charged impurity atom,

$$V = \frac{e}{4\pi r},$$

for which the Fourier-transformed potential is just (ignoring the nonintegrable divergence at $r \to \infty$, for which we need screening)

$$V(q) = \frac{e}{q^2}. \qquad (7.24)$$

We note that we have omitted the ε_0 term from the potential, as we have included it into the total dielectric function below. The resultant screened potential is now just

$$U(q) = \frac{e}{q^2 \varepsilon(q, \omega)}. \tag{7.25}$$

We now want to evaluate the dielectric function in (7.25) for two approximations, both valid at low frequency (the high-frequency limit is just the plasmon excitations). For these, we will first take the limit $q \to 0$, and then return to include the full q dependence. Later, we extend this to include the descreening effects of drift and high scattering rates. For now, we assume also that $\alpha \to 0$.

7.3.1. The Debye Screening Limit

For most transport in semiconductors, we can assume that the distribution function is a nondegenerate Maxwellian in form, especially since the interelectronic scattering serves to drive the system toward the equilibrium Maxwellian. Our main task is to evaluate the summation inherent in the last term of (7.14). We take the low-frequency limit, so that the first two terms combine to just give $\varepsilon(0)$, the low-frequency dielectric function in the absence of free carriers. The denominator is

$$E(k + q) - E(k) \simeq \mathbf{q} \cdot \nabla E(k), \tag{7.26}$$

and the numerator is

$$f(k) - f(k + q) \simeq -\mathbf{q} \cdot \nabla f(k) = -[\mathbf{q} \cdot \nabla E(k)] \frac{\partial f(k)}{\partial E}. \tag{7.27}$$

For a Maxwellian,

$$\frac{\partial f}{\partial E} = -\frac{1}{k_B T} f(E), \tag{7.28}$$

and, since $\sum f(k) = n$, we can write the dielectric function as

$$\varepsilon(q \to 0, \omega = 0) = \varepsilon(0) + \frac{ne^2}{q^2 k_B T}. \tag{7.29}$$

By introducing the inverse square of the Debye length L_D as

$$\lambda^2 = \frac{1}{L_D^2} = \frac{ne^2}{\varepsilon(0) k_B T}, \tag{7.30}$$

we can rewrite the screened potential of (7.25) as

$$U(q) = \frac{e}{\varepsilon(0)[q^2 + \lambda^2]}. \tag{7.31}$$

This can then be inverse-transformed to yield the Debye screened potential

$$U(r) = \frac{e}{4\pi r \varepsilon(0)} \exp(-\lambda r). \tag{7.32}$$

The Debye screened potential is used quite often in the literature to describe the radial variation of the impurity potential. In truth, however, we have made a rather dramatic approximation in assuming that $q \ll k$, but also in assuming that $q \ll \lambda$, although this latter approximation is seldom recognized. For a free-electron concentration of 10^{17} cm^{-3}, the inverse Debye length $\lambda = 7.47 \times 10^5$ cm^{-1} in GaAs at 300 K. This value of λ corresponds to a screening energy of about 6 meV, so the inequalities are actually satisfied only for the fast electrons, which do not scatter well from the Coulomb impurities in any case. What we have left out of the considerations so far is the dynamic character of the dielectric response, and we turn to this now.

7.3.2. Momentum-Dependent Screening

While we will continue to use the low-frequency approximation, we want to now try to evaluate the summation over the free carriers without any limits on the range of q. To proceed, we rewrite the dielectric function as

$$\varepsilon(q, 0) = \varepsilon(0) + \frac{e^2}{q^2} \sum_k f(k) \left\{ \frac{1}{E(k+q) - E(k)} - \frac{1}{E(k) - E(k-q)} \right\}. \tag{7.33}$$

Now,

$$E(k \pm q) - E(k) = \frac{\hbar^2 q^2}{2m} \pm \frac{\hbar^2 kq}{m} \cos \theta. \tag{7.34}$$

We transform the summation over the vector k into an integration as

$$\sum_k A(k) \rightarrow \frac{1}{2\pi^2} \int_0^\infty \int_0^{2\pi} A(k, \theta) k^2 \sin \theta \, d\theta \, dk,$$

where we have already integrated over the azimuthal angle ϕ and included a factor of 2 for spin degeneracy. Incorporating this change, and using (7.34) in (7.33), we can integrate over the angle, yielding

$$\varepsilon(q, 0) = \varepsilon(0) + \frac{me^2}{\hbar^2 q^3} \int_0^\infty f(k) k \ln\left(\left| \frac{k + 2q}{k - 2q} \right| \right) dk. \tag{7.35}$$

To proceed further, the following normalized variables are now introduced:

$$\xi^2 = \frac{\hbar^2 q^2}{8mk_B T}, \qquad x^2 = \frac{\hbar^2 k^2}{2mk_B T}. \tag{7.36}$$

We note that the temperature here is that of the distribution function and represents the *electron* temperature, not that of the lattice. By incorporating the normalization factors on the Maxwellian distribution function, we can now rewrite (7.35) as

$$\varepsilon(q, 0) = \varepsilon(0)\left[1 + \frac{\lambda^2}{q^2} F(\xi)\right], \tag{7.37}$$

where

$$F(\xi) = \frac{1}{\pi^{1/2}\xi} \int_0^\infty x \exp(-x^2) \ln\left(\left|\frac{x + \xi}{x - \xi}\right|\right) dx. \tag{7.38}$$

Hall[7] has shown that we can rewrite $F(\xi)$ as

$$F(\xi) = \frac{1}{\xi} \exp(-\xi^2) \int_0^\xi \exp(t^2)\, dt, \tag{7.39}$$

which is related to an error function of imaginary argument. Actually, except for the $1/\xi$ prefactor, (7.39) is Dawson's integral.[8] Generally, however, $F(\xi)$ is closely related to the plasma dispersion function.

The behavior of $F(\xi)$ is generally not very dramatic. As $q \to 0$ ($\xi \to 0$), $F(\xi) \to 1$, and we obtain the Debye screened behavior. On the other hand, as $q \to \infty$, $F(\xi) \to 0$, and the screening is broken up completely. Thus, for high momentum transfer in the scattering process, the scattering potential is completely descreened. The upshot of this is that the nonlinearity in Coulomb scattering appears once again as a scattering cross section that depends on itself through the momentum transfer $\hbar q$ (and hence through the scattering angle θ). $F(\xi)$ has decreased to a value of 0.5 at $\xi \approx 1.07$, for which $q \approx 1.75 q_T$, where q_T is the thermal wave vector, corresponding to the thermal velocity ($\hbar q_T = m v_T$). For GaAs at 300 K, $q_T \approx 2.61 \times 10^6$ cm^{-1}. This value is only 3.5 times the λ value quoted above at 10^{17} cm^{-3}, and becomes less than the Debye value if the doping is increased a factor of 2. Although the screening is not eliminated in the materials of interest, the strength of the scattering can vary by some 20–30%, depending on the momentum transfer, and must be considered for completeness.

7.4. PLASMON SCATTERING

The scattering from the screened potential of other electrons and from the collective plasmon modes are both part of the total electron–electron interaction among the free carriers. Starting from the noninteracting electron gas, the Coulomb interaction can be treated as a perturbation, and this is the usual case for semiconductors. To go beyond this requires a more formal quantum treatment, which we discuss in a later chapter. It generally is not possible to consider the full Coulomb interaction beyond the lowest order of perturbation theory because of the long range of the potential associated with this interaction. As discussed above, the coupled dielectric function has singularities at the plasmon frequency and at zero frequency, corresponding to the plasmon modes and the Coulomb scattering potential. One can proceed from this observation to split the summation over q in a Fourier transform of the potential

$$V_{ee} = \frac{e^2}{4\pi\varepsilon_0} \sum_{i \neq j} \sum_q \frac{\exp[i\mathbf{q} \cdot (\mathbf{r}_i - \mathbf{r}_j)]}{q^2}$$

into a short-range part, for which $q > q_c$, and a long-range part, for which $q < q_c$. It has been shown that the short-range part of the potential corresponds to the screened Coulomb interaction,[9] which we discussed in the preceding sections. The long-range part, on the other hand, is responsible for the collective oscillations of the electron gas, which describe the motion of the electrons in the field produced by their own Coulomb potential.[10] These collective oscillations are bosons, so their scattering rate can be calculated in much the same manner as that for the longitudinal polar optical phonon. We must use the polar interaction as the collective oscillations are polar in nature. The only difference is that there is now a maximum q ($=q_c$) that can be involved in the scattering.

The electron–plasmon interaction terms are calculated by the normal method with the Fermi golden rule. Under the conditions of a relatively dense electron plasma, we can write the electron–plasmon scattering rate for a free electron at energy E and momentum k as

$$\Gamma(k) = \frac{me^2 \hbar\omega_p}{4\pi\varepsilon_0 \hbar k} \left(N_q \ln\left\{ \frac{q_c}{k} \left[\left(1 + \frac{\hbar\omega_p}{E}\right)^{1/2} - 1 \right]^{-1} \right\} \right.$$
$$\left. + (N_q + 1) \ln\left\{ \frac{q_c}{k} \left[1 - \left(1 - \frac{\hbar\omega_p}{E}\right)^{1/2} \right]^{-1} \right\} u_0(E - \hbar\omega_p) \right), \quad (7.40)$$

where N_q is the Bose–Einstein distribution of the plasmons. In incorporating this scattering rate, we must still determine a value for the cutoff wave vector q_c. Detailed studies of the inverse dielectric function, including the q dependence, give us an answer to this problem. In Figure 7.3, we plot the imaginary part of the inverse dielectric function for the case of silicon. We can see from this figure that the plasmon mode dies out for values of q nearing the Debye length $1/\lambda$. It is from plots such as these that we generally can estimate that the appropriate value for q_c is just this quantity $1/\lambda$. This is the quantity that cuts off the short-range part of the potential as well, so this result is consistent with both ends of the interaction range.

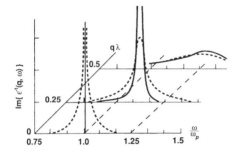

FIGURE 7.3.   The imaginary part of the inverse dielectric function, as a function of frequency, for several values of the scattering momentum q. The two different curves are with and without scattering in the dielectric function.

One important attribute of the electron–plasmon scattering is that the Bose–Einstein distribution for the plasmons does not remain in equilibrium for hot electrons. Studies of transport in high electric fields, carried out by Monte Carlo techniques,[4] show that the plasmon temperature comes into local equilibrium with the electron temperature that describes the free-carrier distribution in the high electric field. This again demonstrates the consistency of the approach, since both distributions are describing properties of the same electrons.

7.5. DESCREENING OF A POTENTIAL

In Section 7.3.2, we encountered the concept that a large momentum transfer (in comparison with the inverse screening length λ) in the scattering process resulted in the scattering being described by the descreened potential. In essence, the large momentum vector q led to a descreened interaction. This large momentum transfer case is only one of a variety of ways in which we must treat the unscreened potential in the scattering problem. In this section, we want to consider this further through a discussion of the dielectric function of the electrons themselves. We will find that descreening is induced by high frequencies, large scattering rates, and drift of the carriers. The latter will be illustrated through the results of ensemble Monte Carlo calculations of the radial distribution function of the electrons, where the potential of interest is the Coulomb interaction between individual pairs of electrons.

We can write the dielectric function in terms of a single distribution function, as we have previously done, as

$$\frac{\varepsilon(q,\omega)}{\varepsilon(0)} = 1 + \frac{e^2}{q^2\varepsilon(0)}\sum_k f(k)\left\{\frac{1}{E(k+q) - E(k) + \hbar\omega - i\hbar\alpha}\right.$$

$$\left. - \frac{1}{E(k) - E(k-q) + \hbar\omega - i\hbar\alpha}\right\}, \quad (7.41)$$

where we have explicitly retained the frequency dependence and the damping terms, and we have assumed the remainder of the dielectric function can be represented by $\varepsilon(0)$. If we need to consider the interaction between the screening term and the lattice terms, we must not use this dielectric constant term, but we will assume for the moment that we are dealing with nonpolar material. In the presence of the drift, we must modify the distribution function to reflect this distortion of the charge distribution. Closed forms for this are difficult to obtain, as this is principally the major task of the entire hot-electron problem. Here, we want to examine a system in which electron–electron scattering is large, and so we can take a drifted Maxwellian distribution function to describe the carriers. Thus,

$$f(k) \sim \exp\left[\frac{-\hbar^2(k - k_d)^2}{2mk_BT_e}\right], \tag{7.42}$$

where k_d is the drift momentum ($=mv_d/\hbar$). The drift effect on the distribution function is no more than a shift of the centroid of the electron cloud in momentum space in the approximation of (7.42). Since we are summing over all of the electrons in (7.41), we incur no error by making the change of variables

$$\mathbf{k} - \mathbf{k}_d \rightarrow \mathbf{k}', \tag{7.43}$$

although we will drop the primes in the treatment below. This change of variables shifts the origins of the energy terms, and moves the dependence upon the drift velocity from the distribution function to the denominator terms. Using the expansions given in (7.34), we can write the dielectric function as

$$\frac{\varepsilon(q, \omega)}{\varepsilon(0)} = 1 + \frac{e^2}{q^2\varepsilon(0)}\sum_k f(k)$$
$$\times \left\{\left[\frac{\hbar^2 kq}{m}\cos\theta + \frac{\hbar^2 q^2}{2m} + \frac{\hbar^2 q}{m}k_d\cos\psi + \hbar\omega - i\hbar\alpha\right]^{-1}\right.$$
$$\left. - \left[\frac{\hbar^2 kq}{m}\cos\theta - \frac{\hbar^2 q^2}{2m} + \frac{\hbar^2 q}{m}k_d\cos\psi + \hbar\omega - i\hbar\alpha\right]^{-1}\right\}. \tag{7.44}$$

In addition to the reduced variables of (7.36), we also introduce the variables

$$\Omega = \frac{\hbar\omega}{4k_BT_e}, \qquad \Gamma = \frac{\hbar\alpha}{4k_BT_e}, \qquad x_d^2 = \frac{\hbar^2}{2mk_BT_e}k_d^2. \tag{7.45}$$

Also as previously, we integrate over the angle θ, ϕ (the azimuthal angle, which is a free angle), and

$$\frac{\varepsilon(q, \omega)}{\varepsilon(0)} = 1 + \frac{\lambda^2}{q^2}F(\xi, x_d, \Omega), \tag{7.46}$$

where now

$$F(\xi, x_d, \Omega) = \frac{1}{2\pi^{1/2}\xi} \int_0^\infty e^{-x^2} x \ln \left\{ \frac{(\xi^2 + x\xi)^2 - (\Omega + x_d\xi \cos \psi + i\Gamma)^2}{(\xi^2 - x\xi)^2 - (\Omega + x_d\xi \cos \psi + i\Gamma)^2} \right\} dx.$$

(7.47)

It is clear from (7.47) that if any of the terms in the second set of parentheses, which are identical in both numerator and denominator, become dominant, the argument of the natural logarithm tends toward unity, and F vanishes. Thus, high-frequency (large Ω), high scattering rates (large Γ), and/or large drift velocities (large x_d) can all work to reduce the effective screening of a potential by the electrons.

In nearly all cases, the presence of a high electric field appears to work to reduce the screening factors, since they are all normalized with the electron temperature. However, we note that all of the terms in both numerator and denominator of the logarithm term in (7.47) have this same normalization. It is the relative strength of the various terms that is important, not the absolute strength. We can illustrate this via a numerical simulation. We take an ensemble of electrons and treat their normal scattering and transport in a high electric field through an ensemble Monte Carlo calculation. The interelectronic Coulomb interaction is retained as a real-space potential, and its effect on the motion of the electrons is computed through a molecular dynamics procedure, in which the local force on each electron, due to the electric field and the repulsion of all other electrons, is calculated each time step of the Monte Carlo process.[11] Once the electrons have reached a stable steady state in which the distribution function is in a local equilibrium, we can calculate the radial distribution function of the electron gas. The strength of this approach comes from the fact that it avoids simplifying assumptions on the form of the electron distribution function and on the form of the dielectric function involved in the screening process. Furthermore, the real-space fluctuations of the carrier gas are naturally included. The radial distribution function gives a representation of the two-particle correlations. In Figure 7.4, we plot the radial distribution function for three different values of the electric field for an electron density of 1×10^{16} cm^{-3} in silicon. At the lowest field, which corresponds to a low-field equilibrium situation, no drift or heating effects are present and $g(r)$ approaches its asymptotic value corresponding to a Debye screened interparticle potential. The equilibrium Debye length $L_D = 1/\lambda$ is shown in the figure.

The behavior of Figure 7.4 is typical for full random-phase-approximation-type calculations for the dielectric function in which reduced screening is often found, as we have discussed. The reduced screening begins already at intermediate electric fields of 5 kV/cm, and $g(r)$ relaxes more rapidly. The analytical details of the competition of the screening and descreening processes is still not fully understood. At the highest electric field, $g(r)$ exhibits a pronounced peak at short distances from the central site, and there might be many rapid oscillations in this region. The inset of Figure 7.4 shows the first Legendre component of a spherical

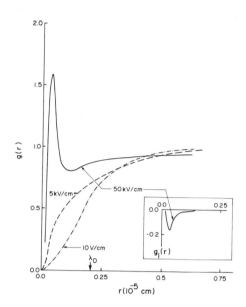

FIGURE 7.4.. The radial distribution function $g(r)$ for several different electric fields. The inset is the first Legendre term in a spherical harmonic expansion, discussed in the text. All curves are for silicon at 77 K with an electron concentration of 10^{16} cm^{-3}.

harmonic expansion of $g(r)$ (also calculated numerically). However, we expect the two-particle correlation to be symmetric with inversion symmetry since the potential is a central-force-type potential. The values obtained in the inset are felt to give a measure of the noise in the calculation in the central region. The reason why rapid oscillation might be expected in this region comes from the integral in (7.47) itself. In the absence of screening, and when the electron cloud is totally uniform, we might expect $g(r)$ to be unity for $r > 0$. Indeed, the descreening behavior in Figure 7.4 is tending toward this condition. However, the integral in (7.47) is related to Fresnel integrals, so that the approach is expected to involve rapid oscillations which eventually will be canceled out by the smoothing effect of fluctuations.

At higher concentrations, the spatial asymmetry of $g(r)$ is greatly reduced, and shifted to much higher values of the electric fields. In Figure 7.5, we plot the radial distribution function for a density of 1×10^{18} cm^{-3} and for the same

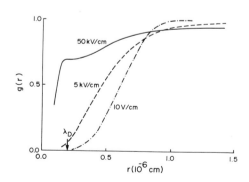

FIGURE 7.5. The radial distribution function $g(r)$ for several different electric fields. All curves are for silicon at 77 K with an electron concentration of 10^{18} cm^{-3}.

values of the electric field as in the previous case. As can be seen by comparison with Figure 7.4, the electron–electron interaction is much stronger at this higher density and reinforces the spatial symmetry. Thus, a much higher electric field, of the order of 100 kV/cm, is needed in order to induce the asymmetric behavior of $g(r)$.

REFERENCES

1. J. M. Ziman, *Principles of the Theory of Solids*, University Press, Cambridge (1964).
2 J. C. Phillips, *Bonds and Bands in Semiconductors*, Academic Press, New York (1973).
3. D. Penn, *Phys. Rev.* **128**, 2093 (1962).
4. P. Lugli and D. K. Ferry, *IEEE Electron Dev. Lett.* **EDL-6**, 25 (1985).
5. D. K. Ferry, *J. Appl. Phys.* **50**, 1422 (1979).
6. M. E. Kim, A. Das, and S. D. Senturia, *Phys. Rev. B* **18**, 6890 (1978).
7. G. L. Hall, *J. Chem. Phys. Solids* **23**, 1147 (1962).
8. M. Abramowitz and I. A. Stegun, *Handbook of Mathematical Functions*, U.S. Department of Commerce, Government Printing Office (1964).
9. A. Huang, *Theoretical Solid State Physics*, Pergamon Press, New York (1972).
10. P. Lugli, *Ph.D. Dissertation*, Colorado State University, unpublished.
11. P. Lugli and D. K. Ferry, *Phys. Rev. Lett.* **56**, 1295 (1986).

8

Lateral Surface Superlattices

As semiconductor technology continues to pursue the scaling down of integrated circuit dimensions into the submicron and ultrasubmicron regimes, many novel and interesting questions will emerge concerning the physics of charged particles in semiconductors. One of the more important topics to be considered is that of carrier confinement in structures that reduce the dimensionality of the system. Notable among these structures are MOS quantized inversion layers discussed in previous chapters and the heterojunction superlattice. In particular, the fabrication of the quantum well superlattice has been possible due to the advent of MBE and MOCVD technology, which we have discussed previously.

The concept of a superlattice, as it is most commonly interpreted, refers to layered structures and was first put forward by Esaki and Tsu.[1] Indeed, as we have tried to convey, the general potential and energy structure has been verified by careful experimental work so that we can now talk of band-gap engineering. We want now to consider lateral superlattices. Lateral superlattices, in which the superstructure lies in a surface or heterostructure layer,[2,3] offer considerable advantages for obtaining superlattice effects in a planar technology.

One reason for considering such a lateral superlattice lies in the growth of the density of devices on an individual integrated circuit chip. The chips are designed in a highly regular manner, with the individual devices—be they RAM cells or active gates in a logic processing section—laid out on a regular square lattice. The interactions between the devices can be expected to lead to rather strong nearest-neighbor interactions. In today's layouts, these extraneous interactions are considered to be *parasitic* interactions which can, for example, be accommodated in the design by extra capacitance assigned to the fan-out line. In the future, however, these parasitic interactions will be the dominant interaction between devices, as this interaction is greatly strengthened as the devices are moved closer together (the strength of this interaction does not scale with the device geometry). Our major rationale in this chapter is then to study the cooperative interactions between a regular array of devicelike structures—our lateral surface superlattice. The test structures that are fabricated are devices which allow us to begin to study these cooperative effects in order to try to understand how they may be usefully used to provide new processing architectures in densely packed arrays of devices.

Consider a periodic array of gates. If the periodicity in the gate array can be fabricated with a spacing small compared with the inelastic mean free path for the carriers, then superlattice effects should manifest themselves in the surface conduction channels. While being a distinct limitation to down-scaling of semiconductor device arrays, the quantum collective effects which arise in the surface superlattices are interesting in their own right and offer new device capabilities.

The lateral superlattice offers conceptual as well as technological advantages over the layered superlattice in terms of achieving the desired quantum well transport effects. In essence, the reason for this is that the minibands in the one-dimensional layered superlattices are not separated by real minigaps, but are "connected" by the two-dimensional transverse continuum of states. In order to create true minigaps, a multidimensional superlattice or quantization is required. This can be achieved by a lateral superlattice imposed in a quantized inversion layer. We shall illustrate how such lateral superlattices can be prepared by a variety of techniques. One formally proposed by Bate[2] is the MOS structure, which is similar to an array of charge-coupled device (CCD) devices, and these superlattices can be fabricated through the use of fine-line lithography.[4] Other approaches use selective area epitaxy, but have not been fabricated to date.[5] Finally, the complement of the MOS structure is a depletion-mode device made in a high-electron-mobility transistor.[6]

8.1. LATERAL SUPERLATTICES

The concept of a lateral superlattice along a surface has considerable advantages, among which is the ability to control the magnitude of the surface potential seen by an inversion layer. The basic structure is shown in Figure 8.1, for an MOS structure. A periodic gate array is placed on the surface by metal liftoff or by etching a grid and metallizing the structure.[6] The top electrode provides energy-gap control without requiring critical alignment of successive levels. A one-dimensional (along the surface) MOS implementation of this structure has been achieved,[4] and the transport studied both along and transverse to the gate stripes. If the periodic gates are biased negatively, the surface potential for

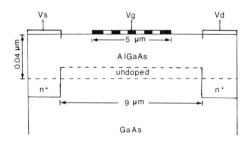

FIGURE 8.1. The Bloch FET in which a two-dimensional superlattice array of dots is placed within the gate oxide. This dot array produces a superlattice potential along the interface at the inversion layer.

electrons decreases under the gate electrodes and, to a lesser extent, in the gaps. Minority carrier generation, injection from an FET source, or optical pumping creates the carriers necessary to maintain the inversion layer under the gaps. Thus, in addition to the normal average surface potential, a periodic superlattice potential is seen by the inversion-layer electrons. The presence of the top electrode allows for critical control of the relative strengths of both the average potential and the superlattice potential.

The effective superlattice potential $U(x, y)$ can be expected to vary along the structure in a form given by

$$U(x, y) = 4U_0 \cos\left(\frac{2\pi x}{d}\right) \cos\left(\frac{2\pi y}{d}\right). \tag{8.1}$$

By introducing a change of variables $u = x + y$, $v = x - y$, with

$$\sqrt{2}\,\frac{\partial}{\partial x} = \frac{\partial}{\partial u} + \frac{\partial}{\partial v}, \qquad \sqrt{2}\,\frac{\partial}{\partial y} = \frac{\partial}{\partial u} - \frac{\partial}{\partial v}, \tag{8.2}$$

the effective two-dimensional Schrödinger equation for the interface electrons in the inversion layer is given by

$$\left\{\frac{d^2}{du^2} + \frac{d^2}{dv^2} + \frac{2m^*}{\hbar^2}\left[E - 2U_0 \cos\left(\frac{2\pi u}{d}\right) - 2U_0 \cos\left(\frac{2\pi v}{d}\right)\right]\right\}\Psi(u, v) = 0. \tag{8.3}$$

This equation is now separable using a product form of the wave function $\Psi_1(u)\Psi_2(v)$, which yields two effective one-dimensional Schrödinger equations of the form

$$\left\{\frac{d^2}{du^2} + \frac{2m^*}{\hbar^2}\left[E - 2U_0 \cos\left(\frac{2\pi u}{d}\right)\right]\right\}\Psi_1(u) = 0. \tag{8.4}$$

Introducing the reduced variables

$$\xi = \frac{gu}{2}, \qquad a = \frac{8m^* E}{\hbar^2 g^2} = \frac{2m^* d^2 E}{\pi^2 \hbar^2}, \qquad q = \frac{2m^* d^2 U_0}{\pi^2 \hbar^2}, \tag{8.5}$$

where $g = 2\pi/d$, we can write (8.4) as

$$\frac{d^2\Psi(\xi)}{d\xi^2} + [a - 2q \cos(2\xi)]\Psi(\xi) = 0. \tag{8.6}$$

This latter form is immediately recognized as the Mathieu equation. For $q = 0$, all values of a (and hence of the energy E) are allowed. However, when $q \neq 0$, gaps open in the spectrum of a. For small q, the lowest gap is centered approximately at the point $a = 1$, and higher gaps occur approximately at $a = 4, 9, \ldots, n$. The general energy structure is shown in Figure 8.2. It is very important to note that the general solution to the Mathieu equation is of Bloch form

$$\Psi(\xi) = \exp(ik\xi)p(\xi), \tag{8.7}$$

where $\rho(\xi)$ has the periodicity of the superlattice potential. This is expected for the periodic potential. For the first minigap to be centered at a particular energy W, we require $a(W) = 1$, and

$$\frac{\hbar^2 \pi^2}{2m^*d} = W, \tag{8.8}$$

or

$$d = \left(\frac{\hbar^2 \pi^2}{2m^*W}\right)^{1/2}. \tag{8.9}$$

In Figure 8.3, we show a plot of the d values required to achieve an energy width of $W = 6k_BT$ in the lowest band. We also show the number of states in the lowest miniband, which is closely related to the surface carrier density required in the inversion layer to completely fill the first miniband. It will be noted that these densities correspond to a relatively strong inversion layer existing at the surface. For these densities, the Fermi level at low temperature will be well into the conduction band.

The first minigap will have a value given approximately by $\Delta_a = a(\Delta) = 1.9q$ for small values of q. For $U = 0.01$ V, we find that $\Delta_a = 2.9k_BT$ at 77 K and

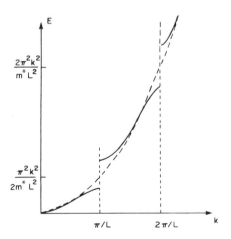

FIGURE 8.2. The general conduction-band energy structure. Gaps are opened at $k = \pi/d$ due to the perturbation of the surface superlattice potential.

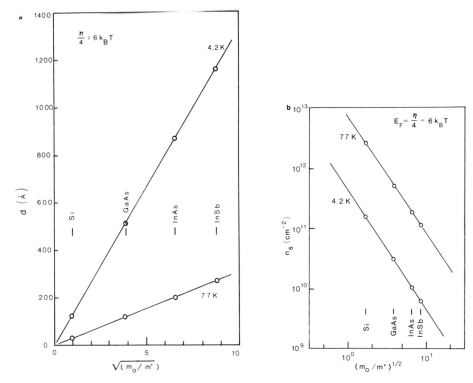

FIGURE 8.3. (a) Value of the gate center-to-center spacing required on the surface superlattice in order to form a first miniband whose width is given by $\eta/4 = 6k_BT$. (b) Inversion density for which the first miniband is completely full. Here, the bandwidth is that given by the curves in part (a).

$52.4k_BT$ at 4.2 K, corresponding to rather large gaps when compared with the miniband energy width. It is clear that relatively small induced superlattice potentials are required to produce the miniband/minigap structure. Indeed, surface band bending corresponding to roughly one trapped electron under each gate could produce sufficient potential to be noticable at 4.2 K in the case of the very light mass InSb. Evidently, from Figure 8.3, lower effective mass material is favored. Whereas Si requires $d = 10$ nm in order to fully produce superlattices, the effects should be observable in InAs and InSb for $d = 50$ nm. One caution, however, must be stated here, and this is that the calculations are done for the size of the bands and gaps and not for whether the effects will be washed out by thermal effects. Rather, the important length is not the wavelength, which enters into the equations leading to Figure 8.3, but the inelastic mean free path which determines the distance over which the electron wave functions remain coherent.

We can probe the dynamics of these superlattices in another manner. The two-dimensional motion of electrons that are subjected to both a two-dimensional periodic potential and a perpendicular magnetic field is a problem that has been studied for a great many years. In general, the solution of this problem is complicated by the fact that there are two characteristic lengths in the problem: one is the period a of the superlattice potential, while the second is the magnetic length $L_m = (\hbar/eB)^{1/2}$. The complete problem can be solved generally in an infinite domain only when these two lengths are related by the ratio of two integers. However, there are two distinct limits in which perturbation theory can be used to obtain solutions. In one limit, the Landau regime, the periodic potential is regarded as a weak perturbation on the usual magnetic Landau-level structure. When the flux coupled through each unit cell of the periodic potential is given by $\phi = ea^2 B/\hbar = p/q$, each Landau level splits into p subbands of equal degeneracy. It has been known for some time that a series of oscillations, periodic in $1/B$, should arise in magnetotransport measurements.[7-9] These oscillations have an appearance similar to, but an origin different than, the Shubnikov–de Haas oscillations. It is easy to think of these oscillations in terms of commensurability of the cyclotron radius with the lattice periodicity, but the physics of the oscillations is related to the Fermi level moving through the split bands of the Landau level. In fact, these predicted oscillations have recently been observed with periodic potentials macroscopically produced by a superlattice, and are now known as Weiss oscillations.[10-12] When the strength of the superlattice periodic potential is increased, the Landau levels are broadened significantly, the subbands merge and the Weiss oscillations, *as well as the normal Shubnikov–de Haas oscillations*, are heavily damped[13] and eventually disappear. This has also been observed experimentally.[14]

In the opposite limit to that above, the magnetic field is treated as a perturbation on the periodic potential. In this regime, known as the Onsager regime, the magnetic transport properties are expected to be periodic in the magnetic flux coupled through each unit cell, i.e., periodic in magnetic field (as opposed to $1/B$ in the previous case).[15-17] While the required magnetic field is unreasonably high in normal semiconductor lattices, it is an observable effect in lateral surface superlattice (LSSL) periodic potentials.[18] Clearly, the observation of these effects, which are linear in the magnetic field, requires the phase coherence length of the electrons to be larger than the superlattice period. On the other hand, the limit being taken here is the tight-binding limit of the superlattice potential, and the transport can be expected to have strong similarities to more localized types of transport.

The structure of the bands can be obtained by solving Schrödinger's equation with the magnetic field as a perturbation. In the absence of the latter field, the energy structure of the superlattice minibands is given by

$$E = E_0[\cos(k_x a) + \cos(k_y a)]. \tag{8.10}$$

With a magnetic field, described in the Landau gauge $\mathbf{A} = (0, Bx, 0)$, the Peierls substitution leads to the equation

$$\left\{ \cos(k_x a) + \cos\left[\left(k_y - \frac{eBx}{\hbar} \right) a \right] \right\} \Psi(x, y) = \frac{E}{2E_0} \Psi(x, y). \tag{8.11}$$

The introduction of the wave function *ansatz* $\Psi(x, y) = g(x)\exp(ik_y y)$ and the substitutions $x = ma$, $y = na$,[17] lead to the iterative equation

$$g(m + 1) + g(m - 1) + 2\cos(2\pi m\alpha - \nu)g(m) = Eg(m), \qquad (8.12)$$

where $\alpha = eBa^2/h$, $\nu = k_y a$, and E ($= E/E_0$) is the reduced energy. This equation is the Harper equation,[16] and the solutions of this equation have been discussed by Hofstadfer[17] in some detail. The energy structure is periodic in α, which means that it is periodic in magnetic field, as this quantity is the ratio of the flux coupled through a unit cell to the quantum unit of flux h/e.

The source of the periodicity in magnetic field in the tight-binding limit can be understood in one sense by its relationship to the Aharonov–Bohm effect. Consider the presence of magnetic translation operators connected with the periodicity of the lattice. In a periodic lattice, it is known that $\Psi(x + a, y) = \exp(ik_x a)\Psi(x, y)$, where $\Psi(x, y)$ is the Bloch function corresponding to the superlattice in the absence of the field. If the magnetic field is normal to the layer, and the vector potential is taken (as above) in the Landau form $\mathbf{A} = (0, Bx, 0)$, the motion of successive translations about a rectangle of unit cells (returning to the original point) leads to

$$T(-na\mathbf{j})T(-ma\mathbf{i})T(na\mathbf{j})T(ma\mathbf{i})$$

$$= T(-na\mathbf{j} - ma\mathbf{i} + na\mathbf{j} + ma\mathbf{i})\exp\left[i\int_0^{2\pi}dy\int_0^{2\pi}dx\frac{eB}{\hbar}\right]$$

$$= \exp(2\pi inm\alpha), \qquad (8.13)$$

where $\mathbf{i}$ and $\mathbf{j}$ are unit vectors in the directions of the LSSL. In fact, the group theoretical arguments for the magnetic translation group have been worked out in some detail.[19] In fact, the source of the periodicity can be understood quite easily with a simple Fermi energy argument. Recall that the periodicity of $1/B$ arises from the Fermi energy being forced down through the Landau levels, and the split bands of the levels. The $1/B$ behavior arises from the increase in the degeneracy of each level with the magnetic field and the spreading apart of the levels in the magnetic field. In the present case, the superlattice potential breaks up the conduction band into a series of minibands of width ΔE. The number of states in each miniband is constant, but as the magnetic field is increased these bands are *depopulated by the magnetic field*. The conductance oscillations arise as the Fermi energy passes through each miniband. In the absence of the periodic potential, the Fermi level is given by $E_{F0} = \pi\hbar^2 n_s/m^*$. This energy range is the range of *allowed states* and will be the sum of the widths of the occupied energy minibands in the presence of the periodic potential. Each periodic potential can accommodate $2/a^2$ electrons, so the number of full and fractionally occupied minibands is just $n_s a^2/2$. In a sense, this number is the *filling factor* for the minibands. Thus, the average miniband width may be found from

$$\Delta E_{\text{av}} = \frac{E_{F0}}{\text{ff}} = \frac{2\pi\hbar^2}{m^*a^2} = \frac{eh}{m^*}\left(\frac{\hbar}{ea^2}\right) = \hbar\omega_c. \qquad (8.14)$$

Here, ω_c is the cyclotron frequency corresponding to the magnetic-field periodicity. The term in parentheses in (8.11) is the flux coupled through each unit cell of the superlattice, and an integer number of flux quanta is coupled through each cell when the Landau level has been swept through a miniband.

8.2. GaAs STRUCTURES

Possible structures discussed for the lateral surface superlattice can be developed. One is fabricated in a GaAs MODFET, while a second can be fabricated through the use of selective area epitaxy. Although the former is more readily fabricated, the latter was suggested earlier in a historical sense. A MODFET is usually a depletion-mode device, in which the gate is used to push electrons out of a channel, much like a MESFET, rather than drawing them into the channel as in a MOSFET. In this case, we need a grid, as shown in Figure 8.4. The grid imposes the periodic potential on the electrons in the channel, and as the device approaches pinchoff the final electrons are left in pockets which are aligned with the holes of the grid. The electrons are thus sitting in small quantum wells induced by the superlattice potential itself. The grid in Figure 8.4 was

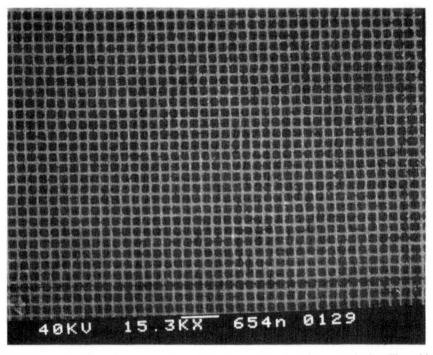

FIGURE 8.4. Electron micrograph of a grid forming a lateral surface superlattice. The grid is fabricated by lifting off gold with a pattern written in single-level PMMA. The lines are approximately 40 nm wide and lie on a 160-nm pitch. From G. Bernstein and D. K. Ferry, *Superlatt. Microstruct.* **2**, 373 (1986).

written by electron-beam lithography using single-level PMMA, and the pattern transferred by normal lift-off processing. The grid shown has 40-nm lines on 160-nm spacing. The grid is 28 periods in the source-drain direction and 170 periods in the transverse direction, so the superlattice gate is approximately 5 mm by 30 mm. (The grid potential can also be made by etching a pattern similar to Figure 8.4 into the GaAlAs, and then coating the surface with a solid metal.[6]) Devices were then fabricated from typical modulation-doped heterostructure material commonly employed for MODFETs. Samples were prepared by molecular-beam epitaxy of a pseudomorphic InGaAs single quantum well structure on an undoped semi-insulating GaAs substrate. The InGaAs layer, 13.5 nm thick with 20% In content, was grown on an undoped GaAs buffer layer (0.5 μm thick on top of a GaAs substrate). An undoped GaAs layer 15.7 nm thick was then grown, followed by a Si-doped (1×10^{18} cm^{-3}) GaAs layer 40 nm thick. The carrier density in the pseudomorphic quantum well was 2×10^{11} cm^{-2} at 5 K. The first step of the processing is mesa isolation, in which a cross structure is defined by photolithography and etched about 200 nm deep. After that, 200-nm-thick AuGe/Ni/Au contacts were deposited by electron-beam evaporation and lift-off. These were alloyed at 450°C for 5 min in forming gas to form the ohmic contacts. The grid gate was patterned by electron-beam lithography and lift-off processing. Finally the bonding pads were made by evaporation and lift-off of 300-nm-thick Cr/Au. The grids are composed of 40-nm lines on a 160-nm pitch. The active area of the device structure is 10 μm $\times$ 20 μm.

In Figure 8.5(a), the source conductance, in which the current is along the long axis of the sample, is shown at 5 K. It is apparent that there are significant fluctuations and a weak periodicity of the conductance that is present in the magnetoconductance. The structure is fully repeatable as long as the sample is maintained at low temperature, but does change somewhat upon heating and recooling of the sample. The applied longitudinal voltage on the sample was only 1 mV over the entire range, so that the amplitude of the fluctuations in conductance is about $0.1 e^2/h$. We have Fourier-transformed the conductance in order to bring out the underlying periodicity, and this is shown in Figure 8.5(b). The dc component has been removed to enhance the signal, but there is still a low-frequency component that arises from the weak magnetoresistance variations in the sample. In this latter figure, we have marked the range expected from estimating the frequency that would arise from the fabricated superlattice periodicity, allowing for the possibility that the actual flux coupled to each well varies due to the finite width of the individual gate lines. A second set of weaker peaks is observed near the second harmonic. Whether these relate to $h/2e$ oscillations seen in weak localization in rings or are simply the second harmonic is not discernible at this time.

In Figure 8.5(b), the dominant peak is approximately 6.9 T^{-1}, which corresponds to a unit cell whose side is 176 nm, while actual scanning electron microscopy measurement of the sample suggests a number closer to 168 nm. Considering the quality of the data, this agreement is quite good. A secondary peak is also observed which lies very close to the first, and within the range of

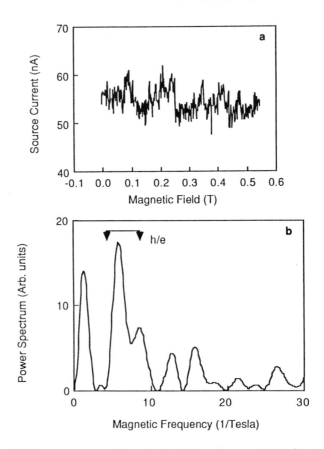

FIGURE 8.5. (a) The source current through the LSSL made on a InGaAs/GaAs high-mobility sample. (b) Fourier transform of the magnetoconductance, showing the peak at the value h/e flux per unit cell of the LSSL.

the spread expected from the fabricated grid. This secondary peak could arise from a slightly different spacing over part of the grid, which could arise from differences across the grid in the linearity of the electron-beam sweep during e-beam lithography.

The source of the conductance fluctuations in the data, and the relatively large amplitude of these fluctuations compared with that expected for universal conductance fluctuations (UCF), is also quite interesting. The UCF is regarded as arising from quantum interference of different modes, or paths, of the electrons as the chemical potential or the magnetic field is varied, so that interference effects appear in the end-to-end conductance, and are related to the Aharonov–Bohm effect. In general, the observations of these effects in the past have been confined to quasi-one-dimensional conductors. The structure we are investigating is considerably larger than the estimate of the inelastic mean free path. UCF is generally found to decay faster than linearly in quasi-one-dimensional wires and a sequence of rings. The inelastic mean free path inferred is such that the amplitude observed for the fluctuations is of the order of magnitude expected from these studies.

In other studies of such superlattices, negative differential conductance (NDC) has been reported once. The striking features of these curves was a strong NDC evident at high reverse-gate voltage. These features were explained as follows: As the device approaches pinchoff, the electrons in the channel are more fully localized in weakly coupled quantum boxes, which are the sites under the holes in the grid. Thus, there is a strong superlattice potential that is strengthened by the localization of the electrons in the boxes. As the channel density is increased, by making the gate potential more positive, the superlattice potential is weakened by the screening of the background charge density. To observe the miniband effects, the electrons must have an inelastic mean free path length that is longer than the period of the surface superlattice. The mobility in these layers was such that we estimate the mean free path to be only about 200 nm at 4.2 K, so that the effect is seen only at these low temperatures. While the experiments are highly suggestive of Bloch oscillations, they have never been repeated. The measurement of Bloch oscillations is quite a difficult task, so that a discussion of transport in such superlattices is given in the next section.

8.3. TRANSPORT EFFECTS

As we have seen, various superlattice structures give rise to energy minibands that vary sinusoidally across the minizone, in at least one dimension, and which have relatively narrow widths. The shape of such bands results in interesting electrical properties. The one-dimensional superlattice is one such structure, and the lateral surface superlattice is another. In this section, we want to now talk about the average velocity and energy of the carriers in the superlattice, and the various transport properties that can occur. We will do this in both the steady-state dc conductivity case and in the small-signal ac conductivity case.

8.3.1. Steady Transport

The average velocity and energy are found by taking the first and second moments of the Boltzmann transport equation with an assumed form for the distribution function. A constant electric field is assumed to be applied in the plane of the sinusoidal bands, while the energy shape in the other two directions is arbitrary. Here, we shall take a simple Wigner function representation as the initial equilibrium distribution. This treatment is valid for a single energy band at low to moderate electric fields for which the sinusoidal band is less than half-filled with carriers. For a band that is more than half-filled, the Pauli exclusion principle must be taken into account. In addition, we shall use the constant relaxation time approximation, as the effects in which we are interested arise from the properties of the energy bands and not from the energy dependences of the scattering rates. In truth, the relaxation time is not strictly constant, since the density of states for cosinusoidal energy bands in two dimensions show strong Van Hove singularities which lead to strong scattering peaks near midband. Nevertheless, we shall see below that the constant relaxation time approximation is not too bad due to the high scattering rates. Lastly, we assume the distribution function is homogeneous in space.

The form of the time-independent, homogeneous Wigner transport equation in the relaxation time approximation is the same as that of the Boltzmann transport equation, and is

$$\frac{eF}{\hbar} \frac{\partial f(\mathbf{k})}{\partial k_z} = -\frac{f(\mathbf{k}) - f_0(\mathbf{k})}{\tau}. \tag{8.15}$$

where the field and the direction of the superlattice are taken to be the z-direction. The quantity $f_0(\mathbf{k})$ is found from the equilibrium quantum density distribution by using the Hamiltonian equivalence principle, followed by a Wigner transformation, which leads to the Bloch form

$$f_0(\mathbf{k}) = \exp[\beta\varepsilon - \beta\varepsilon \cos(k_z d)] f_0(k_x, k_y), \tag{8.16}$$

in which the energy is

$$E = \varepsilon - \varepsilon \cos(dk_z), \tag{8.17}$$

ε is the half-width of the energy band, and d is related to the periodic spacing of the superlattice.

Taking the first moment of the velocity in the z-direction and the second moment (the total energy) results in the equations

$$\frac{eF}{\hbar} \int\int_{-L}^{L} v_z \frac{\partial f(\mathbf{k})}{\partial k_z} d^3k = -\Gamma\langle v_z \rangle, \tag{8.18}$$

$$\frac{eF}{\hbar} \int\int_{-L}^{L} E \frac{\partial f(\mathbf{k})}{\partial k_z} d^3k = \Gamma n[\langle E \rangle - \langle E_0 \rangle], \tag{8.19}$$

where $\Gamma = 1/\tau$, $L = \pi/d$, and $\langle S \rangle$ is defined as the average of the quantity S. Here, E_0 is the equilibrium energy without an applied field.

To proceed, we must now introduce the analytical expressions for the velocity and the energy. For these, we use the assumed energy-band shape

$$E = \varepsilon - \varepsilon \cos(dk_z) + E(k_x, k_y), \tag{8.20}$$

$$v = \frac{\varepsilon d}{\hbar} \sin(dk_z). \tag{8.21}$$

The left-hand sides of both moment equations may be integrated by parts in the z-direction, and the fact that both the energy and the distribution function are periodic in $2L$ and the velocity vanishes at $L, -L$, to find that

$$\frac{eFd^2}{\hbar^2} \langle \varepsilon \cos(dk_z) \rangle = \Gamma \langle v \rangle, \tag{8.22}$$

$$-eF\langle v_z \rangle = \Gamma[\langle E \rangle - \langle E_0 \rangle]. \tag{8.23}$$

The first bracketed average in (8.22) can be replaced, using (8.20) by

$$\langle \varepsilon - E + E(k_x, k_y) \rangle = \varepsilon - \langle E \rangle + \langle E_t \rangle \quad [E_t = E(k_x, k_y)].$$

Solving these equations simultaneously gives the expressions for the velocity and energy as functions of the field as

$$\langle v_z \rangle = [\varepsilon + \langle E_t \rangle - \langle E_0 \rangle] \frac{eFd^2 \tau / \hbar^2}{1 + \omega_B^2 \tau^2}, \tag{8.24}$$

$$\langle E \rangle = [\langle E_0 \rangle + (\varepsilon + \langle E_t \rangle) \omega_B^2 \tau^2] \frac{1}{1 + \omega_B^2 \tau^2}. \tag{8.25}$$

where $\omega_B = eFd/\hbar$ is the Bloch frequency. Thus, we see that as the field increases the average energy rises from its equilibrium value $\langle E_0 \rangle$ ($= 3k_B T/2$) to the half-band energy plus the average transverse energy, and the velocity behaves in a corresponding manner.

The velocity has the same field dependence in (8.21) as that obtained earlier by Lebwohl and Tsu,[20] except for the energy prefactor in front of the expression. The difference is caused by the different equilibrium distribution function chosen. In this latter work, the authors assumed the distribution was a zero-temperature Fermi–Dirac. Here, on the other hand, the distribution is a real temperature one that includes the details of the band shape. Note that the velocity shows a negative differential conductivity that sets in above $\omega_B \tau = 1$. Monte Carlo calculations have been performed as well, in which the exact details of the energy-dependent scattering processes were included.[21] The general shape of the velocity curves of (8.24) are found there as well. In fact, for the same set of material (band) parameters, the curves are very close together, which justifies our earlier assumption of a constant relaxation time. The details of the scattering processes just do not make a large difference in the present circumstances and the important aspect for negative differential conductivity is the band shape itself.

8.3.2. High-Frequency Response

While the above analysis provides limits within which Bloch oscillations may be seen, it does not provide an existence proof on their presence. We want now to consider several possible routes, besides the normally considered potential negative differential conductivity, by which Bloch oscillations may be experimentally verified. We pursue this by looking at a number of consequences of superimposing a small ac signal on top of the applied dc bias electric field. Here, we will seek first the ac component of the velocity and, hence, of the mobility that relates to the time varying frequency ω at which the applied field is oscillating. To achieve this, we must add the time derivative term to (8.12). We then assume that $\langle E \rangle = \langle E \rangle_0 + E_1$, $\langle v_z \rangle = \langle v_z \rangle_0 + v_1$. Then, (8.22) and (8.23), for the ac terms, become

$$(i\omega - \Gamma)v_1 = \frac{eF_1 d^2}{\hbar^2} \langle \varepsilon \cos(k_z d) \rangle, \tag{8.26}$$

$$(i\omega - \Gamma)E_1 = eF_1 \langle v_z \rangle_0 + eF_0 v_1. \tag{8.27}$$

from which the resulting ac mobility (the real part of the velocity response to F_1) is given by

$$\mu(\omega) = \mu_0 \frac{(1 - \omega_B^2 \tau^2)[1 - (\omega_B^2 - \omega^2)\tau^2]}{[1 + (\omega_B^2 - \omega^2)\tau^2]^2 + 4\omega^2 \tau^2}, \tag{8.28}$$

where $\mu_0 = e\tau/m$ is the low-field dc mobility. In Figure 8.6, we plot this mobility as a function of frequency for a case in which the dc field lies in the negative differential conductivity regime ($\omega_B \tau = 2$). It is important to note that there is no peak in the conductivity at the Bloch frequency. Rather there is just a falloff in the negative conductivity to positive values. The lack of a resonance at the Bloch frequency suggests, but does not prove, that the Bloch oscillations are not radiative. In fact, what is probably meant by this result is that the individual Bloch electrons oscillate, but that their phases add incoherently, so that no

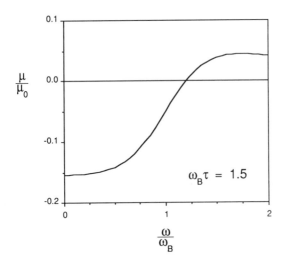

FIGURE 8.6. The small signal ac mobility (the real part of the complex mobility) that arises when a small ac electric field is superimposed on the dc electric field. The dc field is large enough to have the device in the negative differential conductivity regime with $\omega_B \tau = 2$. The curve is calculated in the relaxation-time approximation, but agrees well with calculations using an ensemble Monte Carlo approach.

coherent radiation is produced. This result is somewhat reinforced by Monte Carlo calculations of the velocity correlation function, which show that the noise spectra does exhibit a peak at the Bloch frequency.[22] This noise spectra is shown in Figure 8.7, for a GaAs-based simulation. In principle, both the ac conductivity and the spectral density of the noise can be measured, although this is difficult for frequencies close to the Bloch frequency as it lies in the far-infrared portion of the spectrum. On the other hand, measurements of the noise emission with, say, an FTIR system can in principle provide evidence of the Bloch oscillations.

The above analysis indicates that it may be difficult to couple directly to the amplitude variations of the Bloch oscillations of the electrons themselves. However, it may be possible to couple to the *phase* of the oscillations. This limitation on amplitude arises due to the fact that the velocity amplitude is limited by the band structure itself, but the phase is not so constrained and may be the preferred coupling scheme. In addition, we note that coupling to the phase has a direct analogy to the flux in the Josephson tunnel junctions and the resulting ac Josephson effect in which steps are produced in the current-voltage characteristics by self-rectification of the ac signal. This result is a direct consequence of the cosinusoidal nature of the energy bands. We illustrate this by treating a single electron confined to the cosinusoidal band described by (8.20), but ignoring the transverse energy for the moment. Under the influence of both a dc electric field F_0 and an ac electric field F_1, the time variation of the momentum wave vector, in the absence of scattering, is given by

$$k(t) = k + \frac{eF_0 t}{\hbar} + \frac{eF_1 t}{\hbar} \sin(\omega t), \tag{8.29}$$

and the corresponding velocity is just

$$v(t) = v_0 \sin\left[\omega_B t + k_0 d + \frac{\omega_B F_1}{\omega F_0} \sin(\omega t)\right], \tag{8.30}$$

FIGURE 8.7. The noise spectrum calculated using an ensemble Monte Carlo approach. The spectrum is calculated from the Fourier transform of the velocity autocorrelation function for the electrons. The peaks occur at the Bloch frequency at each value of the electric field (the parameters are those for GaAs).

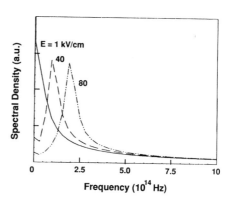

Spectral Density (a.u.)

E = 1 kV/cm

40

80

0   2.5   5.0   7.5   10

Frequency (10^{14} Hz)

where $v_0 = \varepsilon d / \hbar$. This expression can be written as[22]

$$v = v_0 \sum_n J_n(\zeta)\{[\sin(\theta_B + n\theta) + (-1)^n \sin(\theta_B - n\theta)] \cos(\lambda)$$

$$+ [\cos(\theta_B + n\theta) + (-1)^n \sin(\theta_B - n\theta)] \sin(\lambda)\}, \qquad (8.31)$$

where $\zeta = \omega F_1 / \omega_B F_0$, $\theta = \omega t$, $\theta_B = \omega_B t$, $\lambda = k_0 d$, and J_n is the Bessel function of order n. A dc component of the velocity, and hence of the current, occurs when $\omega_B = \pm n\omega$. For a fixed ω, we change ω_B by changing the dc electric field, and we then expect to see resonance effects at the critical multiples of the ac frequency. The occurrence of such structures would be an unambiguous demonstration of the existence of the Bloch oscillations, but we caution that these will occur only in the negative differential conductivity region of the dc characteristics. The resonances above arise because the ac signal is inducing transitions, either emission or absorption, across the individual ladder states of the Stark ladder that coexists with the Bloch oscillations. The presence of superlattice implies that the Stark ladder is not a real ladder but a virtual one, and the ac effect is coupled to tunneling of the electron from one well to the next. In this case, the electron tunnels through to the adjacent well, emits a photon corresponding to the ac frequency, and drops in energy to a new state described by $E - eF_0 d(\omega - \omega_B)$. The absorption response corresponds to first absorbing the photon and then tunneling to the adjacent well. There is a preferred direction for each of these tunneling processes, as determined by the symmetry breaking of the applied electric field. It is evident from this that the resonances are very closely tied to the concept of sequential resonant tunneling in superlattices themselves.[22,23] The importance of the virtual nature of the Stark ladder was only recently clarified.[24] At the highest electric fields, the band nature is completely destroyed, leaving just the virtual Stark ladder of states, one state in each quantum well of the superlattice, which can be probed by the optical limit of the above equations. In this case, the Stark ladder has been observed in such measurements.[25]

REFERENCES

1. L. Esaki and R. Tsu, *IBM J. Res. Develop.* **14**, 61 (1970).
2. R. T. Bate, *Bull. Am. Phys. Soc.* **22**, 407 (1977).
3. D. K. Ferry, *Phys. Stat. Solids* (*b*) **106**, 63 (1981).
4. A. C. Warren, D. A. Antoniadis, H. I. Smith, and J. Melngailis, in: *IEEE Electron Devices Meeting Technical Digest*, p. 866, IEEE Press, New York (1984). See also the work of M. Wassermeier, H. Pohlmann, and J. P. Kotthaus, in: *18th Int. Conf. on the Physics of Semiconductors* (O. Engström, ed.), World Scientific Press, Singapore, Vol. 1, p. 441 (1987). This latter work clearly shows the presence of the minibands through measurements of the density of states in a high magnetic field.
5. G. J. Iafrate, D. K. Ferry, and R. K. Reich, *Surf. Sci.* **113**, 485 (1982).
6. G. Bernstein and D. K. Ferry, *Superlatt. Microstruct.* **2**, 373 (1986); U.S. Patent 4,872,038.
7. A. B. Pippard, *Phil. Trans. Roy. Soc. London* **A68**, 317 (1964).
8. M. Ya. Azbel, *J. Exp. Theor. Phys.* **46**, 929 (1964) [transl., *Sov. Phys. JETP* **19**, 634 (1964)].
9. A. Rauh, G. H. Wannier, and G. Obermair, *Phys. Stat. Solids* (*b*) **63**, 215 (1974).

10. D. Weiss, K. von Klitzing, K. Ploog, and G. Weimann, in: *High Magnetic Fields in Semiconductor Physics II* (G. Landwehr, ed.), pp. 357–365, Springer-Verlag, Heidelberg (1988).

11. R. R. Gerhardts, D. Weiss, and K. von Klitzing, *Phys. Rev. Lett.* **62**, 1173 (1989).

12. R. W. Winkler, J. P. Kotthaus, and K. Ploog, *Phys. Rev. Lett.* **62**, 1177 (1989).

13. H. J. Schellnhuber, and G. M. Obermair, *Phys. Rev. Lett.* **45**, 276 (1980).

14. P. Beeton, E. S. Alves, M. Hennini, L. Eaves, P. C. Main, O. H. Hughes, G. A. Toombs, S. P. Beaumont, and C. D. W. Wilkinson, *Proc. Symp. on New Phenomena in Mesoscopic Systems,* Kona, Hawaii, Jpn. Soc. Promotion Sci., unpublished (1989); E. Paris, J. Ma, A. M. Kriman, D. K. Ferry, and E. Barbier, *J. Phys. Cond. Matter,* in press.

15. M. Ya. Azbel, *J. Exp. Theor. Phys.* **44**, 980 (1963) [transl., *Sov. Phys. JETP* **17**, 665 (1963)].

16. P. G. Harper, *Proc. Phys. Soc., London,* **A68**, 874 (1955).

17. D. R. Hofstadter, *Phys. Rev. B* **14**, 2239 (1976).

18. D. K. Ferry, G. Bernstein, R. Puechner, J. Ma, A. M. Kriman, R. Mezenner, W. P. Liu, G. N. Maracas, and R. Chamberlin, in: *High Magnetic Fields in Semiconductor Physics II* (G. Landwehr, ed.), pp. 344–352, Springer-Verlag, Heidelberg (1988).

19. J. Zak, *Phys. Rev.* **134**, A1602, A1607 (1964).

20. P. A. Lebwohl and R. Tsu, *J. Appl. Phys.* **41**, 2664 (1970).

21. R. K. Reich, R. O. Grondin, and D. K. Ferry, *Phys. Rev. B* **27**, 3483 (1983).

22. R. O. Grondin, W. Porod, J. Ho, D. K. Ferry, and G. J. Iafrate, *Superlatt. Microstruct.* **1**, 183 (1985).

23. R. F. Kazarinov and R. A. Suris, *Sov. Phys. Semicond.* **5**, 707 (1971).

24. J. B. Krieger and G. J. Iafrate, *Phys. Rev. B* **33**, 5494 (1986).

25. E. E. Mendez, F. Agullo-Rueda, and J. M. Hong, *Phys. Rev. Lett.* **60**, 2426 (1988).

9

Quantum Transport in Small Structures

Nearly all of the transport that has been dealt with in treating devices has its conceptual basis in the Boltzmann transport equation. It is perhaps worthwhile at this point to actually summarize the various approximations and limitations that impact this equation. In general, transport processes are viewed on a coarse-grained time scale $t \gg \tau_c$, τ_m, etc., so that many completed, independent collisions occur in the passage of a carrier through the system. In addition, each collision is treated as an irreversible process which is completed prior to the next one, and is (a) local in space (collision spheres do not overlap in space), (b) local in time (instantaneous collisions), (c) independent of any driving fields or other scattering processes (no multiple scattering effects and no field acceleration during the collision), and (d) at low frequency.

Each of the above assumptions provides a factor that is neglected in the Boltzmann transport equation, and these neglected factors provide warning signs for the onset of failure of this semiclassical approach. These factors, which can be expected to arise in nanometer-scale devices, may be summarized as follows: (1) nonlocality of the scattering processes, both in space and time, (2) strong driving forces, (3) strong scattering, (4) dense systems, (5) small systems, and (6) nonclassical influence of the driving fields. Each collision is actually extended in both space and time. If the spatial scale or temporal scale (given by the wave vector $\mathbf{q}$ and frequency ω) approach the microscopic scale of the scattering interaction, then one collision cannot be completed before the onset of another one. Normally, we use the Boltzmann equation only when $\omega\tau < 1$ and $qL < 1$, so that many independent scattering events can occur within one cycle of the driving forces. If these assumptions are violated, appreciable quantum effects can occur due to the incomplete scattering. Notable among these are the loss of irreversibility and the onset of multiple scattering—the scattering from two or more centers simultaneously. This also occurs with strong driving forces, where the carriers can be accelerated during the collision itself, an *intracollisional field effect.* Here, the field and scattering terms actually interfere with one another.

If strong scattering exists in the system, it is no longer acceptable to treat the scattering as a weak perturbation, and polaronic effects can occur. In dense systems, many-body effects must be considered and the Markovian approximations of the Boltzmann equation are no longer acceptable. Then, correlation between the electrons is a significant effect. In small systems, size quantization (discussed previously) and surface-limited transport become important. Eventually, spatial variations on the scale of the de Broglie wavelength can be expected to occur in the conduction channels (and not merely perpendicular to them), as already occurs in transport normal to the layers of a superlattice. Nonclassical effects of the driving fields are evident in Landau quantization in high magnetic fields and in Stark ladder effects in high electric fields; in either case the Boltzmann equation breaks down.

When the Boltzmann equation breaks down, we must turn to more fundamental transport equations. Although the ensemble Monte Carlo transport technique overcomes some of the non-Markovian limits of the Boltzmann equation, it is still based in the semiclassical world where most of the above limitations still apply. We must turn to the quantum transport formalism, in which all of the above processes can be included in a proper fashion (if the equations can then be solved).

In this chapter, we want to discuss the general realm of the quantum transport problem, although the treatment will not be extensive or complete. We first discuss the Fermi golden rule, as it is the principle method by which scattering rates are calculated for the semiclassical treatment generally used to determine the relaxation times for Boltzmann transport and for Monte Carlo techniques. We show that strong scattering and the presence of an electric field drastically modify the normal concepts of energy conservation. Then, transport through thin oxides is discussed as an example of the first corrections that arise. We then turn to resonant tunneling structures and quantum fluctuation (interference) effects that have been observed in devices. This is used as a basis to develop a set of moment equations similar to those developed earlier for use in device modeling. In the final two sections, we treat an irreversible equation of motion for the density matrix, which has similarities to the Boltzmann equation, and for its cousin, the Wigner function.

9.1. THE GENERAL PROBLEM

Much of the rationale for beginning to consider quantum transport in very small semiconductor devices can be understood merely by reconsidering first-order, time-dependent perturbation theory that gives us the Fermi golden rule used in the calculation of scattering rates by various electron–phonon interactions. From this, we can examine the various approximations and their breakdown that leads us to the necessity for a more complete treatment.

9.1.1. Fermi Golden Rule

Generally, we must begin with the basic formulation of quantum mechanics. Here, we use the Hamiltonian form, in which the basic equation is just given by

$$H\Psi = E\Psi = i\hbar \frac{\partial \Psi}{\partial t}, \tag{9.1}$$

where H is the total Hamiltonian, E is the total energy, and $\Psi(x)$ is the wave function of the system. We normally write the Hamiltonian as

$$H = H_0 + H_1, \tag{9.2}$$

where H_0 is that part of the Hamiltonian that is continuously applied, while H_1 is the perturbing part of the Hamiltonian. We will later include an electric field by incorporating it within H_0. We proceed by selecting a basis set of functions $\{\psi_n\}$, for which

$$H_0\psi_n = E_n\psi_n \tag{9.3}$$

and

$$\Psi(x, t) = \sum_n c_n \exp\left(\frac{-iE_nt}{\hbar}\right)\psi_n(x). \tag{9.4}$$

Here, the time-varying coefficient c_n is

$$c_n = (\psi_n, \Psi(t_0)) \exp\left(\frac{iE_nt_0}{\hbar}\right). \tag{9.5}$$

We now need to solve the problem in which the perturbing potential is added. Here, we consider that the perturbing potential produces a slow change in the occupancy of the various basis states, and that this is reflected by a time varying c_n. Using (9.4) in (9.1), we can solve for the time variation of these factors as

$$i\hbar \frac{\partial c_n(t)}{\partial t} = \sum_k V_{nk}c_k \exp(i\omega_{nk}t), \tag{9.6}$$

where

$$V_{nk} = \int dx\, \psi_n^* H_1 \psi_k = \langle n|V|k \rangle \tag{9.7}$$

and

$$\omega_{nk} = \frac{E_n - E_k}{\hbar}. \tag{9.8}$$

In general, we assume that at $t = 0$ only a single state c_s is occupied, while all the others are completely empty. Thus,

$$c_s(0) = 1 \quad \text{and} \quad c_k(0) = 0 \qquad \text{for } k \neq s. \tag{9.9}$$

Moreover, we assume that the scattering out of state s is quite small, so that $c_s(t) = 1$ for all time. This is one of the major assumptions that is violated in heavy scattering regimes, and we will return in the next subsection to remove it. Using the initial conditions given in (9.9), we can now rewrite (9.6) as

$$i\hbar \frac{\partial c_k}{\partial t} = V_{ks} \exp(i\omega_{ks}t), \tag{9.6a}$$

and

$$c_k(t) = -\frac{i}{\hbar} \int_0^t V_{ks} \exp(i\omega_{ks}t') \, dt' = \frac{V_{ks}}{\hbar \omega_{ks}} [1 - \exp(i\omega_{ks}t)]. \tag{9.10}$$

The probability that the state k has some occupancy is then just

$$p_k(t) = |c_k(t)|^2 = 2 \frac{|V_{ks}|^2}{\hbar^2 \omega_{ks}^2} [1 - \cos(\omega_{ks}t)], \tag{9.11}$$

and the transition probability is

$$w_k = \frac{d}{dt} p_k(t) = \frac{2}{\hbar^2} |V_{ks}|^2 \frac{\sin(\omega_{ks}t)}{\omega_{ks}}. \tag{9.12}$$

The second major approximation now appears, and that is that we take the limit as $t \to \infty$, so that the last term is zero (it actually oscillates very rapidly, having a zero average value) unless $\omega_{ks} = 0$. Thus, we can finally write the Fermi golden rule transition probability as

$$w_k = \frac{2\pi}{\hbar^2} |V_{ks}|^2 \delta(\omega_{ks}) = \frac{2\pi}{\hbar} |V_{ks}|^2 \delta(E_k - E_s). \tag{9.13}$$

Thus, if we are dealing with time scales of the order of $1/\omega_{ks}$, we must be concerned that the energy-conserving delta function is not fully formed, and that the energy (or momentum) transferred is not exactly the phonon value, or that the collision is not completed. (We note here that the energies $E_{k,s}$ involve the complete energy of a basis state, hence involve both the electron and the phonon energies.) In GaAs, the LO phonon energy is 36 meV, and corresponds to a radian frequency of 5.46×10^{13} Hz and this requires $t \gg 1.8 \times 10^{-14}$ s. On the picosecond time scale, this would seem to be satisfied. On the other hand, velocity overshoot in Si occurs on the 0.1-ps time scale. Since the phonon energies are not drastically different (they are actually lower in energy), we should be concerned about the validity of (9.13) in calculations of the scattering rates. The extreme case is that of SiO_2 which, with its low mobility of 20 cm^2/V-s and effective mass of $0.5m_0$, has an average relaxation time of 5.7×10^{-15} s. This is to be compared with $1/\omega$ of 4.3×10^{-15} and 10.3×10^{-15} s for the 150-meV and 63-meV LO phonons, respectively. In this latter case, (9.13) certainly cannot be used, since almost certainly both critical approximations have been violated. We return to this in the next section, where we can evaluate the impact of the failure of (9.13).

9.1.2. Initial-State Decay

We now want to repeat the derivation of the transition probability, but relax the first approximation; that is, we will no longer assume that $c_s(t) = 1$. Rather, the scattering out of state s gradually reduces the probability $|c_s|^2$ for occupancy, and this effect will in turn appear in the transition rates. Thus, we replace (9.6a) with

$$i\hbar \frac{dc_k(t)}{dt} = V_{ks}c_s(t) \exp(i\omega_{ks}t). \tag{9.14}$$

This is easily solved to yield

$$c_k(t) = -\frac{i}{\hbar} V_{ks} \int_0^t c_s(t') \exp(i\omega_{ks}t') \, dt'. \tag{9.15}$$

Throughout this section, we assume that $V_{ss} = V_{kk} = 0$, although the treatment is easily extended to include these terms. We must now use (9.15) in the equivalent expression for c_s, as

$$i\hbar \frac{dc_s}{dt} = \sum_{k \neq s} V_{sk}c_k(t) \exp(i\omega_{sk}t)$$

$$= \sum_{k \neq s} |V_{ks}|^2 \int_0^t c_s(t') \exp[i\omega_{ks}(t' - t)] \, dt'. \tag{9.16}$$

To proceed further, we Laplace-transform (9.16) to obtain $C_s(z)$, where z is the transform variable, as

$$C(z) = \int_0^\infty c(t) \exp(-zt)\, dt.$$

Then

$$zC_s(z) - 1 = -\frac{1}{\hbar^2} \sum_{k \neq s} |V_{ks}|^2 \frac{C_s(z)}{z + i\omega_{ks}}. \tag{9.17}$$

Before proceeding, we want to simplify the right-hand side of (9.17). We will eventually retransform the result, but in this term we are still interested in the longer time results. Thus, we are only going to relax the first approximation, that of very weak scattering. In the (relatively) long-time limit, we consider that z can be replaced by a factor η, which is small in comparison with ω_{ks}. We then use the expansion

$$\lim_{\eta \to 0}(\omega - i\eta)^{-1} = \omega^{-1} + i\delta(\omega).$$

The right-hand side is then

$$C_s(z)i\Sigma = C_s(z)\left\{\frac{i}{\hbar^2} \sum_{k \neq s} \frac{|V_{ks}|^2}{\omega_{ks}} - \frac{\pi}{\hbar^2} \sum_{k \neq s} |V_{ks}|^2 \delta(\omega_{ks})\right\}. \tag{9.18}$$

Here, we have defined Σ as the *self-energy* of the state s. Its real part describes a shift ΔE in the energy E_s, while the imaginary part $(= \Gamma)$ describes the decay of the state s due to the total scattering out to states k (we note that the units of Σ are s^{-1}, so we must be careful to incorporate the factor of h in its definition in terms of energies). Thus,

$$C_s(z) = \frac{1}{z - i\Sigma},$$

or

$$c_s(t) = \exp\left[-\frac{\Gamma}{\hbar}t - \frac{i}{\hbar}\Delta E t\right]. \tag{9.19}$$

We can now use (9.19) in (9.15) to find the rate of change of the population in c_k:

$$c_k(t) = -\frac{i}{\hbar} \int_0^t V_{ks} \exp\left[i\left(\omega_{ks} - \frac{\Delta E}{\hbar}\right)t' - \frac{\Gamma t'}{\hbar}\right] dt' \tag{9.20}$$

and

$$|c_k|^2 = \frac{1}{\hbar^2}|V_{ks}|^2 \left| \int_0^t \exp\left[\frac{i}{\hbar}(E_k - E_s - \Delta E)t' - \frac{\Gamma}{\hbar}t'\right] dt' \right|^2. \qquad (9.21)$$

From (9.21), we can now define the transition probability as

$$w_k = \frac{d}{dt}|c_k|^2 = \frac{2\pi}{\hbar}|V_{ks}|^2 g(E_{ks}), \qquad (9.22)$$

where $g(x)$ is a spectral function that constrains the argument x. In the previous subsection, $g(x)$ was found to be a delta function that ensured energy conservation. In our case, we take the limit $t \to \infty$ after taking the derivative, and

$$\begin{aligned} g(E_{ks}) &= \frac{1}{2\pi\hbar}\frac{d}{dt}\left| \int_0^t \exp\left[i\left(\omega_{ks} - \frac{\Delta E}{\hbar}\right)t' - \left(\frac{\Gamma}{\hbar}\right)t'\right] dt' \right|^2 \\ &= \frac{1}{\pi\hbar}\operatorname{Re}\int_0^t \exp\left[i\left(\omega_{ks} - \frac{\Delta E}{\hbar}\right)t' - \left(\frac{\Gamma}{\hbar}\right)t'\right] dt' \\ &= \frac{1}{\pi}\frac{\Gamma}{\Gamma^2 + (\hbar\omega_{ks} - \Delta E)^2}. \end{aligned} \qquad (9.23)$$

In this last equation, we have taken one form of the derivative, although others can be used. This particular one happens to give the result, in the second line, that agrees with the more exact Green's function approach for the joint spectral density function. We note that the delta function from the earlier treatment has been broadened into a Lorentzian line shape. While the transition probability has a peak rate at $E_k = E_s + \Delta E$, there is no real requirement that this energy conservation be satisfied during the process. The reason for this is that the state energies are really defined at $t = 0$ and are changing slightly during the process (ΔE). Thus, for transitions which occur at $t > 0$, the energies are different and the Lorentzian line shape reflects this change.

So far, we have relaxed only the one approximation and allowed the initial state to decay through state interaction. We could have relaxed the second approximation and treated very short times in evaluating (9.17). To do so, however, we would have had to really limit ourselves to just two states for simplicity. The response is much more complicated, with very rapid amplitude oscillations as the transition rate builds up. This is beyond our current treatment, however, and we forego further discussion of this problem.

9.1.3. Field Interactions

We now want to include the electric field to show that it further distorts the joint spectral density, primarily due to an intracollisional field effect. This latter arises because we are taking the collision to have a real, nonzero duration, which replaces the energy-conserving delta function by the Lorentzian line. Because the collision now takes a small amount of time to occur, it is possible for the electron to be accelerated during the collision itself, rather than just between the collisions.[1,2] To achieve this, we will incorporate the electric field within the vector potential notation. This choice is merely the selection of a proper gauge with which to pursue the interaction. By the choice of placing the field within the vector potential, we can retain our earlier formalism by just modifying the term H_0. The total Hamiltonian is then

$$H = \frac{(\mathbf{p} + e\mathbf{F}t)^2}{2m} + H_1.$$

(9.24)

Here, we have assumed that we are in the quasi-particle approximation of nearly free electrons dressed with an effective mass. We note that at any instant of time

$$p(t) = p(t_0) - \int_{t_0}^{t} eF\, dt'.$$

(9.25)

More interestingly, we find that

$$\omega_{ks}(t) = \omega_{ks}(t_0) + \frac{e\mathbf{F}}{m\hbar} \cdot [\mathbf{p}_k(t_0) - \mathbf{p}_s(t_0)](t - t_0),$$

(9.26)

where the values at t_0 correspond to the onset of the collision process. For simplicity, we take $t_0 = 0$ in the following, just as was done in the previous two subsections. We also denote the bracketed difference in momenta as the scattering wave vector $\mathbf{q}$. We can then immediately utilize the results in (9.21) to write

$$|c_k|^2 = \frac{|V_{ks}|^2}{\hbar^2} \left| \int_0^t \exp\left\{ -\frac{\Gamma t'}{\hbar} + i\left[\left(\omega_{ks}^0 - \frac{\Delta E}{\hbar} \right) t' + \frac{et'^2}{m} \mathbf{F} \cdot \mathbf{q} \right] \right\} dt' \right|^2,$$

(9.27)

where we now have to note that the matrix elements are computed with different wave functions. The wave functions in a constant applied electric field are Airy functions rather than plane waves. However, it has been found that this produces only a small change in the matrix elements that breaks their symmetry properties by admixing parts of all plane waves in the system. For our purposes, the change is quite small except at extraordinarily large values of the applied field. Corresponding to (9.23), we now find the new spectral density is

$$g(E_{ks}) = \frac{1}{\pi\hbar} \text{Re} \int_0^{\infty} \exp\left\{ -\frac{\Gamma}{\hbar} t' + i[\Omega_0 t' + \Omega_1 t'^2] \right\} dt'.$$

(9.28)

Again, we have taken the limit $t \to \infty$ after performing the derivative and used a form that agrees with that obtained by Green's function techniques. We have also dropped the vector notation, as its usage is now obvious, and defined the reduced quantities $\Omega_0 = \omega_{ks}^0 - \Delta E/\hbar$ and $\Omega_1 = eFq/m$. The general solution now involves a combination of complicated Fresnel integrals. If the field is too large, then the field effectively wipes out the scattering process itself. For fields which are not too large, the integral can be evaluated by a long, asymptotic approximation involving a contour integration, with the contour taken along the path of steepest descent in a saddle-point evaluation. This yields the approximation

$$g(\Omega_0, \Omega_1) = \frac{1}{\pi} \frac{\Gamma(\hbar^2\Omega_0^2 + \Gamma^2) + \hbar^3\Omega_0\Omega_1}{(\Omega_0^2\hbar^2 + \Gamma^2)^2 + \hbar^3\Omega_1(\hbar\Omega_1 + \Omega_0\Gamma)}. \tag{9.29}$$

In the absence of the field term ($\Omega_1 = 0$), (9.29) reduces exactly to the Lorentzian line of (9.23). The peak does not occur at $\Omega_0 = 0$, as is the case for (9.23), but rather occurs at

$$\Omega_{0,\max} \sim \frac{\hbar\Omega_1}{2\Gamma}, \tag{9.30}$$

which has a sign corresponding to the vector product of the field and the momentum change. The form of (9.30) also clearly shows that increased scattering lowers the effective field shift, so that the intracollisional field effect is a delicate interplay between field acceleration and a scattering duration, depending on the actual scattering rates. While there are other subtle, and small, effects of the field, such as the change in the matrix elements, the major effect is the shift in the resonance energy corresponding to the peak of the Lorentzian line shape of the joint spectral density.

The role of the energy shift in (9.30) depends on whether the particle is gaining or losing energy to the phonon. If the electron is scattered against the field by the emission of a phonon, the field acts to lower the amount of energy lost to the phonon by reaccelerating the electron during the collision. On the other hand, if the momentum change is along the field direction, so that the particle is being accelerated in the same direction as the phonon wave vector, then the energy lost to the phonon is increased. For phonon absorption, the role is played out in a similar fashion. If the acceleration is in the same direction as the phonon momentum, the energy gained from the phonon is increased by the field, and conversely. By the energy gained or lost, we are referring to this quantity as the electron sees it. The energy change in the phonon is still just the mode energy, but the electron sees a modified energy by virtue of the field acceleration during the collision.

From the discussion in this section, we can summarize a number of the reasons for more properly treating the quantum transport in a semiconductor that go beyond just the quantizing effects of small structures. To properly incorporate the presence of dense carrier plasmas, rapid scattering processes, and the joint presence of both the field and the scatterers, we need to greatly modify the normal semiclassical treatment of transport that is commonly used. A proper quantum transport treatment will accomplish these goals, but the development of such treatments for the fully nonequilibrium, open system that devices represent is still in its infancy.

9.2. TRANSPORT IN SILICON DIOXIDE

Silicon dioxide is the standard insulator in Si MOS technology, both for the gate oxide and as a field oxide. Thus, it is quite important to understand its properties. For example, submicron VLSI typically employs gate oxides in the 20 nm or less thickness range, and this results in an electric field on the order of MV/cm across the oxide for modest gate voltages. Under these enormous fields, the electron distribution function may become unstable if the energy which is gained from the field can no longer be efficiently relaxed to the lattice. This situation is referred to as "runaway." Clearly, it is of great technological interest to understand under which conditions runaway and, as a consequence, breakdown can be expected to occur.

Until very recently, it was thought that the dominant scattering mechanism for energy loss to the lattice in SiO_2 was primarily due to the emission of longitudinal optical phonons, of which there are two primary modes at 0.063 and 0.153 eV. This emission was thought to be efficient enough to prevent runaway for fields less than about 7 MV/cm in thick oxides and greater in thinner oxides. In this picture stable distributions exist for the electrons up to the runaway field, beyond which the electrons move to high energy and are contained by impact ionization. The stable distribution would then have an average energy of only about 0.15 eV, the higher optical phonon energy.

Recently, however, DiMaria and co-workers[3-5] observed experimentally that the electronic distribution is stable at a much higher average energy, typically a few electron volts, and that this stability exists to much higher fields than earlier thought. In fact, they find no evidence for impact ionization well beyond 10 MV/cm. These new results cannot be explained in terms of the earlier model. More recent Monte Carlo studies, however, have shown that the electronic distribution is stabilized by the polar phonons only up to about 1-2 MV/cm, and runaway already occurs for fields in the 2-3 MV/cm. The stabilization of the electronic distribution at the higher energy level must then be due to the onset of a previously unsuspected scattering process that becomes large at high energies. Fischetti[6] has suggested that this extra scattering mechanism is nonpolar acoustic umklapp processes, which become quite efficient at these high energies. In these processes, electrons which approach a significant fraction of the zone boundary momentum can scatter from a phonon which transfers the momentum to the second zone, creating a type of Bragg reflection for the electron. By using a Monte Carlo model, he was able to demonstrate that this process is able to stabilize the electrons near the experimentally observed energies over a wide range of fields. However, in his treatment, he assumed the electronic structure was characterized by the use of a simple parabolic band for the electrons (as is usually the case). On the other hand, theoretical studies of the electronic structure of SiO_2 suggest that this structure is quite complicated.[7]

Porod and Ferry took a different approach,[8] in which the existence of satellite valleys of the conduction band was taken into account. The increased scattering strength arising from the onset of intervalley scattering also results in a stable distribution function and average energy quite similar to that found experimentally over a wide range of fields.

Both of these approaches essentially formulate a new scattering process which incorporates a high density of final states, thus producing a high scattering rate. In this regard, their effect is similar in that a high q phonon is involved in scattering to this large density of states. The resulting effect in the model Monte Carlo calculations is thus quite similar, and it is probably not possible to pick which of the two mechanisms (or both?) is truly responsible for stabilizing the distribution function. What is important here, however, is that the scattering rates become quite high for the energies of interest here. Thus, the quantum mechanical effects cannot be ignored in the calculations.

The main results of Section 9.1 are that the energy-conserving delta function in Fermi's golden rule has to be replaced by a distribution of final states, roughly distributed in a Lorentzian fashion. The width of this distribution of final states is determined both by the natural linewidth of the interacting electron system, the collisional broadening, and by the acceleration of carriers during the finite collision duration, the intracollisional field effect. These results have been established both by the simple approach of Section 9.1 and by more complete studies by a variety of analytical techniques.

Porod and Ferry[9] have included the low-order quantum corrections into a Monte Carlo calculation by incorporating the results of Section 9.1. In the Monte Carlo simulation, the final energy is chosen according to the Lorentzian distribution (using the Monte Carlo selection technique). Once the energy is known, the standard techniques select the final momentum vector with the proper distribution, again according to standard techniques. This scheme no longer conserves energy for an individual scattering process. However, for an ensemble, the total energy is conserved because of cancellation of positive and negative contributions.

The result of collisional broadening for the average energy is shown in Figure 9.1 (curve a). The broken curve is that obtained for the classical Monte Carlo approach, which obviously fits the average energy data much better. The net effect of the collisional broadening is a marked increase in the average energy for electrons below 3 MV/cm, and a decrease above this field. Curve b shows the added effect of the intracollisional field effect, which was implemented by an additional, field-dependent broadening of the final state and a shift of the resonance. Even in this case, the results are affected only slightly.

A different approach was followed by Fischetti et al.[10] In this latter case, the scattering rates for the high fields were calculated as the imaginary part of the complex-valued self-energy usually evaluated by a Green's function analysis

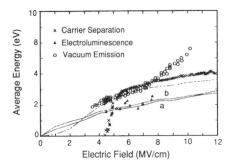

FIGURE 9.1. Calculated average energies are compared with experimental data (Refs. 3–5). A value of 2.5 eV has been used for the offset between the central and satellite valleys. The broken curve is for the classical Monte Carlo model, the solid lines (a) incorporates just collisional broadening, and (b) incorporates the additional intracollisional field effect.

in quantum transport. In order to make these self-consistent calculations manageable, only one effective scattering mechanism was utilized for the high-energy range. It was found that the resulting scattering rates do not significantly differ from the classical ones, and that the results were essentially the same as those shown in Figure 9.1, except that the crossover point in energy moved slightly higher to 3.5 MV/cm. In both cases, the results show that the inclusion of the quantum mechanical effects do not change the result that the distributions are stable. However, the general variation of the energy differs from that found experimentally (and found from the semiclassical approach as well). It should, however, be pointed out that the Monte Carlo approach to transport modeling relies on a basically semiclassical path integral, and the quantum approaches discussed here are only first steps at getting the proper modification to handle quantum transport. There is still much to be done, and this work is ongoing.

9.3. RESONANT TUNNELING

One of the advantages of constrained-dimensionality devices is the ability of applying textbook quantum mechanics to their understanding and still achieving a reasonable degree of success in this task. Perhaps the classic example of this is tunneling, particularly in the case of the resonant tunneling device. The concept of tunnel diodes goes back several decades, and is implemented in very heavily doped *p-n* junction diodes. In this case the tunneling is through the forbidden-gap region and involves electrons making transitions from the conduction band to the valence band, and vice versa. In the present context, however, we are concerned with fabricating tunnel barriers by band-gap engineering. Thus, we can separate two GaAs regions by a thin barrier region of GaAlAs, and the tunnel barrier is formed by the conduction- (and valence-) band discontinuity. In this sense, the barrier formed is a textbook example, and the results on tunneling current can be calculated in a straightforward manner.

Interest arises in the ability to combine two or more barriers, and to sandwich thin GaAs regions between the barriers. Here, the small-gap GaAs layers are actually quantum wells, weakly coupled to one another. This now sets up the concept of resonant tunneling, in which the tunneling probability is quite low except for the resonant energies of the quantum levels in the individual wells. As we will see, this opens the possibility of negative differential conductivity in the device characteristics. In fact, the advent of MBE (and MOCVD) allows one to engineer an entire range of tunneling and quantum structures within more normal semiconductor devices in each case producing novel or enhanced performance characteristics.

9.3.1. Tunneling Probabilities

The tunneling probability has been calculated in many introductory texts for simple cases, either as rectangular barriers or in the WKB approximation, often called the quasi-classical approximation. In actual devices, the barrier is quite often distorted by the applied bias. In this section, we want to introduce the tunneling probability, and calculate it in a number of cases. We will begin with a simple single barrier before moving to the more complicated resonant tunneling structure. Although the double-barrier structure is quite old in concept, it has become a very popular device recently due to measurements showing definite peaks in conductivity (and negative differential conductance) and very high frequency behavior.[11] The key factor in this recent success appears to be much better material quality. However, the investigators have spent time in optimizing the devices for peak performance. The bias upsets any prior symmetry in the two tunneling barriers and changes the conditions for resonance. Consequently, optimization must be based on exact knowledge of the germane peak and the overall structure.

The general treatment for rectangular barriers to be followed is that of most textbooks, although we modify this approach through the use of a transfer matrix formulation. Consider, for example, the rectangular barrier of Figure 9.2. The points A and D represent sites just outside the barrier in which the wave function is allowed to propagate. Points B and C, on the other hand, are points just inside the barrier in which the wave function is strongly attenuated and nonpropagating. The transfer-matrix approach sets down a matrix of coefficients that describes the change in the wave functions at the barrier edge and through the barrier (or the propagating region). We differ from some earlier treatments here, by noting

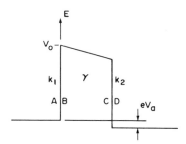

FIGURE 9.2. The rectangular single-barrier tunneling structure.

that it is possible to have different barrier heights on the two sides of the barrier, as shown in the figure. For this reason, we will write the equations entirely in terms of the wave vectors themselves rather than in terms of the potentials. We shall assume that the top row of the matrices refers to the forward wave, propagating in the positive x-direction, and the bottom row refers to the backward wave, propagating in the negative x-direction.

To begin, we must write the wave at C in terms of the wave existing at D. Primarily the transfer matrix at this point just expresses the phase shifts encountered by the waves at the boundary. This matrix for $C(D)$ we will call $2\mathbf{M}_0$. It is given by

$$\left[\begin{matrix} \left(1 - \dfrac{ik_2}{\gamma}\right) & \left(1 + \dfrac{ik_2}{\gamma}\right) \\ \left(1 + \dfrac{ik_2}{\gamma}\right) & \left(1 - \dfrac{ik_2}{\gamma}\right) \end{matrix}\right], \tag{9.31}$$

where $k_2 = (2mE/\hbar^2)^{1/2}$ is the propagating wave vector to the right of the barrier and $\gamma = [2m(V_0 - E)/\hbar^2]^{1/2}$ is the attenuation constant within the barrier. Within the barrier, we need to now write the wave functions at B in terms of those at C. We call this propagation matrix $\mathbf{M}_\gamma$, and it is

$$\left[\begin{matrix} \exp(\gamma a) & 0 \\ 0 & \exp(-\gamma a) \end{matrix}\right] \tag{9.32}$$

for a constant γ. We remark at this point that (9.32) is an approximation if γ varies with position in the barrier even if γa is replaced by an averaging integral. Rather, the proper treatment is to integrate the incremental exponential operators, which is not the same as the exponential of the integrated operator. In this case, (9.32) is not strictly diagonal anymore. However, we will continue to use the form (9.32), as it is a reasonable approximation that actually improves upon the usual WKB approximation.

Finally, we need a matching matrix for A in terms of B, given by $2\mathbf{M}_i$ as

$$\left[\begin{matrix} \left(1 + \dfrac{i\gamma}{k_1}\right) & \left(1 - \dfrac{i\gamma}{k_1}\right) \\ \left(1 - \dfrac{i\gamma}{k_1}\right) & \left(1 + \dfrac{i\gamma}{k_1}\right) \end{matrix}\right]. \tag{9.33}$$

The total transfer matrix is now $\mathbf{M}_i\mathbf{M}_\gamma\mathbf{M}_0 = \mathbf{M}$. If we assume that there is no backward wave at D, then the tunneling coefficient for the wave function is just $1/M_{11}$, and the probability current is then $1/|M_{11}|^2$. The term M_{11} is just

$$\frac{1}{2}\left[\left(1 + \frac{k_2}{k_1}\right)\cosh(\gamma a) + 2i\left\{\frac{\gamma}{k_1} - \frac{k_2}{\gamma}\right\}\sinh(\gamma a)\right].$$

This leads to the tunneling probability

$$D = \frac{4k_1 k_2}{(k_1 + k_2)^2 + W(k_1, k_2, \gamma) \sinh^2(\gamma a)},$$ (9.34a)

where

$$W = \frac{(\gamma^2 + k_1^2)(\gamma^2 + k_2^2)}{\gamma^2}.$$ (9.34b)

One thing we note immediately is that the tunneling probability is symmetrical in k_1 and k_2. This symmetry result is independent of any distortion of the barrier by the field, and is simply a statement that the tunneling barrier is reciprocal and, hence, a linear "resistor." Generally, we may approximate the tunneling probability by using the fact that γa is a large number. Hence,

$$D \simeq \left[\frac{16 k_1 k_2}{W(k_1, k_2, \gamma)} \right] \exp(-2\gamma a).$$ (9.35)

Equation (9.35) provides us with the basis for using a WKB-type formula for a general barrier, in which we replace the term γa by

$$\gamma a \to \int_0^a \gamma(x)\, dx = \left(\frac{2m}{\hbar^2} \right)^{1/2} \left(\frac{2}{3eF} \right) \{ (E_B - E)^{3/2} - (E_B - eFa - E)^{3/2} \},$$ (9.36)

where we have assumed that the energy of the barrier top varies linearly with an applied electric field F. Here, E_B is the barrier height (in absolute energy) at $x = 0$, and E is the energy of the tunneling particle.

What we note from the exponential behavior is that the tunneling carriers will be those of highest energy on either side of the barrier, and the current calculated in the Landauer formula (discussed later) is a sum over these carriers. If the field is small, $F \ll (E_B - E)/ea$, expansion of (5.18) just yields the result γa, and the role of the field in the barrier can generally be ignored. For $E_B - E = 0.3$ eV and $a = 3$ nm, we require $F \ll 10^6$ V/cm. While this seems like a substantial field, we note that this is just the field required to lower the barrier on the right-hand side to the energy level E. In fact, these energies and fields can be expected to be routinely encountered in tunneling problems, and the expansion discussed above cannot be utilized. This means that the field induces a significant change in the quantity in (9.36), which is reflected exponentially in the tunneling probability and, hence, in the current. The second term in an expansion of (9.36) is linear in the field, which means that the current will rise exponentially with the field, as $D = D_0 \exp(F/F_0)$, where D_0 is the tunneling probability in the absence of the field and F_0 is a normalization field obtained by the numerical factors in (9.36).

The resonant tunneling structure, for general barriers, is shown in Figure 9.3. Here, a quantum well is sandwiched between two tunneling barriers. In its most general, and current, implementation, the tunneling barriers are GaAlAs, whose composition is adjusted to give a barrier height of about 0.3 V. The quantum well and the two end, or cladding, layers are then usually pure GaAs. In a simple case, the resonant levels correspond to the bound states of the quantum well. For electrons, incident with this energy, the tunneling probability is considerably larger than that expected from the two barriers. Indeed, in the WKB approximation, with equal barriers and no bias, the tunneling probability can approach unity. In actual experiments, this has not been found. While the measurements have been carried out on realistic structures, where the barriers are distorted by the applied bias, the assumed theory mentioned above requires pure symmetry. This can lead to a drastic drop in the effective tunneling probability for the structure, even on resonance.[12] Moreover, the resonances themselves will be shifted by the applied bias.[13] In Figure 9.4, we illustrate a typical i-v curve obtained for a structure with uniform thickness of barriers and well of 5 nm.[11]

A second important point that must be made with regard to the actual experiments that have been carried out is that tunneling can be a time-dependent problem. As in any resonant cavity problem in electromagnetics, or quantum mechanics, the population of the resonant level will build up with a characteristic time constant related to the Q of the cavity. This latter quantity is also related to the natural line width of the resonance itself, and this has been suggested as being quite small in resonant tunneling devices. However, in the current-carrying case, when the device is under bias, we must remember that it is the *loaded Q*, not the natural or unloaded value, that must be used in computing the time required for the buildup of the resonant level population. The physical device itself, when placed under bias, is an open system. Such systems have not been treated with much consideration in the quantum literature. However, the presence of the bias and external loads will certainly broaden the resonance beyond its natural, or unloaded, value. In general, this time constant must be found by self-consistent solutions of the bias distribution, local carrier density, and current flow. Not much progress has been made in this direction, but several approaches are discussed in the following sections.

The general case, shown in Figure 9.3, for two rectangular potential barriers is the basis for our approach. The tunneling matrix **M** for one of these barriers was developed earlier and is given in (9.34). Each of the two barriers has its own

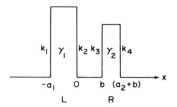

FIGURE 9.3. The double-barrier resonant tunneling structure.

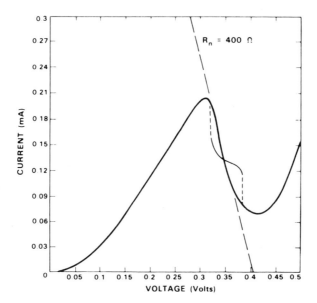

FIGURE 9.4. Experimental curves of the tunneling current through the resonant structure. After Sollner *et al.*[11] The measurements were at 100 K, and the dark line is measured for a stable device, while the light line was measured during oscillations at 4 GHz. The difference is thought to be due to self-detection.

matrix. We will denote the matrix for the left-hand barrier in Figure 9.3 by $\mathbf{M}_L$ and will also take the tunneling probability for this barrier alone to be $D_L = (k_2/k_1)|M_{L,11}|^{-2}$. This matrix is written in terms of k_1 and k_2 as the propagating wave vectors at the left- and right-hand edges, respectively. In a similar fashion, we may write the tunneling probability of the right-hand barrier in Figure 9.3 as $D_R = (k_4/k_3)|M_{R,11}|^{-2}$, where $\mathbf{M}_R$ is written in terms of k_3 and k_4 as the propagating wave vectors at the left- and right-hand edges of the barrier, respectively. For a flat-bottom potential well, as shown in the figure, we have $k_2 = k_3$, but retain the difference for generality under applied bias to be treated below.

The overall tunneling probability can be calculated by connecting the two barrier layers through the equation

$$\mathbf{M}_T = \mathbf{M}_L \mathbf{U} \mathbf{M}_R, \tag{9.37}$$

where $\mathbf{U}$ is a connection matrix describing the propagation through the quantum well. This latter matrix is

$$\begin{bmatrix} \exp(-ikb) & 0 \\ 0 & \exp(ikb) \end{bmatrix}, \tag{9.38}$$

where b is the well thickness and k is an average propagating wave vector (we will have to replace the quantity kb by an integral over the region when we apply bias, in keeping with the approximation discussed following (9.32), but for the moment $k = k_2 = k_3$). There are, in addition to (9.37), two propagator matrices which account for the shift in origins of the left and right barriers with respect to the origin assumed in (9.33). However, these matrices are left- and right-multipliers of (9.37), are diagonal as in (9.38), have the same form as (9.38), so that when we take the magnitude squared of $\mathbf{M}_{11}$ they do not play any role. We can therefore ignore them. Finally, we find that the tunneling probability for the entire structure is now given by

$$D = \frac{k_4}{k_1} |\mathbf{M}_{T,11}|^{-2} = \frac{k_4}{k_1} |M_{L,11} M_{R,11} \exp(-ikb) + M_{L,12} M_{R,21} \exp(ikb)|^{-2}. \quad (9.39)$$

We can now expand the denominator term to give

$$D^{-1} = \frac{k_1}{k_4} \{ |M_{L,11}|^2 |M_{R,11}|^2 + |M_{L,12}|^2 |M_{R,21}|^2$$

$$+ 2 \operatorname{Re}[M_{L,11}^* M_{R,11}^* M_{L,12} M_{R,21}] \cos(2kb)$$

$$- 2 \operatorname{Im}[M_{L,11}^* M_{R,11}^* M_{L,12} M_{R,21}] \sin(2kb) \}. \quad (9.40)$$

It is immediately noted that these quantities can be expressed in terms of the tunneling probabilities for the left and right barriers alone, which means that we can express the result by the properties of these barriers and the quantum well separately. We also note that, because $M_{11} = M_{22}^*$ and $M_{12} = M_{21}^*$ for each of the barriers, the net tunneling at this point should be symmetrical. This symmetry cannot be broken by the applied bias, and we must be careful to distinguish between symmetry of the tunneling coefficient and symmetry in the current flow for two different directions of the applied bias. This latter result is different and nonsymmetric current flow can arise from different positions of the resonant level for the two directions of applied bias. The asymmetry in current for reverse bias can also arise from nonlinear effects associated with charge transfer in the barriers, which we do not treat in this approach. We return to this point later.

We can examine the denominator in some detail to determine the appropriate terms. We want to examine the largest terms in the denominator, as these will create the smallest tunneling coefficient, as they dominate the overall value of D. We can pursue this somewhat by recognizing that

$$|M_{L,11}|^2 - |M_{L,12}|^2 = \frac{k_2 \gamma_1}{k_1 \gamma_2}, \quad (9.41a)$$

$$|M_{R,11}|^2 - |M_{R,12}|^2 = \frac{k_4 \gamma_3}{k_3 \gamma_4}, \quad (9.41b)$$

and that we can rewrite a number of the terms as functions of the various "reflection angles" as

$$|M_{11}|^2 = W_1[\cosh(2\overline{\gamma a}_1) - \cos(2\phi_2 - 2\phi_1)], \tag{9.42}$$

$$M_{11}^* M_{12} = -W_1[\cosh(2\overline{\gamma a}_1 - 2i\phi_2) - \cos(2\phi_1)], \tag{9.43}$$

$$M_{11}^* M_{12}^* = W_1[\cosh(2\overline{\gamma a}_1 - 2i\phi_1) - \cos(2\phi_2)], \tag{9.44}$$

where

$$W_1 = \frac{(k_2^2 + \gamma_2^2)(k_1^2 + \gamma_1^2)}{8k_1^2\gamma_2^2}, \tag{9.45}$$

$$\phi_1 = \arctan\left(\frac{\gamma_1}{k_1}\right), \tag{9.46}$$

$$\phi_2 = \arctan\left(\frac{\gamma_2}{k_2}\right). \tag{9.47}$$

The quantity $\overline{\gamma a}_1$ is the integrated average over the barrier, as discussed above. These values are for the left barrier. For the right barrier, we obtain the equivalent of (9.45)–(9.47) by making the changes $\gamma_1 \to \gamma_3$, $\gamma_2 \to \gamma_4$, $W_1 \to W_2$, $k_1 \to k_3$, $k_2 \to k_4$, $\phi_1 \to \phi_3$, and $\phi_2 \to \phi_4$. In evaluating the lead terms of the denominator, we need only use the small-tunneling approximation of (9.35). Then the leading term of the denominator is

$$2W_1 W_2 \exp(2\overline{\gamma a}_1 + 2\overline{\gamma a}_2)[1 - \cos(2kb - 2\phi_2 - 2\phi_3)], \tag{9.48}$$

and

$$D \simeq D_L D_R \tag{9.49}$$

reflects the fact that we really have two tunneling barriers in series. In this case, there is no coherence in the tunneling wave function. However, this term vanishes at resonance, a condition given by the bracketed term in (9.48); that is, this leading term disappears on resonance when

$$kb = \arctan\left(\frac{\gamma_1}{k_2}\right) + \arctan\left(\frac{\gamma_2}{k_3}\right) + n\pi, \tag{9.50}$$

where we assume that the integer n is not zero. On resonance, the next leading terms in the series of the denominator are terms of the order of

$$D_L^{-1} + D_R^{-1}, \tag{9.51}$$

but the coefficient of these terms also involves the same bracketed angle-dependent terms of (9.48). Thus, these terms also vanish on resonance. In (9.48) and (9.51), the form of these terms is not significantly changed by having different barriers. On the other hand, the next leading term, which really governs the tunneling coefficient on resonance, is one that does not appear in the equal barrier case. This next term is

$$C_{RL} \sinh(2\overline{\gamma a}_1 - 2\overline{\gamma a}_2), \tag{9.52}$$

where

$$C_{RL} = \frac{W_1 W_2}{2} \cos(2kb + 2\phi_3 - 2\phi_2), \tag{9.53}$$

which does not vanish. Thus, on resonance, the tunneling coefficient is

$$D_{res} \simeq \frac{D_{min}}{D_{max}}, \tag{9.54}$$

where D_{min} is the smaller of D_L or D_R, and D_{max} is the larger of the two. It may readily be shown that in the case of identical barriers, this quantity goes to unity exactly.

The results obtained above hold even when the incident energy range is quite narrow. We must remember, however, that the tunneling current is not given by D alone but is a summation over all of the states that are allowed to tunnel. In the resonant tunneling device, current begins to flow when the bias is such that the resonant level is brought down to the top of the occupied states on the left cladding layer (assuming the bias is such that the right layer is lowered with respect to the left). The position of the resonance itself is affected by the bias, since this affects the values of the parameters in (9.50). In any case, current flows as long as the resonance energy is sweeping through the occupied states of the left cladding layer. The peak is reached when this energy is reduced to the conduction-band edge on the left. Then the current is reduced to the $D_L D_R$ value. One should be careful in estimating the valley current, however, as inelastic processes, transport through a second resonance, and/or emission over one or the other of the barriers can all contribute to the total current.

We also note that in the absence of resonance, the wave function decays continuously from the left to the right of the barrier (assuming the bias is as discussed above). On resonance, however, the wave function has a local maximum in the quantum well. Carrier retention in the well is a dynamic process, while the equations developed above are for the static case. Thus, the number of particles within the well and the net current flowing through the well will be time-dependent quantities. This requires that the transient solutions be obtained, which is a somewhat more difficult project. This is especially true, as the bias will force carriers through the barrier in a driven mode. This is a reflection of the fact that it is the loaded-Q of the tunnel structure, and not the unloaded-Q, that is important in determining the time-dependent quantities. There is an important feedback mechanism at work as well, as the local shape of the barriers is modified when the number of particles trapped within the well varies. This also affects the resonance time evolution.

In Figure 9.5, we show a resonant tunneling barrier with an applied bias. We have assumed that the bias is linearly dropped across the tunnel barriers and the well in the structure, and that the two barriers have equal thicknesses. Then

$$D_L \simeq \exp\left[-\frac{4(2m)^{1/2}}{3\hbar eF}\{\Phi_0^{3/2} - \Phi_1^{3/2}\}\right], \qquad (9.55)$$

$$D_R \simeq \exp\left[-\frac{4(2m)^{1/2}}{3\hbar eF}\{\Phi_2^{3/2} - \Phi_3^{3/2}\}\right], \qquad (9.56)$$

and

$$\frac{D_{\min}}{D_{\max}} \simeq \exp\left[-\frac{e(2m)^{1/2}}{3\hbar eF}\{\Phi_0^{3/2} - \Phi_1^{3/2} + \Phi_2^{3/2} - \Phi_3^{3/2}\}\right], \qquad (9.57)$$

where

$$\begin{aligned}
\Phi_0 &= eV_{\text{barrier}} - E, \\
\Phi_1 &= \Phi_0 - eFa, \\
\Phi_2 &= \Phi_0 - eF(a + b), \\
\Phi_3 &= \Phi_0 - eF(2a + b).
\end{aligned} \qquad (9.58)$$

We also note that the applied voltage is given by $V_a = F(2a + b)$. In (9.57), the energy is referenced to the left-hand conduction-band edge, and so the peak in

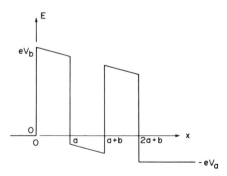

FIGURE 9.5. Shift in the double-barrier structure under applied bias. Here it is assumed that
the bias is linearly distributed across the barriers.

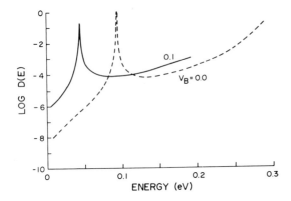

FIGURE 9.6. The tunneling probability as a function of energy for the structure of Figure 5.6, for 0 and 0.1 V of applied bias.

the current is obtained when $E = 0$. Then resonance is obtained at this same energy and requires

$$\frac{2(2m)^{1/2}}{3\hbar eF}\{(eFa)^{3/2} - (eFa + eFb)^{3/2}\}$$

$$= \tan^{-1}\left\{\frac{\Phi_0 - eFa}{eFa}\right\}^{1/2} + \tan^{-1}\left\{\frac{\Phi_0 - eFa - eFb}{eF(a + b)}\right\}^{1/2}. \qquad (9.59)$$

Ricco and Azbel[13] have evaluated this expression for the parameters of Ref. 11 and obtained a bias voltage of 0.21 V, which agrees well with the experimental data. In Figure 9.6, we plot the net tunneling coefficient for the case of barriers of 5 nm and a quantum well of 5 nm, assuming the barrier height is 0.3 eV, in the absence of an applied bias and with a bias of 0.1 V applied. In Figure 9.7, we plot the resonance energy in this well with applied bias. Here, we find that the peak occurs slightly lower than the Ricco and Azbel value, and is found at about 0.19 V.

9.3.2. Pseudodevice Calculations

In calculating the current through a quantum structure such as a resonant tunneling transistor, we must use a fully quantum transport scheme if we are to

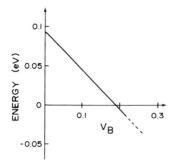

FIGURE 9.7. The shift of the resonant level as a function of applied bias for the structure of Figure 5.6. The energy is measured from the conduction-band edge on the left of the structure, as indicated in Figure 5.6. The peak of the tunneling current through the structure occurs as the resonance level passes through zero energy.

accurately reflect the microdynamics of the structure. Yet, the understanding of quantum transport applied to far-from-equilibrium, and open, systems like semiconductor devices is still in the primitive state. Consequently, this device—the resonant tunneling diode—has become the "fruit fly" for quantum studies of device dynamics. In detail, we still can do no better than the Landauer formula. On the other hand, it is to be hoped that the experimental results will actually push the theoretical field forward toward understanding the details of the quantum interference that can go on in this device.

Most quantum transport is developed in terms of either the wave function for a discrete quantum state, or in terms of the density matrix for the mixed system. The density matrix is a nonlocal function in either the momentum or the position representation, and differs from the normal representations used for classical transport. For example, we introduced the distribution function in earlier chapters. This distribution function was a function of the six dimensions of momentum and position, even though we treated only a one-dimensional (in position) device. The normal quantum distribution is the density matrix and does not have this mixed property of describing both position and momentum. On the other hand, we can go to a different, but still proper, quantum description through the introduction of a Wigner distribution function. This latter distribution is written in a mixed representation of both position and momentum, at the expense of introducing quantum interference through negative values and at the expense of introducing nonlocal terms in the Hamiltonian for the time evolution of the function itself.[14]

Although the Wigner function is normally written in terms of all of the generalized coordinates, we will treat only a single spatial coordinate and a single momentum coordinate. Then, we can write it as

$$f_W(x, p) = (2\pi\hbar)^{-1} \int_{-\infty}^{\infty} dy\, \psi^*\left(x + \frac{y}{2}\right) \psi\left(x - \frac{y}{2}\right) \exp\left(\frac{iyp}{\hbar}\right), \qquad (9.60)$$

in terms of a general wave function $\psi(x)$, and

$$f_W(x, p) = (2\pi\hbar)^{-1} \int_{-\infty}^{\infty} dy\, \rho\left(x + \frac{y}{2}, x - \frac{y}{2}\right) \exp\left(\frac{iyp}{\hbar}\right), \qquad (9.61)$$

in terms of the density matrix $\rho(x, x')$. The Wigner function has the proper properties as well. For example, if we integrate f_W over all momenta, we obtain the probability density function, and if we integrate over all position we obtain the momentum probability density, as

$$\int f_W(x, p)\, dp = \psi^*(x)\psi(x) \qquad (9.62)$$

and

$$\int f_W(x, p)\, dx = \phi^*(p)\phi(p), \qquad (9.63)$$

where

$$\phi(p) = (2\pi\hbar)^{-1} \int \psi(x) \exp\left(-\frac{ipx}{\hbar}\right) dx \qquad (9.64)$$

is the momentum wave function.

It follows immediately from (9.61) that the expectation value of an observable $W(x, p)$ is

$$\langle W \rangle = \int\int W(x, p) f_W(x, p) \, dx \, dp, \qquad (9.65)$$

which is analogous to the classical expression for the average value. Herein lies the interesting aspect of the Wigner function; the result of (9.65) suggests that it is possible to transfer many of the results of classical transport theory into quantum transport theory by simply replacing the classical distribution function by the Wigner distribution function. However, unlike the nonlocal density matrix, the Wigner distribution function itself cannot be viewed as the quantum analog in a simple sense since it is in general not a positive definite quantity. Indeed, general potentials complicate the problem due to the fact that quantum interference appears in the Wigner function through negative regions whose extent in phase space is such that the uncertainty principle is fulfilled. This is to say, the negative regions are sufficiently large that measurements cannot be made in small parts of the phase space that would violate the Heisenberg principle.

The time evolution equation for the Wigner function can be found by substituting the Schrödinger equation into (9.60), integrating the momentum functions by parts, and obtaining

$$\frac{\partial}{\partial t} f_W(x, p) + \frac{p}{m^*}\frac{\partial}{\partial x} f_W(x, p) = \frac{1}{\pi\hbar^2}\int ds \, dP \, K(s, P) f_W(x, p + P, t), \qquad (9.66)$$

where

$$K(s, P) = \left[V\left(x + \frac{s\hbar}{2}\right) - V\left(x - \frac{s\hbar}{2}\right)\right]\sin(sP). \qquad (9.67)$$

In fact, this equation is not adequate to fully determine the Wigner function, since it does not define the initial condition. It only defines the manner in which the initial function is propagated forward in time, just as the Schrödinger equation does the same for the wave function. One often sees (9.66) with $K(s, P)$ expanded in low-order terms of a Taylor series; the classical equation contains only the first two such terms. However, it is through $K(s, P)$ that the nonlocal behavior characteristic of quantum mechanics is evident. (There would be an equivalent form for the kinetic energy if we did not assume parabolic energy bands!) Especially in the case of the tunneling structures fabricated by molecular-beam epitaxy, the potential barriers are quite sharp, and the Taylor series must retain an almost infinite number of terms.

In Figure 9.8, we illustrate the use of the Wigner distribution by calculating f_W for a narrow Gaussian wave packet, and then propagating it through a double-barrier structure. The occupation in the resonant level and the nonpositive portions of the distribution are clearly seen in this figure. It is calculations of propagation such as this that have been used to estimate tunneling times.[15,16]

There are a number of problems that must be faced when we get down to actually simulating real devices. The first is the initial distribution function $f_W(x, p)$ that exists within the device. While one might initially expect this to be just a normal classical one, this is incorrect because of the sharp potential barriers that can exist. We illustrate this with the double-barrier tunneling structure. We can calculate the initial distribution from the density matrix for the equilibrium situation following the property

$$\rho(x, x') \sim \exp(-\beta \hat{H}), \tag{9.68}$$

in which $\hat{H}$ is the total Hamiltonian operator that includes the details of the potentials. Both the boundary conditions and the basis set for the expansion must be selected prior to evaluating the density matrix. Once a basis set is selected, the density matrix may be evaluated from the usual expansion

$$\rho(x, x') = \langle x| \left[1 - \frac{\beta \hat{H}}{n} \right]^n |x'\rangle \tag{9.69}$$

in the limit as n goes to infinity. This density matrix can then be transformed

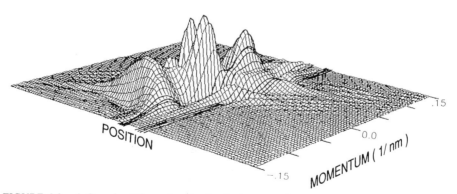

FIGURE 9.8. A Gaussian Wigner function impinging upon the structure of Figure 9.3. Reflection and tunneling are evident. The center of the momentum of this packet is at 3.0×10^8 m^{-1}.

with the Wigner transform (9.68) to yield the Wigner distribution function for the device. In Figure 9.9, a Wigner distribution calculated with periodic (but equilibrium) boundary conditions and with a basis set of plane waves weighted by the distribution function for the position representation eigenfunctions. The number was determined by the grid size, but the finite number of these introduces some problems. We see from this figure that the distribution is much broader in momentum range near the barrier than one would expect from just classical considerations. This is because the higher momentum states have a shorter wavelength and therefore penetrate much closer to the barrier. The impact of this on tunneling currents has not been fully assessed yet, but is thought to relate strongly to some zero-bias anomaly in tunneling structures. What can also be determined from Figure 9.9 is the *lack* of sharp resonance levels between the barriers, in distinction to Figure 9.8. This failure is thought to be due to the fact that the ground state in the quantum well is made up of oppositely propagating plane waves which in fact create a standing wave with zero wave vector.

A second problem is the proper boundary conditions that must be introduced into the quantum treatment. As indicated above, this consideration must enter the problem from the very first step. In classical devices, we just use "ohmic" contacts generated by high doping densities (in principle), and do not entertain further complications. In quantum systems, on the other hand, there is no description as to what constitutes an "ohmic" contact. Periodic boundary conditions introduce anomalous quantization levels within the system, and these should be avoided as well. The assumption of some arbitrary equilibrium distribution has also been made; i.e., one can assume that with no applied bias the incoming electrons (positive k at the left boundary and negative k at the right boundary) derive from equilibrium Maxwellians. However, this infers a randomization of

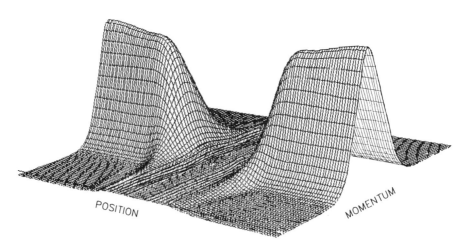

FIGURE 9.9.   The initial Wigner distribution function of a resonant tunneling diode for a degenerate electron gas. The dark band at the right is negative valued. This distribution assumed a boundary condition in which the gradient of the density is zero at the boundary.

electrons at the boundary that introduces dissipation into the system.[17,18] Thus, the boundary condition for the equilibrium system, as well as for the time-developing system, is a major concern for the quantum problem.

Still another problem arises with the incorporation of dissipation within the quantum system, as well as at the boundaries. While the introduction of a simple relaxation time approximation is possible,[16] this avoids the problem that one wants to address with quantum transport in the first place—extended, nonlocal scattering in a dense, many-body system. The problems associated with quantum transport in a general homogeneous system have only now begun to be solved in a general way, without incorporating the details of specific scattering processes. This remains a complicated problem under very active investigation, but the solution to quantum devices will not be achieved until this problem is solved. The problem is complicated in the device arena because of the possibility that individual device dimensions will be of the size of the electron wavelength. This complicates the scattering problems, and the redistribution of charge accompanying the self-consistent solutions of the Poisson equation for the potential still further complicates the problem.[16]

In Figure 9.10, we show the current-voltage curve calculated for the system of this section. Two curves are shown: one for no dissipation and one for dissipation within the relaxation time approximation. In the latter case, the peak is shifted to higher voltage and occurs very nearly at the value predicted from the transfer matrix approach discussed earlier. The kink in the falling part of the current in the negative resistance regime is due to charge storage in the resonant level itself.

In summary, while we can talk in quite specific detail about the nature of boundary condition matching and tunneling probabilities and current formulations, the general problem of actual quantum device simulation is far from being

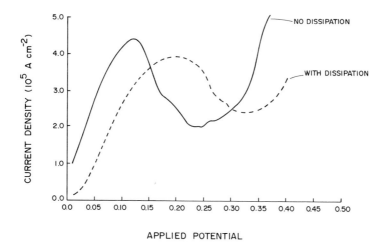

FIGURE 9.10. The current–voltage curve calculated by the Wigner approach for a resonant tunneling structure with 0.3-eV barriers 5 nm thick and a well that is also 5 nm thick.

solved. We tend to forget about the many decades that have gone into solving the classical device problem (to the level at which it is currently understood). Extrapolating from this, it will be quite some time until we achieve a comparable understanding of the quantum transport in devices.

9.4. LINEAR RESPONSE THEORY

In much of semiconductor transport theory, it is our main aim to calculate the response of the distribution of electrons, within our device or bulk semiconductor sample, to an applied perturbation. While this perturbation is usually an electric field, it may also be a magnetic field, temperature gradient, density gradient, pressure, or any combination of these generalized forces. Energy from these forces is coupled to the electrons and subsequently decays to the lattice via interaction with the phonons. In small semiconductor devices, we would like to know the entire time dependence of the appropriate interactions, and usually we use some form of kinetic theory based on the Boltzmann transport equation, which is not valid itself on the short time scales. The response is directly related to the nature of the scattering of the electrons by the lattice, and is often characterized in terms of relaxation times, such as the momentum relaxation time and the energy relaxation time. In turn, these averaged quantities are the macroscopic effects of the microscopic fluctuations introduced by the scattering processes themselves. As a consequence, it is possible to relate directly the averaged response to the spectrum of the fluctuations themselves—the traditional fluctuation-dissipation theorem.

While a kinetic theory is usually the basis of transport theory, there is an alternative approach tied rather directly to the use of Langevin equations to describe the transport of carriers. The latter has been known in statistical physics for a long time, but it is only in the past few decades that a formal response theory, pioneered by Kubo,[19] has grown up to utilize this approach. In this chapter, it is our goal to introduce the Kubo formalism through linear response theory, to introduce the connection with Green's functions, and to give a concrete (although semiclassical) example for electron transport.

The approach to be used here is primarily that of linear response theory. On the other hand, when we treat semiconductor transport, we are primarily dealing with concepts such as hot electrons and velocity saturation. It is important to note that these ideas do not arise from deviations from formal linear response theory, but from the idea that the system itself evolves to a far-from-equilibrium state for which the normal ideas of *stationarity* do not hold. It is more important to incorporate changes arising from the evolution of the distribution function itself (energy relaxation) than to address higher-order terms in response theory. Accompanying this nonstationary behavior is the need to recognize that most of the response functions will be two-time functions. Nevertheless, our introductory treatment will primarily deal with these functions in a single time-variable formalism, although no properties requiring stationarity will be utilized. The explicit two-time nature will be discussed at the end of the treatment in terms of results that have appeared in the literature.

In this treatment, we desire to find the response of the coupled electron–phonon system to a time-dependent perturbation by calculating to lowest order the change induced in the density matrix. Thus, we write the Hamiltonian as

$$H = H_0 - A^+ f(t), \tag{9.70}$$

where H_0 includes the electron, lattice, and interaction terms, and $f(t)$ is a c-number variable which describes the strength and time variation of the perturbation whose coupling to the system is described by the operator A^+. As usual, the time variation of the density matrix is given by the Liouville equation

$$i\hbar \frac{\partial \rho(t)}{\partial t} = [H(t), \rho(t)]. \tag{9.71}$$

In the spirit of the linear response approach, we write $\rho = \rho_0 + \delta\rho$, where ρ_0 describes the system prior to the application of the perturbation and is thus a system in which the electrons and phonons are in equilibrium with each other. Then, the linearized equation for $\delta\rho$ is

$$i\hbar \frac{\partial \delta\rho(t)}{\partial t} = [H_0, \delta\rho(t)] + f(t)[\rho_0, A^+]. \tag{9.72}$$

We may solve for $\delta\rho$ as

$$\delta\rho(t) = -\frac{i}{\hbar} \int_{-\infty}^{t} dt' f(t')[\rho_0, A^+(t'-t)]. \tag{9.73}$$

To find the expectation value of a dynamical operator P, we then can use (9.73) as

$$\langle P(t) \rangle = \text{Tr}\{\delta\rho(t)P\} = \int_{-\infty}^{t} dt' f(t') K(t-t'), \tag{9.74}$$

where we have introduced the linear response function

$$K(t) = -\frac{i}{\hbar} \text{Tr}\{[\rho_0, A^+(-t)]P\} = \frac{i}{\hbar}\langle[P, A^+(-t)]\rangle = \frac{i}{\hbar}\langle[P(t), A^+]\rangle. \tag{9.75}$$

In (9.75), the cyclic property of the trace operation has been used to rearrange the various operators. The response function $K(t)$ describes a causal effect induced by the application of the perturbation. Thus, the range of integration in (9.75) is from the time t' of application of the perturbation up to the observation time t.

Suppose we want to calculate the velocity response to a steady electric field turned on at $t = 0$. Then $f(t) = eFu(t)$, where $u(t)$ is the Heaviside function, and $A^+ = r$, the position operator. The response is then

$$\langle v(t) \rangle = \frac{eF}{m} \int_0^t K_v(t') \, dt', \tag{9.76}$$

where

$$K_v(t) = \frac{i}{\hbar} \langle [p(t), r] \rangle, \tag{9.77}$$

and p is the momentum operator. Let us now look at a simple semiclassical example.

9.4.1. The Langevin Equation

The motion of the carriers in an applied electric field is a balance of the acceleration process due to the field and the relaxation process due to scattering. However, the latter is a retarded process, due to its dependence on other degrees of freedom, which is approximated by a history dependence. The retardation effect modifies the usual form of a Newton's equation by the introduction of a convolution process, as

$$\frac{dv}{dt} = \frac{eF}{m} - \int_0^t v(t')\gamma(t - t') \, dt' + W(t), \tag{9.78}$$

where the relationship, between the total momentum relaxation rate Γ_m and γ,

$$\Gamma_m = \int_0^\infty \gamma(t') \, dt' \tag{9.79}$$

is required for consistency with the Boltzmann result for the long-time limit, and $W(t)$ is a random force. In fact, (9.78) is just the well-known retarded Langevin equation. (To achieve this result from the Boltzmann transport equation, one must first overcome the well-known limitations of this latter equation on the short time scale.) While we write this stochastic equation for a classical variable, the result of (9.78) can be derived quite generally for quantum mechanical systems.[20-22]

The function γ is proportional to the force–force correlation function introduced in basic statistical mechanics. In equilibrium systems, this correlation function is often taken to be a δ-function in time, as the random forces at two different times are assumed to be uncorrelated. This is the Markovian approximation, which is a poor approximation, especially far from equilibrium. The proper correlation function reflects the upper frequency limit at which the ensemble can respond to fluctuations. This is related to the energy relaxation time in our problem, because this is the time scale of the evolution of the ensemble (distribution function). The relaxation rate, described by γ, is an average which describes the decay of momentum (and energy) due to the random scattering events themselves. The random force, arising from the fluctuations of the random scattering forces about the average decay, vanishes in an ensemble average.

We can now readily solve (9.78) by Laplace transformation. We introduce a new function $X(t)$, defined by its Laplace transformation as

$$X(s) = [s + \gamma(s)]^{-1}. \tag{9.80}$$

Laplace-transforming (9.78) and using (9.80), we find then that the time evolution of the ensemble averaged velocity is given by

$$\langle v(t) \rangle = \langle v(0) \rangle X(t) + \frac{eF}{m} \int_0^t X(t')\, dt', \tag{9.81}$$

which is a general expression of the evolution of the velocity under the influence of a high electric field. In general, $v(0)$ is taken to be zero. It may easily be shown, in keeping with the Kubo formula, that $X(t)$ is related to the normalized velocity autocorrelation function for the electrons. If we multiply (9.78) by $v(0)$ and then take the ensemble average, we find that

$$X(t) = \frac{\langle v(t)v(0) \rangle}{\langle v^2(0) \rangle}. \tag{9.82}$$

If $\langle v(0) \rangle = 0$, then we can relate $X(t)$ to $K(t)$ and also to the velocity autocorrelation function, although $K(t)$ certainly does not bear a strong resemblance to this latter function. Had we defined the perturbation term as a frequency-dependent quantity as, e.g., $evF(\omega)/\omega$, then $K(\omega)$ would appear as a velocity–velocity autocorrelation function more in keeping with $X(s)$. However, it is important to note that the appearance of $X(t)$ as an autocorrelation function for the velocity is only true in this special case of response originating in the relaxed initial state and does not carry over to the case of a small perturbation around the far-from-equilibrium steady state. To be able to treat this latter case, we must develop the treatment more fully.

9.4.2. Relaxation and Green's Functions

If we now look at a system that relaxes toward equilibrium, we can introduce another function, the relaxation function. This new function describes the change in the expectation value of a dynamical operator after the external perturbation has been turned off. An explicit form can be obtained by letting $f(t) = \exp(\eta t)$ for $t \leq 0$ and zero for $t > 0$. The change in the operator P is then

$$\langle P \rangle = \int_{-\infty}^{0} dt' \exp(\eta t') K(t - t') = \int_{t}^{\infty} dt' \exp[\eta(t - t')] K(t'). \qquad (9.83)$$

By letting $\eta \to 0$, we can define the relaxation function as

$$R(t) = \int_{t}^{\infty} dt' \, K(t'), \qquad \frac{\partial R(t)}{\partial t} = -K(t). \qquad (9.84)$$

One can easily verify by direct differentiation that the relaxation function is a quantum correlation function, given by[23]

$$R(t) = \int_{0}^{\beta} d\tau \, \langle A^{+}(-i\hbar\tau) P(t) \rangle - \beta \langle A^{+} \rangle \langle P \rangle. \qquad (9.85)$$

It is a peculiarity of the semiclassical case, discussed in the previous section, that the functional forms that arise in the initial velocity response form are such that $R(t)$ and $K(t)$ differ only in the constant term (which is zero for this case anyway) and in the normalization. However, the use of $R(t)$, which is a proper correlation function, brings the semiclassical case into line with the more rigorous quantum treatment.

Another function of interest is the Green's function. Here, our response function $K(t)$ and relaxation function $R(t)$ describe the response of the electron system to an external perturbation. However, this is just the terminology used for describing the causal, or retarded, Green's function $G_r(t)$. Consistently, the relationship between these functions is defined to be

$$G_r(t) = -u(t) K(t), \qquad (9.86)$$

where once again $u(t)$ is the Heaviside function. From this definition, it is clear that we can describe the response of the system to a perturbation by use of the retarded Green's function (and which also relates to the equilibrium case where $K(t)$ is an even function of time). Calculation of the Green's function for the case of a full electron–phonon system in which all many-body effects are included can be a formidable task.

We can use the relationship between $K(t)$ and $R(t)$ to finally get a form for the quantum velocity response that agrees with the classical one. From an earlier section, we have

$$K(t) = -\frac{\partial}{\partial t} \int_{0}^{\beta} d\tau \, \langle r(-i\tau\hbar) p(t) \rangle = -\int_{0}^{\beta} d\tau \left\langle r(-i\hbar\tau) \frac{\partial p(t)}{\partial t} \right\rangle$$

$$= \frac{i}{\hbar} \int_{0}^{\beta} d\tau \left\langle \frac{\partial}{\partial \tau} r(-i\hbar\tau) p(t) \right\rangle = \int_{0}^{\beta} d\tau \, \langle v(-i\hbar\tau) p(t) \rangle, \qquad (9.87)$$

which, with the normalization introduced by the β integration, is the same as that of the semiclassical approach.

9.4.3. Extension to Two-Time Functions

In the far-from-equilibrium state that describes the hot-electron problem arising in semiconductor devices, the correlation functions, response functions, and relaxation functions are proper two-time variable functions. One explicit reason for this lies in the fact that the inverse temperature β is itself a time-evolving function. Moreover, the ensemble used in the averages (the density matrix) can be a time-evolving function also. In the above treatments, we used ρ_0 as the equilibrium density matrix. This is a severe limitation on the linear response formalism, as the deviation term can be quite large. As a result, more modern approaches try to utilize as much of the complete, time-evolving density matrix as possible. One common approach is to introduce a *quasi-equilibrium* density matrix, which is parameterized.[24-26] The parameters have the same form as the equilibrium density matrix, but are allowed to evolve in time. The philosophy is the same as that of the drifted Maxwellian distribution function in the Boltzmann equation approach to classical transport. Then, the linear response formalism is used to describe the deviations around this quasi-equilibrium density matrix, and a set of moment equations, such as the retarded Langevin equation for momentum, is developed to describe the temporal variation of the parameters. As a consequence, a family of correlation functions is obtained describing the fluctuations affecting each of the parameters. For example, the momentum is described by the velocity–velocity correlation function, the temporal evolution of β is described by an energy–energy fluctuation function (which becomes a δ-function in equilibrium), etc.

Determination of the correlation functions is a complicated process, especially in the two-time variable form required in devices far from equilibrium. At least in the weak scattering limit, in which the electron–phonon interaction is kept to second order in the matrix elements, these functions can be evaluated with the quasi-equilibrium formulation.[25-27] The results are effectively the kernal of the Fermi golden rule for scattering induced relaxation of the variable as determined in semiclassical transport theory. For effects beyond this order, it is not clear if any simplification is possible, or whether the full complexity of the Green's function approach is necessary. However, even in this latter approach (which differs only in the fact that no reference is made to the Langevin equations), the use of the quasi-equilibrium form of the density matrix is often employed.

Finally, we want to exhibit the relevance of this approach by an example that describes fluctuations in small devices. These fluctuations are related to quantum interference phenomena within the devices themselves and appear as current fluctuations, that have in turn been suggested as the fundamental basis of $1/f$ noise. We can rewrite the response function $K(t)$ for the velocity correlations in terms of the currents $ev(t)$ and obtain the conductivity in the frequency domain as

$$\sigma(\omega) = K(\omega) = m \int_0^\infty e^{-i\omega t} \, dt \int_0^\beta d\tau \, \langle e^2 v(-i\hbar\tau) v(t) \rangle, \qquad (9.88)$$

where the conductivity must be summed over the transverse dimensions of the region of interest. Our interest here is a small localized transport region that is situated between two metallic conducting regions. As we go toward low temperature, the imaginary time integration really goes into a summation over all occupied states, and the time integration gives rise to energy-conserving delta functions on the individual state energies (two because of the two different time variables). Finally, by inserting a complete set of basis states, in order to evaluate the trace operation with a density matrix formed by the basis states, we find[28]

$$\sigma(\omega) = \frac{\pi}{\omega L^2} m \int dE' \sum_{\alpha,\beta} \left| \int_0^L J_{\alpha\beta}(z) \, dz \right|^2 \delta(E' - \omega - E_\beta)\delta(E' - E_\alpha), \qquad (9.89)$$

where $J_{\alpha\beta}$ is the matrix element between basis states ϕ_α and ϕ_β, each of which have been properly normalized in the transverse directions. We can see this further by expanding the current as

$$J_{\alpha\beta}(z) - J_{\alpha\beta}(z') = i\omega \int_{z'}^z dz'' \int d\rho \, \phi_\alpha^*(\mathbf{r})\phi_\beta(\mathbf{r}), \qquad (9.90)$$

where $\mathbf{r} = \rho + z\mathbf{a}_z$ is the total position vector. As the frequency goes to zero, the current becomes spatially uniform, and these terms lead to the cancellation of the leading frequency factor as well.

One can now introduce Green's functions and proceed formally, but the form of the above equations really give us a different approach. In the localized regime, current flows by carriers hopping along an irregular path from one metallic region to the other. Each of these paths will be composed of a current operator made up of the various basis sets. We can rearrange the sum over the current matrix elements by *a sum over the various paths themselves*. If we have a series of channels going into the localized region, denoted by the index j, and a series of channels coming out of the region, denoted by the index j', we can then write the wave function part as a propagator or Green's function. Then (9.89) can be written as[29]

$$\sigma(0) = \frac{e^2}{4\pi} \sum_{jj'} m v_j v_{j'} [|G_{jj'}(z, z')|^2 + |G_{jj'}(z', z)|^2], \qquad (9.91)$$

and the causal solution is always taken for the Green's function. The Green's function form can easily be related to the tunneling, or transition matrix, form and

$$\sigma(0) = \frac{e^2}{2\pi\hbar} \mathrm{Tr}(D^+ D). \qquad (9.92)$$

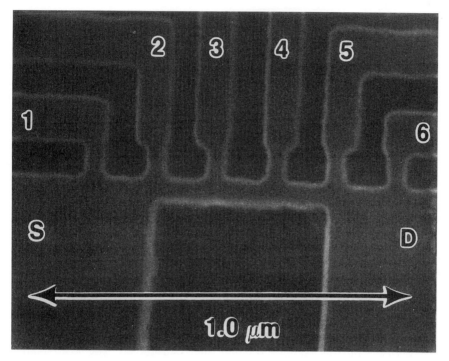

FIGURE 9.11. Top view of submicron region of the gate pattern that defines a narrow-channel MOSFET from source S to drain D, probed by voltage measurements on the sidearms 1–6. After Skocpol.[30]

The tunneling coefficients are dimensionless, as discussed in the previous sections. This particular form is essentially the Landauer formula, but its current importance lies in the fact that any change in the conduction through the localized region by a single path causes a change in the conductance of the order of $e^2/\hbar$. Any changes of the chemical potential, by the gate voltage for example, will

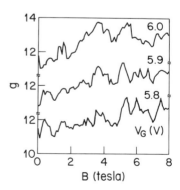

FIGURE 9.12. Dimensionless conductance vs. magnetic field at three gate voltages, for the inversion-layer segment indicated in Figure 9.11, showing aperiodic conductance variations of order e^2/h. After Skocpol et al.[38]

cause a rearrangement of the current paths and can lead to observable changes in the conductance of the device.

In Figure 9.11, we show a long, slender MOSFET fabricated at Bell Labs.[30] The conductance through the active inversion channel is shown in Figure 9.12 for small variations in the gate voltage and an applied magnetic field, although the latter is not necessary to observe the quantum fluctuations. The figures are labeled in units of e^2/h. While these fluctuations look as if they are random, they are actually correlated quite well over narrow ranges of the two applied fields. In these experiments, the temperature is quite low, so that the inelastic mean free path is larger than the transverse dimensions of the sample. However, as device sizes shrink, we can begin to expect to see these effects at higher temperatures as well.

9.5. THE DENSITY MATRIX EQUATION OF MOTION

The general scope of Boltzmann transport theory, as developed previously, is a nonideal theory that fails on the short time and space scales. Yet, it has its virtues, which lay in its conceptual and mathematical simplicity. In addition, it tends to work far better than one might expect from its origins in the classical theory of dilute gases. Quantum transport theory, on the other hand, enjoys no such status. Rather, it is quite often neither conceptually nor mathematically simple. The resulting equations, particularly those utilizing the Green's function approach, are complicated and difficult to apply. Yet, after considerable labor, the result often reduces to that obtained from the Boltzmann picture, usually because of the approximations that have had to be applied to achieve analytical results. Still, quantum transport theory is necessary to explain how the Boltzmann picture can arise from the underlying framework of quantum statistical mechanics. Our interest here, though, is in developing the necessary novel concepts and transport kinetics for genuine quantum transport phenomena.

Our particular interest is in submicron semiconductor devices. Here, the key quantum aspects are primarily the correction terms for the Boltzmann equation, and we shall pursue only those approaches to quantum transport that yield a kinetic equation of this form. We begin, in this section, to discuss the development of the appropriate kinetic equations for transport in the quantum picture by treating the quantum kinetic equation, derived for the density matrix. The correction terms that arise deal with temporal retardation, the intracollisional field effect, and higher-order scattering dynamics. In the next section, we convert this quantum kinetic equation into one for the Wigner distribution function, which exists in both momentum and position coordinates and is more easily related to the Boltzmann distribution function.

The quantum kinetic equation we wish to derive describes the evolution of the density matrix in applied fields and spatial gradients. However, the primary form we wish to deal with is that for a homogeneous system, so that we shall deal only with the applied potentials. We shall approach this with the concepts of projection operators, in particular the set of commutator-generating super-operators. These operators are tetradic operators that lie in a Hilbert space of operators. The density matrix ρ at time t has a temporal evolution described by the Liouville equation

$$\frac{\partial \rho}{\partial t} = -\frac{i}{\hbar} \hat{H}\rho = -\frac{i}{\hbar}[H, \rho], \tag{9.93}$$

where the "hat" denotes a commutator-generating superoperator. We can illustrate this operator's effects by considering the matrix element of the product as

$$\langle m|\hat{H}\rho|n\rangle = \sum_{rs} (H_{mr}\rho_{rn} - \rho_{ms}H_{sn}) = \sum_{rs} (H_{mr}\delta_{ns} - H_{sn}\delta_{mr})\rho_{rs} = \sum_{rs} \hat{H}_{mnrs}\rho_{rs}. \tag{9.94}$$

Thus, the matrix for the superoperator is a tetradic, or fourth-rank, tensor. If the operator itself is diagonal, then

$$\hat{H}_{mnrs} = (H_{mm} - H_{nn})\delta_{mr}\delta_{ns}. \tag{9.95}$$

For any function of a diagonal superoperator $\hat{A}$,

$$\langle m|f(\hat{A})B|n\rangle = f(A_{mm} - A_{nn})B_{mn}, \tag{9.96}$$

and, in particular,

$$\langle m|(\hat{A} - z)^{-1}B|n\rangle = (A_{mm} - A_{nn} - z)^{-1}B_{mn}, \tag{9.97a}$$

$$\exp(i\hat{A})B = \exp(iA)B\exp(-iA). \tag{9.97b}$$

The latter expression is particularly important due to the compact notation that can be introduced for expressions such as the kinetic equation we will develop.

In the following, we want to consider a system composed of electrons and phonons, which is describable by the Hamiltonian

$$H = H_0 + H_F + H_L + H_{eL}. \tag{9.98}$$

Here, H_0 is the unperturbed Hamiltonian for the electrons, in the effective mass approximation, and has the eigenvalues and eigenstates specified by

$$H_0|n\rangle = E_n|n\rangle. \tag{9.99}$$

We remark that this Hamiltonian can also include the full electron–electron interaction, with all of its appropriate self-energy shifts, so that the solutions specified in (9.99) already are appropriate to the full many-body problem. We will not deal with this specifically, but note its occurrence at this point, so that the electronic part of the total density matrix is the effective single-electron form. The term H_F represents the driving fields, which for an electric field uniform in space can be represented as

$$H_F = -e\mathbf{F} \cdot \mathbf{x}. \tag{9.100}$$

Finally, the term H_L represents the lattice degrees of freedom, and H_{eL} represents the interaction between the electrons and the phonons.

We Laplace-transform the Liouville equation, and thus obtain the structurally simpler form

$$\left(s + \frac{i\hat{H}}{\hbar}\right)\rho(s) = \rho(0). \tag{9.101}$$

We assume here that H is not specifically time-dependent, although this restriction can be readily removed by working in the time domain exclusively. In general, one could write $iH/\hbar$ as iL, where L is the Liouville operator of classical mechanics and make an approximate connection with the classical world. This correspondence is in fact well established.

What we desire is the diagonal part of ρ, as it is just these terms which enter the normal transport equations.[31] The projection operator we want to use effectively projects out the one-electron density matrix from the total one, and results in a summation over phonon variables. The leading term, in parentheses in (9.101), is a super-resolvent which produces the projection operator expansion[32]

$$\begin{aligned}
\hat{R}(s) &= (i\hbar s - \hat{H}) \\
&= \{1 + \hat{Q}\hat{H}\hat{P}(i\hbar s - \hat{Q}\hat{H}\hat{Q})^{-1}\} \\
&\quad \times (i\hbar s - \hat{P}\hat{H}\hat{P} - \hat{C})^{-1}\{\hat{P} + \hat{P}\hat{H}\hat{P}(i\hbar s - \hat{Q}\hat{H}\hat{Q})^{-1}\hat{Q}\} \\
&\quad + (i\hbar s - \hat{Q}\hat{H}\hat{Q})^{-1}\hat{Q}, \tag{9.102}
\end{aligned}$$

where

$$\hat{Q} = 1 - \hat{P} \quad \text{and} \quad \hat{C} = \hat{P}\hat{H}\hat{Q}(i\hbar s - \hat{Q}\hat{H}\hat{Q})^{-1}\hat{Q}\hat{H}\hat{P}$$

is the "collision superoperator." The projection operator used here has the additional property, for the density matrix, that $(\hat{P}\rho)_{mn} = \rho_{mm}\delta_{mn}$. Moreover, it commutes with the resolvant operator $\hat{R}$, $\hat{P}\hat{R} = \hat{R}\hat{P}$. Thus, the projected electronic density matrix ρ_1 is now given by ($\hat{P}\hat{Q} = 0$)

$$-\frac{i}{\hbar}\rho_1(s) = \hat{P}\rho(s) = (i\hbar s - \hat{P}\hat{H}\hat{P} - \hat{C})^{-1}\{\hat{P} + \hat{P}\hat{H}\hat{P}(i\hbar s - \hat{Q}\hat{H}\hat{Q})^{-1}\hat{Q}\}\rho(0),$$

which is readily converted to

$$\begin{aligned}
\frac{\partial \rho_1(t)}{\partial t} &= -\frac{i}{\hbar}\hat{P}\hat{H}\hat{P}\rho_1(t) - \frac{i}{\hbar}\hat{P}\hat{H}\hat{Q}\exp\left(-\frac{it\hat{Q}\hat{H}\hat{Q}}{\hbar}\right)\hat{Q}\rho(0) \\
&\quad -\frac{1}{\hbar^2}\int_0^t dt'\,\hat{K}(t')\rho_1(t - t'), \tag{9.103}
\end{aligned}$$

where $\hat{K}(t')$ is the inverse transform of $\hat{C}$. The second term on the right is a residual memory of the initial state of the system. Generally, for long times, we do not have to concern ourselves with $\rho(0)$, and the resulting equation yields a kinetic equation for just the diagonal part of $\rho_1(t)$ without resort to the random phase approximation. We note that the basis functions are not true eigenfunctions of the total H due to the terms in the field and the electron–phonon interaction. The term H_F operates on the diagonal part as well, and

$$\frac{i}{\hbar} \hat{P}[H_F, \rho_1] = e\mathbf{F} \cdot \frac{\partial \rho_1}{\partial \mathbf{p}}. \tag{9.104}$$

Using this expression and denoting the coefficient of $\hat{Q}\rho(0)$ by $\hat{D}(t)$, we have

$$\frac{\partial \rho_1}{\partial t} + e\mathbf{F} \cdot \frac{\partial \rho_1}{\partial \mathbf{p}} = -\frac{1}{\hbar^2} \int_0^t dt' \, \hat{K}(t - t')\rho_1(t') - \frac{i}{\hbar} \hat{D}(t)\hat{Q}\rho(0). \tag{9.105}$$

The similarities between $K(t)$ and $D(t)$ are quite obvious, and both have their roots in the electron–phonon interaction. This form of the kinetic equation was apparently first derived by Barker,[33] but its form is quite analogous to the Prigogine-Resibois equation,[34] and is also in a form quite similar to that obtained by Levinson,[35] which is the form we shall pursue.

We first turn our attention to the exponential term in $K(t)$, which may be expanded as follows:

$$\hat{Q} \exp\left(-\frac{it\hat{Q}\hat{H}\hat{Q}}{\hbar}\right) \hat{Q} = \hat{Q}\left\{1 - \frac{it\hat{Q}\hat{H}\hat{Q}}{\hbar} - \left(\frac{t}{\hbar}\right)^2 \hat{Q}\hat{H}\hat{Q}\hat{H}\hat{Q} + \cdots\right\}\hat{Q}$$

$$= \hat{Q} \exp\left(-\frac{it\hat{H}\hat{Q}}{\hbar}\right) \hat{Q}, \tag{9.106}$$

where the effect of the trailing Q is on trailing functionals. The importance of this form is that the full Hamiltonian arises in the propagator, which in turn generates the full perturbation series if needed. Further, the leading (and trailing) terms are

$$(\hat{P}\hat{H}\hat{X})_{nnmm} = \sum_{rs} P_{nnrs}\left(\sum_{pq} H_{rspq}X_{pqmm}\right) = \sum_{pq} H_{nnpq}X_{pqmm}$$

$$= \sum_{pq} (H_0 + H_F)_{nnpq}X_{pqmm} + \sum_{pq} H_{eL,nnpq}X_{pqmm}. \tag{9.107}$$

Since the first term involves diagonal terms in H, it vanishes due to the properties of the tetradic formed from diagonal operator matrices. A similar result arises from examining the elements of $\hat{Q}\hat{H}$. Hence, $\hat{K}(t)$ is in general a function that involves only the electron–phonon interaction H_{eL}, except in the propagator exponential. As a consequence, we can rewrite $\hat{K}(t)$ as

$$\hat{K}(t) = \hat{P}\hat{H}_{eL}\hat{Q} \exp\left[-\frac{i}{\hbar}(\hat{H}_0 + \hat{H}_F + \hat{H}_{eL})\hat{Q}\right]\hat{Q}\hat{H}_{eL}\hat{P}, \tag{9.108}$$

which generates a power series in the interaction potential. Clearly, the lowest-order term arises from neglecting the interaction term in the exponential, which gives rise to the Fermi golden rule studied in Section 9.1. However, the presence of this term in the resolvent implies that a higher-order series should be considered—indeed must be considered—in many applications. Several approximations have been made to take this sum into account properly, and many of these appear in the literature.

We note also that the initial projection operator in $\hat{K}(t)$ and $\hat{D}(t)$ also includes a trace over the phonon variables and modes. This separate notation has been suppressed in the above, but it is the sum over modes that gives, in the Fourier-transformed case, the summation over the phonon wave vectors necessary to produce the scattering probabilities.

By and large, one finds that the self-energy corrections that arise from higher-order scattering processes are not as large as the corrections that arise from the presence of high scattering rates and the intracollisional field effect. In these cases, it is a fair assumption to ignore the term in the interaction in the exponent of (9.108). In essence, this is tantamount to claiming that lowest-order perturbation theory is adequate. This is a serious limitation if not valid, but for the moment we pursue this approach. If we also neglect for the moment the field term, the exponential term then becomes

$$\left[\exp\left(-\frac{it\hat{H}_0}{\hbar}\right)\right]_{mnm'n'} = \exp\left[-\frac{it(E_m - E_n)}{\hbar}\right]\delta_{mm'}\delta_{nm'}. \qquad (9.109)$$

Then,

$$\begin{aligned}
K_{nnmm}(t) &= \sum_{pqrs} \exp\left[-\frac{it(E_p - E_q)}{\hbar}\right]\delta_{pr}\delta_{qs}H_{eL,rsmm}H_{eL,nnpq} \\
&= \sum_{pqrs} (H_{eL,np}\delta_{nq} - H_{eL,qn}\delta_{pn}) \\
&\quad \times \exp\left[-\frac{it(E_p - E_q)}{\hbar}\right](H_{eL,pm}\delta_{qm} - H_{eL,mq}\delta_{pm}) \\
&= -2|H_{eL,nm}|^2 \cos\left[\frac{(E_m - E_n)t}{\hbar}\right], \qquad (9.110)
\end{aligned}$$

which, with the time integration in the kinetic equation, produces the Fermi golden rule terms used in Section 9.1. Here we have used the fact that H_{eL} has only off-diagonal terms, although we could have retained the diagonal terms to get the state broadening as was done in that section. It is clear that we can retain all of the proper terms to reachieve the state broadening, the intracollisional field effect, and multiple scattering. Moreover, the form of the kinetic equation (9.105) illustrates that the breakdown of the Boltzmann equation arises when the time evolution of the density operator, through the Hilbert space described by the basis states, occurs with frequency components that are comparable to the reciprocal of the scattering rate. On this time scale, (9.105) describes the convolution of this evolution with the scattering functions.

The memory term $\hat{D}(t)$ accounts for two effects in its lowest-order expansion terms. The first is the source of the random force that arises in Langevin equation approaches,[36] while the next term involves the screening of the driving field by the collisions.[32] Neither of these terms are of importance in semiconductors utilized for submicron devices, although the latter one has been utilized to describe localization phenomena in amorphous semiconductors. We will neglect these terms in further discussions.

9.6. THE WIGNER TRANSFORM AND EQUATION OF MOTION

In (9.61), we introduced the Wigner transform of the density matrix, termed the Wigner distribution function. The Wigner distribution becomes important when the physical problem is one that is better understood in terms of a phase-space representation that contains both momentum and position, as is the case for the classical Boltzmann distribution function. This is not the case with the normal density matrix, but the Wigner distribution attempts to present an analogy between quantum and classical phase space for statistical mechanics. Since the statistical picture in phase space is well understood, indeed uses the Boltzmann equation, for classical mechanics, transforming to a similar picture in quantum statistical mechanics allows the physical picture of a problem to be better understood. Unfortunately, position and momentum do not commute in quantum mechanics, and the two cannot simultaneously be measured to great accuracy in phase space. This appears in the Wigner picture by negative values of the distribution function.

Formerly, the Boltzmann transport equation can be thought of as being derivable from the diagonal elements of the density matrix equation of motion, providing we take the semiclassical limit. Unfortunately, this does not fully incorporate the positional variations. The Wigner distribution is thought to be a more proper approach to obtaining a semiclassical distribution in phase space. There is a difficulty in realizing a Wigner distribution as a proper probability distribution. It has regions of negative probability, as discussed above, which relate to the uncertainty principle. This negative probability is definitely not a characteristic of a true probability density function. However, if the function is smoothed over a size corresponding to the uncertainty principle, tantamount to coarse-graining, the resulting distribution is positive definite as required.

The Wigner representation puts the phase-space formulation of quantum mechanics on a firmer basis.[37] Quantum mechanics is known to be a mainly statistical theory by nature, as indicated by the Heisenberg uncertainty principle. Dynamical parameters are treated as statistical quantities, even in the case of isolated atomic structures. Thus, it seems to follow directly that the dynamic variables of observables given by operators should be adapted to a statistical theory. And yet, we would prefer the phase-space formulation, where trajectories are governed by the laws of motion, and the statistical aspects arise from an undetermined initial state of the system. It is fundamental then that a Wigner representation is the proper approach to obtain this technique, and is then useful to relate to the semiclassical Boltzmann distribution.

Wigner was the first to introduce a method of calculating expectation values in a phase-space format.[14] If the expectation value of an operator is expanded in powers of $\hbar$, the zero term is then the classical value. Higher powers can be described as higher-order quantum corrections. If the system is almost classical, the series converges rapidly. However, in the case of sharp potentials, which give rise to a series of resonant energies or quantum reflections, this series does not converge at all. This is the reflection of the fact that the nonlocal behavior of potentials, in which the potential is felt for a distance of one or several wavelengths of the particle, has no correspondence in classical physics. This nonlocality clearly is reflected in (9.66). Yet, as is indicated in this equation, the phase-space formulation of the Wigner distribution still incorporates these quantum effects quite precisely and still yields the proper quantum mechanics.

In this section, we want to transform the density matrix's kinetic equation developed in the previous section. In this way, we develop a Fredholm integral equation for the Wigner distribution in the presence of the electron–phonon interaction and an applied electric field. We use only the second-order terms, as used in the Fermi golden rule, but the technique can readily be extended to include all orders of scattering. As in the previous section, we treat the one electron distribution function, within the quasi-particle approximation, as we are interested in the interactions between the field and the scatterers. Thus, the ground state should be considered to be the proper many-body ground state, which already incorporates the interparticle interactions and correlations. However, this approach neglects the modification of this many-body state by the evolving distribution and is incomplete in that regard.

The basic kinetic equation for the density matrix is given by (9.105). In this equation, it is assumed that the field and the scattering are turned on at $t = 0$. The more normal expression takes the lower limit of the convolution integral as $t \to -\infty$ and adiabatically switches on both of these quantities. We do not lose generality by going over to this more usual notation, but we must understand that a convergence factor may be necessary in the integral. We can rewrite this equation, neglecting the memory of the initial state, as[35]

$$\frac{\partial \rho_1}{\partial t} + e\mathbf{E} \cdot \frac{\partial \rho_1}{\partial \mathbf{p}} = I(\rho_1 | t), \tag{9.111}$$

with

$$I(\rho_1 | t) = -\frac{1}{\hbar^2} \int_{-\infty}^{t} dt' \, \mathrm{Tr}\{[H_{eL}, U(t, t')[H_{eL}(t - t'), \rho_1 \sigma] U^+(t, t')]\}.$$

In the latter expression, we have unfolded the commutator generating super-operators, σ is the density matrix of the particular phonon mode that results from the projection, and

$$U(t, t') = \exp\left\{-\frac{iH(t - t')}{\hbar}\right\} \qquad (9.112)$$

is the normal propagator expression. The trace is over the phonon modes of the lattice, and its result, along with the phonon second-quantization operators in H_{eL}, produces

$$\text{Tr}\{H_{eL} \cdots \sigma\} \to \exp(\pm i\Omega t)(N + \tfrac{1}{2} \pm \tfrac{1}{2})$$

and the corresponding creation or annihilation operator (a_q^+ or a_q) for the phonon shift between the new and old electron states. We can separate I into two terms, corresponding to the emission or absorption of phonons, as

$$I(\rho_1|t) = I^{(+)}(\rho_1|t) + I^{(-)}(\rho_1|t),$$

where the first term is

$$I^{(+)}(\rho_1|t) = (N + 1)\frac{|c_s|^2}{\hbar^2} \int_{-\infty}^{t} dt'\, \exp[-i\Omega(t - t')]$$

$$\times [U(t, t')a_q^+\rho_1(t')U^+(t, t'), a_q] + \text{Herm. conj.} \qquad (9.113)$$

The second term represents the absorption of a phonon and is obtained by the replacement of $N + 1$, Ω, a_q^+, a_q by N, $-\Omega$, a_q, a_q^+, respectively. In the semi-classical limit, and even here, the two terms of the commutator in (9.113) correspond to the inscattering and outscattering processes of the collision term in the Boltzmann equation. Therefore, we can write this factor as

$$I^{(+)}(\rho_1|t) = -A^{(+)}(\rho_1|t) + B^{(+)}(\rho_1|t), \qquad (9.114)$$

where the analog of the outscattering term (due to phonon emission) is

$$A^{(+)}(\rho_1|t) = (N + 1)\frac{|c_s|^2}{\hbar^2} \int_{-\infty}^{t} dt'\, \exp[-i\Omega(t - t')]a_q U(t, t')a_q^+\rho_1(t')U^+(t, t')$$

$$+ \text{Herm. adjoint,} \qquad (9.115)$$

and the inscattering term is

$$B^{(+)}(\rho_1|t) = (N + 1)\frac{|c_s|^2}{\hbar^2} \int_{-\infty}^{t} dt'\, \exp[-i\Omega(t - t')]U(t, t')a_q^+\rho_1(t')U^+(t, t')a_q$$

$$+ \text{Herm. adjoint.} \qquad (9.116)$$

The corresponding terms for inscattering and outscattering by absorption can be easily written in analogy with these two.

In order to solve for the Wigner equation of motion, we must now Wigner-transform each of the terms that make up the scattering terms and the streaming terms on the left of (9.111). The time-derivative term is straightforward, while the field term is just the leading term of the potential expansion in (9.66). If the field is not constant, then this term is much more complicated and reflects the full nonlocal behavior of the driving potentials. This will be the case in a semiconductor device, for example, in which the actual potential must be solved in a self-consistent manner. The major difference from the treatment in the earlier section lies in the scattering terms on the right-hand side of (9.111). For simplicity, we treat only the term $A^{(+)}$ describing the outscattering of an electron by the emission of a phonon. All other terms and their Hermitian adjoints follow from a similar method of evaluation and result in similar expressions. To evaluate this term, we project it onto a position representation and then Wigner-transform the result according to (9.60). This gives

$$A_W^{(+)}(\rho_1|t) = \int_{-\infty}^{\infty} dz\, e^{ipz/\hbar} \int_{-\infty}^{t} dt' \left\langle x - \frac{z}{2} \left| a_q U(t,t') a_q^+ \rho_1(t') U^+(t,t') \right| x + \frac{z}{2} \right\rangle$$

$$+ \text{ Herm. adj.} \tag{9.117}$$

In order to simplify this, we introduce a spectral operator defined through

$$R(x,p) = \int d\Gamma\, e^{i\Gamma x/\hbar} \left| p - \frac{\Gamma}{2} \right\rangle \left\langle p + \frac{\Gamma}{2} \right|.$$

This operator has the Wigner representation

$$R_W(x,p) = (2\pi)^3 \delta(x - x') \delta(p - p').$$

Thus, $R(x,p)$ maps into the Dirac delta function density in phase space. It should be remembered that this is just a convenient heuristic device and with little meaning. On the other hand, it allows us to shift the various arguments within the expectation value on the right of (9.117) and to develop the Wigner operators for each of the factors in turn. The various properties that can be attributed to $R(x,p)$ can be summarized as follows:

$$\int dx\, dp\, R(x,p) = 1,$$

$$\text{Tr}\{R(x,p)\} = 1,$$

$$A_W(x,p) = \text{Tr}\{A R(x,p)\},$$

$$\text{Tr}\{R(x,p) R(x',p')\} = \delta(x - x') \delta(p - p').$$

This latter can be used to prove the important identity for an operator A:

$$A = \int dx\, dp\, \text{Tr}\{AR(x, p)\}R(x, p).$$

With the aid of the operator $R(x, p)$ and the above operator identity, we can expand the collision operator on the density matrix in terms of the Wigner function, which is a transform of the density operator itself. Then, (9.117) takes the resulting form (for the moment, we let $\hbar = 1$)

$$A_W^{(+)}(f_W \mid t) = (N + 1)|c_s|^2 \int_{-\infty}^{t} dt' \int_{-\infty}^{\infty} dz\, e^{ipz}\, e^{i\Omega(t'-t)}$$

$$\cdot \int_{-\infty}^{\infty} dx_n\, dp_n\, dz_n\, f_W(p_4, x_4, t') \left\langle x - \frac{z}{2} \middle| a_q \middle| x_1 + \frac{z_1}{2} \right\rangle \exp(ip_1 z_1)$$

$$\cdot \left\langle x_1 - \frac{z_1}{2} \middle| U(t', t) \middle| x_2 + \frac{z_2}{2} \right\rangle \exp(ip_2 z_2) \left\langle x_2 - \frac{z_2}{2} \middle| a_q^+ \middle| x_3 + \frac{z_3}{2} \right\rangle$$

$$\cdot \exp(ip_3 z_3) \left\langle x_3 - \frac{z_3}{2} \middle| x_4 + \frac{z_4}{2} \right\rangle \exp(ip_4 z_4)$$

$$\cdot \left\langle x_4 - \frac{z_4}{2} \middle| U^+(t, t') \middle| x + \frac{z}{2} \right\rangle + \text{Herm. adj.}, \qquad (9.118)$$

where the contraction

$$dx_n\, dp_n\, dz_n = dx_1\, dx_2\, dx_3\, dx_4\, dp_1\, dp_2\, dp_3\, dp_4\, dz_1\, dz_2\, dz_3\, dz_4$$

has been used. This form can be reduced and simplified by using the identities

$$\langle x | x' \rangle = \delta(x - x'),$$

$$\int dp\, e^{ipx} = \delta(x),$$

$$\int dx\, e^{ipx} = \delta(p),$$

and the useful property on the generation of the phonon wave vector by the creation and annihilation operators

$$\langle x | a_k | x' \rangle = \langle x | e^{iqx} | x' \rangle = e^{iqx}\delta(x - x'),$$

where q is the phonon wave vector. In addition, we use the properties of the translation operators

$$e^{ipz}|x\rangle = |x + z\rangle, \qquad e^{ixq}|p\rangle = |p + q\rangle,$$

and finally make a change of variables $x_4 - x = x'$, $x_4 + x = x''$. Finally, we find that (after reintroduction of $\hbar$)

$$A_W^{(+)}(f_W|t) = (N + 1)\frac{|c_s|^2}{\hbar^2}\int_{-\infty}^{t} dt' \int_{-\infty}^{\infty} dz\, dx'\, dx''\, dp_4 \exp[i\Omega(t' - t)]$$

$$\cdot f_W\left(p_4, x' + \frac{x''}{2}, t'\right) \exp\left\{\frac{i(p - p_4 - q)x'' + i(p - q)z}{\hbar}\right\}$$

$$\cdot \left\langle -\frac{z}{2}\left|U(t, t')\right|\frac{x}{2}\right\rangle \exp\left[\frac{i(p_4 + q)x'}{\hbar}\right]\left\langle -\frac{x'}{2}\left|U^+(t, t')\right|\frac{z}{2}\right\rangle$$

$$+ \text{Herm. adj.} \tag{9.119}$$

For this expression to be reduced further, an assumption of translational invariance must be invoked. Even in this form, however, certain trends can be shown. The electron will traverse the trajectory given by the propagator U, then it will be scattered. It then drifts under the second propagator and is scattered by the second interaction which finally produces the second-order interaction. During the scattering, the motion of the particle is influenced by the full Hamiltonian (for higher-order scattering) including the applied potentials. In the absence of an electric field, the contribution of the momentum-dependent exponentials reduces to a Dirac delta function in both momentum and energy, corresponding to the integrations in momentum and time, respectively. In the presence of the applied fields, however, the delta functions are smoothed and broadened in keeping with the discussion in Section 9.1. This derivation incorporates the full intracollisional field effects arising from the interference between the scattering and acceleration terms in the Hamiltonian. The collision takes a nonzero amount of time, and the particle can be accelerated during this time. This causes an additional broadening and a shift in the "resonance" accounting for energy conservation, as discussed above. The spectral function that results from (9.119) is not quite the same as that obtained in Section 9.1.3, as might be expected, but the general form for the second-order interaction is quite similar. In the translationally invariant system, we can finally write the scattering terms as

$$I(f_W|k, t) = \int_{-\infty}^{t} dt' \int dk'\, f_W(k', t')I(k, k'|t, t'), \tag{9.120}$$

where the scattering kernel $I(k, k' | t, t')$ is made up of the four terms[35]

$$A_W^{(+)}(k, k' | t, t') = \int dq\, W_q^{(+)} \Delta_q \left(k - \frac{q}{2}, k' - \frac{q}{2} \Big| t, t' \right), \tag{9.121a}$$

$$A_W^{(-)}(k, k' | t, t') = \int dq\, W_q^{(-)} \Delta_q \left(k + \frac{q}{2}, k' + \frac{q}{2} \Big| t, t' \right), \tag{9.121b}$$

$$B_W^{(+)}(k, k' | t, t') = \int dq\, W_q^{(+)} \Delta_q \left(k + \frac{q}{2}, k' - \frac{q}{2} \Big| t, t' \right), \tag{9.121c}$$

$$B_W^{(-)}(k, k' | t, t') = \int dq\, W_q^{(-)} \Delta_q \left(k - \frac{q}{2}, k' + \frac{q}{2} \Big| t, t' \right), \tag{9.121d}$$

with

$$W_q^{(\pm)} = \frac{2\pi}{\hbar} |c_s|^2 (N + \tfrac{1}{2} \pm \tfrac{1}{2}), \tag{9.121e}$$

and the spectral operator is

$$\Delta_q(k, k' | t, t') = \frac{1}{2\pi\hbar} \int dx \int dx'\, \delta[p - p' - eE(t - t')]$$

$$\cdot \exp\left\{ -\frac{i}{\hbar} \left[\hbar\Omega(t - t') - \frac{q}{2}(x - x') + \frac{p + p'}{2}(x + x') \right] \right\}$$

$$\cdot \left\langle \frac{x}{2} \Big| U(t, t') \Big| -\frac{x}{2} \right\rangle \left\langle \frac{x'}{2} \Big| U^+(t, t') \Big| -\frac{x'}{2} \right\rangle + \text{c.c.} \tag{9.121f}$$

It can be seen here how the spectral densities also incorporate the spatial nonlocality of the potentials into the response. The presence of the accelerating fields reflect the influence these fields have on the act of scattering, since the scattering no longer takes place between plane-wave states appropriate to free electrons.

REFERENCES

1. J. R. Barker, *J. Phys. C* **6**, 2663 (1973).
2. K. K. Thornber, *Solid-State Electron.* **21**, 259 (1978).
3. T. N. Theis, J. R. Kirtley, D. J. DiMaria, and D. W. Dong, *Phys. Rev. Lett.* **50**, 750 (1983).
4. T. N. Theis, D. J. DiMaria, J. R. Kirtley, and D. W. Dong, *Phys. Rev. Lett.* **53**, 1445 (1984).
5. D. J. DiMaria, T. N. Theis, J. R. Kirtley, F. L. Pesavento, D. W. Dong, and S. D. Brorson, *J. Appl. Phys.* **57**, 1214 (1985).
6. M. V. Fischetti, *Phys. Rev. Lett.* **53**, 1755 (1984).
7. For example, see *The Physics of SiO₂ and Its Interfaces* (S. T. Pantelides, ed.), Pergamon Press, New York (1978).
8. W. Porod and D. K. Ferry, *Phys. Rev. Lett.* **54**, 1189 (1985).
9. W. Porod and D. K. Ferry, *Physica* **134B**, 137 (1985).
10. M. V. Fischetti, D. J. DiMaria, S. D. Brorson, T. N. Theis, and J. R. Kirtley, *Phys. Rev. B* **31**, 8124 (1985).

11. T. L. L. G. Sollner, W. D. Goodhue, P. E. Tannenwald, C. D. Parker, and D. D. Peck, *Appl. Phys. Lett.* **43**, 588 (1983); **45**, 1319 (1984).

12. E. O. Kane, in: *Semiconductors and Semimetals*, Vol. 1, p. 75 (R. H. Willardson and A. C. Beer, eds.), Academic Press, New York (1966).

13. B. Ricco and M. Ya. Azbel, *Phys. Rev. B* **29**, 1970 (1984).

14. E. Wigner, *Phys. Rev.* **40**, 749 (1932).

15. J. R. Barker, *Physica* **134B**, 22 (1985).

16. U. Ravaioli, M. A. Osman, W. Pötz, N. Kluksdahl, and D. K. Ferry, *Physica* **134B**, 36 (1985); N. C. Kluksdahl, A. M. Kriman, C. Ringhofer, and D. K. Ferry, *Solid-State Electron.* **31**, 743 (1988); N. C. Kluksdahl, A. M. Kriman, and D. K. Ferry, *Superlatt. Microstruct.* **4**, 127 (1988); N. C. Kluksdahl, A. M. Kriman, D. K. Ferry, and C. Ringhofer, *Phys. Rev. B* **39**, 7720 (1989); N. C. Kluksdahl, A. M. Kriman, and D. K. Ferry, *Solid-State Electron.* **32**, 1273 (1989).

17. M. Büttiker, *Phys. Rev. B* **33**, 3020 (1986).

18. W. Frensley, *Phys. Rev. Lett.* **57**, 2853 (1986); *Solid-State Electron.* **31**, 739 (1988).

19. R. Kubo, *J. Phys. Soc. Jpn.* **12**, 570 (1957).

20. H. Mori, *Prog. Theor. Phys.* **33**, 423 (1965).

21. J. J. Niez and D. K. Ferry, *Phys. Rev. B* **28**, 889 (1983).

22. N. Hashitsume, M. Mori, and T. Takahashi, *J. Phys. Soc. Jpn.* **55**, 1887 (1986).

23. S. W. Lovesey, *Condensed Matter Physics: Dynamic Correlations*, Benjamin/Cummings, Reading, MA (1980).

24. D. N. Zubarev, *Nonequilibrium Statistical Mechanics*, Consultants Bureau, New York (1974).

25. V. P. Kalashnikov, *Physica* **48**, 93 (1970).

26. D. K. Ferry, *J. Phys. Colloq.* **42**, C7-253 (1981).

27. J. J. Niez, K. S. Yi, and D. K. Ferry, *Phys. Rev. B* **28**, 1988 (1983).

28. P. A. Lee and D. S. Fisher, *Phys. Rev. Lett.* **47**, 882 (1981).

29. D. S. Fisher and P. A. Lee, *Phys. Rev. B* **23**, 6851 (1981).

30. W. Skocpol, in: *Physics and Fabrication of Microstructures* (M. Kelly and C. Weissbuch, eds.), Springer-Verlag, Berlin (1986).

31. W. Kohn and J. M. Luttinger, *Phys. Rev.* **108**, 590 (1957).

32. J. R. Barker, in: *Physics of Nonlinear Transport in Semiconductors* (D. K. Ferry, J. R. Barker, and C. Jacobini, eds.), Plenum Press, New York (1979).

33. J. R. Barker, *Solid-State Electron.* **21**, 197 (1978).

34. R. Balescu, *Equilibrium and Nonequilibrium Statistical Mechanics*, Wiley, New York (1975).

35. I. B. Levinson, *Sov. Phys. JETP* **30**, 362 (1970).

36. N. Pottier, *Physica* **A117**, 243 (1983).

37. J. E. Moyal, *Proc. Cambridge Phil. Soc.* **45**, 99 (1947).

38. W. J. Skocpol, P. M. Mankiewich, R. E. Howard, L. D. Jackel, D. M. Tennant, and A. D. Stone, *Phys. Rev. Lett.* **56**, 2865 (1986).

10

Noise in Submicron Devices

10.1. INTRODUCTION

The essential problem in noise theory is relating local, microscopically produced fluctuations at an interior point to the observable, fluctuations in the macroscopic terminal currents and voltages. A variety of schemes for this have been developed, and a good review can be found in van Vliet et al.[1] The oldest scheme is the "salami" method, in which the device is subdivided into slices. Each slice possesses a local noise source, and a noise voltage appears across the slice as a result. The total mean square noise voltage is the sum of the mean square noise voltages of the individual slices. This summation process ignores correlations between slice noise voltages. In more sophisticated variants, a tensor Green's function is defined, which relates fluctuations in field at point x to fluctuations in current or carrier density at point x'. The terminal noise voltage then is an integral over the product of the Green's function and the noise source. The terminal mean square noise voltage is then the *mean* of the product of two such integrals. This mean involves a two-point spatial correlation function for the noise source. Although such spatial correlations of the microscopic fluctuations do exist, they are generally ignored.

In this chapter we will usually be more interested in the noise source modeling than we are in the development of the Green's functions. We will start with a discussion of how velocity fluctuations lead directly to both noise and diffusion. The role of intervalley scattering in this process will be discussed. We will discover that the spectral density of the velocity fluctuations is a crucial function. Our discussion will then proceed to present a simple but incorrect model of the spectrum. Monte Carlo estimation of the spectral density will then be discussed, and more correct results will be presented. Some simple forms used in FET models for describing the magnitude of the velocity fluctuations will be discussed next. We then will consider the spatial correlations mentioned above. While, as we already have noted, little attention will be paid to the estimation of the pertinent Green's functions, we review some models in which nonstationary transport effects were included in the modeling process.

Statistical fluctuations in the injection of carriers into a region of interest for shot noise will then be addressed. We are particularly interested in Poisson point process models and transit-time effects. We will end with a discussion of certain correlations which occur in time-periodic systems such as heterodyne receivers and oscillators, with their most thorough exploration having been with regard to shot noise in mixers. We argue that similar correlations are expected in velocity fluctuation or thermal noise as well.

10.2. VELOCITY FLUCTUATION NOISE SOURCES

10.2.1. Velocity Fluctuations and Noise Spectral Densities

An electron moves through a semiconductor with velocity $v_x(t)$, which can be viewed as a stochastic process with some nonzero mean $\overline{v_x(t)}$. (Unless otherwise specified, all averages are ensemble averages.) Then the random fluctuations can be modeled by $\Delta v_x(t) = v_x(t) - \overline{v_x}$, whose spectral density is [2]

$$S_{\Delta v_x}(\omega) = 2 \int_{-\infty}^{\infty} \overline{\Delta v_x(t)\, \Delta v_x(t+s)} \exp(j\omega s)\, ds \qquad (10.1)$$

$$= 4\,\mathrm{Re}[D(\omega)], \qquad (10.2)$$

where $D(\omega)$ is the complex-valued frequency-dependent diffusion coefficient defined by

$$D(\omega) = \int_0^{\infty} \overline{\Delta v_x(t)\, \Delta v_x(t+s)} \exp(-j\omega s)\, ds. \qquad (10.3)$$

The factor of 2 in (10.1) essentially converts the double sided transform into a single-sided transform. Therefore, the frequency used on the left-hand side of that equation actually should be restricted to nonnegative values in its application. The velocity fluctuation noise current created by $\Delta v_x(t)$ of a single electron located somewhere inside a region of length x_d and cross-sectional area A is

$$\Delta i(t) = \frac{e\Delta v_x(t)}{x_d}. \qquad (10.4)$$

It has spectral density

$$S_{\Delta i}(\omega) = \frac{e^2}{x_d^2} S_{\Delta v_x}(\omega) = \frac{4e^2\,\mathrm{Re}[D(\omega)]}{x_d^2}. \qquad (10.5)$$

If the carrier density is $n(x)$, then the total noise-current spectral density is

$$S_i(\omega) = \int_0^{x_d} \frac{4e^2 n(x) \, \text{Re}[D(\omega, x)] A}{x_d^2} \, dx \qquad (10.6)$$

$$= \frac{4e^2 A}{x_d^2} \int_0^{x_d} n(x) \, \text{Re}[D(\omega, x)] \, dx, \qquad (10.7)$$

where we are including the possibility that our diffusion coefficient also varies as we move through the region.

10.2.2. Velocity Fluctuations and Diffusion

Now assume a collection of particles having velocities $v_x(t)$ are all located at $x = 0$ at time $t = 0$. Their mean square position fluctuation at time T then is

$$\overline{\Delta x_j^2} = \int_0^T ds \int_{-s}^{T-s} \overline{\Delta v_x(s) \, \Delta v_x(s+w)} \, dw. \qquad (10.8)$$

If T is sufficiently long, (i.e., much longer than the time frame over which the $\Delta v_x(t)$ velocity is expected to be correlated with the velocity distribution at $t = 0$, then

$$\overline{\Delta x_j^2} \cong \int_0^t du \int_{-\infty}^{\infty} \overline{\Delta v_x(u) \, \Delta v_x(u+w)} \, dw \cong \int_0^t du \, [2D_0] = 2D_0 t. \qquad (10.9)$$

Therefore (10.3) reduces to the Einstein formula in the steady state, even in a nonequilibrium setting.

An example of what is meant by long time frames is a situation where phonon scattering randomizes the velocity with an associated mean free time τ. Then

$$\overline{\Delta v_x(s) \, \Delta v_x(s+t)} = \langle \Delta v_x(s)^2 \, e^{-t/\tau} \rangle_E. \qquad (10.10)$$

Here τ is a function of energy only and $\langle \ \rangle_E$ denotes an expectation or averaging over energy. The low-frequency diffusion coefficient then is

$$D = \int_0^{\infty} \langle \Delta v_x^2 \, e^{-t/\tau} \rangle_E \, dt. \qquad (10.11)$$

Interchanging the order at integration and averaging yields

$$D = \langle \Delta v_x^2 \tau \rangle_E. \qquad (10.12)$$

For carriers of isotropic effective mass, $\langle v^2 \rangle = 3 \langle \Delta v_x^2 \rangle$, $3kT = m \langle v^2 \rangle$, and therefore

$$D = \frac{kT}{e} \mu, \qquad (10.13)$$

where

$$\mu = \frac{e\langle v^2 \tau \rangle}{\langle v^2 \rangle m}. \tag{10.14}$$

Therefore (10.3) is consistent with the Einstein relation for at least one important case in the near-equilibrium setting. These points were also discussed in Chapter 4.

As can be seen, the velocity fluctuation component Δv leads to two measurable physical effects: diffusion and noise. This common physical origin explains why the label "diffusion noise" is often applied to velocity fluctuation spectrum.[3,4] By measuring the low-frequency portion of this noise spectrum, we can measure diffusion coefficients;[5] i.e.,

$$D = \tfrac{1}{2} S_{\Delta v_x}(0). \tag{10.15}$$

10.2.3. Connection with Thermal Noise

The connection between velocity fluctuations and thermal noise is established by considering the case of thermal equilibrium. In (10.7) we let $n(x) = n$ and $\mathrm{Re}[D(f, x)] = D_0$. Then

$$S_i(f) = \frac{4e^2 A n D_0}{x_d}. \tag{10.16}$$

Now we use the Einstein relation (thus restricting ourselves to an equilibrium case) and obtain

$$S_i(f) = \frac{4eAnkT\mu_0}{x_d} = \frac{4kT\sigma A}{x_d}, \tag{10.17}$$

where σ is the conductivity. Equation (10.17) is the expression for thermal noise. Therefore thermal noise is the diffusion noise observed from a sample in a new thermal equilibrium state.

10.2.4. Diffusion Noise in Nonequivalent Valleys

A useful division of the total spectrum in multivalley semiconductors was suggested by Shockley et al.[3] and Price.[4] In a two-valley semiconductor, three components may be considered: one due to electron motion in the first valley, one due to electron motion in the second valley, and a "partition noise" spectrum due to the random partition of the electrons between the two valleys. Here we will follow the analysis of Shockley et al.[3] The subscripts c and s denote the central and satellite valleys respectively, ξ is the fraction of the carriers found in the central valley, v_c and v_s are the mean (drift) velocities, and Δv_c and Δv_s

are the velocity fluctuation components. Under steady-state conditions it is required that

$$\frac{\xi}{\tau_{cs}} = \frac{1-\xi}{\tau_{sc}}, \tag{10.18}$$

where τ_{cs}^{-1} is the central to satellite transition rate of central valley electrons and τ_{sc}^{-1} is the satellite-to-central transition rate for satellite valley electrons.

Equation (10.18) can be rewritten as

$$\xi = \frac{\tau_{cs}}{\tau_{cs} + \tau_{sc}}. \tag{10.19}$$

An effective decay constant for any disturbance from the steady-state values can be derived. Let

$$\frac{dn_c}{dt} = -\frac{n_c}{\tau_{cs}} + \frac{n_s}{\tau_{sc}} = \left(\frac{-\xi}{\tau_{cs}} + \frac{1-\xi}{\tau_{sc}}\right)n \tag{10.20}$$

$$\frac{dn_s}{dt} = \frac{n_c}{\tau_{cs}} - \frac{n_s}{\tau_{sc}} = \left(\frac{\xi}{\tau_{cs}} - \frac{1-\xi}{\tau_{sc}}\right)n \tag{10.21}$$

$$\frac{d}{dt}\left(\frac{n_c}{\tau_{cs}} - \frac{n_s}{\tau_{sc}}\right) = -\left(\frac{1}{\tau}\right)\left(\frac{n_c}{\tau_{cs}} - \frac{n_s}{\tau_{sc}}\right), \tag{10.22}$$

where

$$\frac{1}{\tau} = \frac{1}{\tau_{cs}} + \frac{1}{\tau_{sc}}. \tag{10.23}$$

Following Shockley et al., the diffusion coefficient then is

$$D(\Delta v_r, \omega) = \frac{\xi(1-\xi)\Delta^2\tau}{1 + (\omega\tau)^2} + \xi D(\Delta v_c, \omega) + (1-\xi)D(\Delta v_s, \omega), \tag{10.24}$$

where Δv_r is the "average" velocity fluctuation, Δv_c is central valley electron velocity fluctuation, and Δv_s is the satellite valley electron velocity fluctuation. The term Δ is a volume element large enough to allow a carrier to make many transitions between the valleys. The terms $D(\Delta v_c, \omega)$ and $D(\Delta v_s, \omega)$ are the central and satellite valley diffusion coefficients and are defined by the real part of (10.3). This expression is then used in the diffusion–impedance field noise formula, which is[3]

$$S(\delta V_n, \omega) = \int |\nabla Z_{vr}|^2 4e^2 D(\Delta v_r, \omega)n\, d(\text{vol}), \tag{10.25}$$

where $S(\delta V_1, \omega)$ is the diffusion noise-voltage spectral density, n is the carrier concentration, and ∇Z_{vr} is the impedance field vector.[3] The impedance field vector essentially is an example of the Green's functions alluded to earlier. It transforms a local noise current source, the diffusion or velocity fluctuation noise, into a terminal noise voltage. We again refer the reader to the discussion by van Vliet et al.[1] for more details.

10.3. THE SPECTRAL DENSITY FUNCTIONS

10.3.1. Physical Background

The preceding discussions clearly show that an important issue is the spectral density of the velocity fluctuations. Much is known about such spectra as a result of Monte Carlo estimation of their form. They do not have the simple form expected, that of a flat behavior out to some knee frequency, above which a declining spectral density is seen. Instead, a sharp peak is seen roughly near a terahertz for the velocity fluctuations parallel to the applied field. The simple model is seen for the fluctuations perpendicular to the field. The simpler thinking includes the existence of scattering but ignores the field acceleration between scatterings, and it is this field acceleration that creates the peak. There are some practical difficulties faced in spectral estimates, however, and in this section we discuss the estimation of the velocity fluctuation spectrum in multivalley semiconductors in the presence of a constant electric field.

We start by doing the simple analysis. Assume that

$$\overline{\Delta v_x(t)\,\Delta v_x(t+s)} = \overline{\Delta v_x^2(s)\,e^{-t/\tau}}. \tag{10.26}$$

This assumption, while intuitively appealing, we will discover later is wrong. However, if we make this assumption, then

$$D(\omega) = \int_0^\infty \overline{\Delta v_x^2(s)\,e^{-t/\tau}}\,e^{-j\omega t}\,dt. \tag{10.27}$$

Interchanging order of integration and expectation operations and then integrating yields

$$D(\omega) = \left\langle \Delta v_x^2(s)\,\frac{\tau}{1+j\omega\tau} \right\rangle, \tag{10.28}$$

where the brackets and the bar both denote expectation. For constant τ, van der Ziel and van Vliet[6] have shown that a high-frequency Einstein relation exists between the real part of D and the corresponding complex-valued mobility's real part. Equation (10.28) suggests a very crude fudge for high-frequency diffusion coefficients, which is

$$D(\omega) = \frac{D_{dc}}{1+(\omega\tau)^2}. \tag{10.29}$$

Another approach uses a nonconstant τ but a constant mean free path L. Then

$$\tau = \frac{L}{v},\tag{10.30}$$

and by equating our Δv_x^2 to $v^2/3$, where v is the "thermal" velocity of the electrons, (10.28) can be rewritten as

$$D(\omega) = \left\langle \frac{v^2}{3} \frac{\tau - j\omega\tau^2}{1 + (\omega\tau)^2} \right\rangle \tag{10.31}$$

$$= \frac{L}{3} \left\langle \frac{v - j\omega L}{1 + (\omega L/v)^2} \right\rangle. \tag{10.32}$$

The real part of $D(\omega)$ is then

$$\mathrm{Re}[D] = \frac{L}{3} \left\langle \frac{v}{1 + (\omega L/v)^2} \right\rangle. \tag{10.33}$$

At high fields, where inelastic phonon scattering dominates, a critical frequency dependent on L can be defined as

$$\omega_c = \left(\frac{\langle v \rangle}{L} \right). \tag{10.34}$$

Following Moll,[7] we express $\langle v \rangle$ in terms of the electron temperature T_e as

$$\langle v \rangle = \sqrt{\frac{8kT_e}{\pi m}}. \tag{10.35}$$

Therefore

$$\omega_c = \frac{1}{L} \left(\frac{8kT_e}{\pi m} \right)^{1/2}. \tag{10.36}$$

We express kT_e as

$$kT_e = \frac{e^2 L^2 F^2}{3 E_{op}}, \tag{10.37}$$

where F is the electric field and E_{op} is the optical phonon energy. Then

$$\omega_c = \frac{1}{L} eLF \left(\frac{8}{3\pi m E_{op}} \right)^{1/2} \tag{10.38}$$

$$= eF \left(\frac{8}{3\pi m E_{op}} \right)^{1/2}. \tag{10.39}$$

The simple-minded conclusion is that the velocity fluctuation spectrum is flat out until a cutoff frequency determined by an energy balance between the electric field and the inelastic scattering process. However, as we note in the following discussion, this is incorrect. Here, it was implicitly assumed that only scattering events change the velocity fluctuations. What has been ignored is that the field acceleration of the carriers between scattering events also changes the velocity fluctuations for the velocity component parallel to the applied field.

10.3.2. Mathematical Background

In the notation used in this section, the velocity of a single electron as a function of time, as discussed at the beginning of Section 10.2.1, is represented by a random process which we call the velocity process. The velocity process is

$$v(t) = E[v] + \Delta v(t) \tag{10.40}$$

where $E[v]$ is the expected or mean value of v, the velocity of a single electron, and Δv is a random fluctuation about this mean. We assume that this process is wide-sense stationary. This requires that the mean is constant and therefore we are restricted to the case of steady-state transport in a constant electric field. This, however, also requires that the fluctuations $\Delta v(t)$ be represented by a wide-sense stationary process whose spectrum we will estimate.

Since we are also considering multivalley semiconductors, we model the valley occupied by a single electron by a second, wide-sense stationary, random process which we call the valley process. The valley process is developed by assigning the numerical identifier 1 to the central valley and 2 to satellite valleys. The valley random variable $\text{val}(t)$ is assigned the identifier of the valley occupied by the electron at time t.

By using Monte Carlo techniques in conjunction with various statistical methods, we can obtain the autocovariance and spectrum of the velocity and valley random processes. The autocovariance of the velocity process is denoted by

$$K_v(t_1, t_1 + \tau)$$
$$= E(\{v(t_1) - E[v(t_1)]\}\{v(t_1 + \tau) - E[v(t_1 + \tau)]\}). \tag{10.41}$$

Due to our assumption of wide-sense stationarity this also is the autocovariance of the velocity fluctuation process. The autocovariance of the valley process is

$$K_{\text{val}}(t_1, t_1 + \tau)$$
$$= E\{\langle \text{val}(t_1) - E[\text{val}(t_1)]\rangle\langle \text{val}(t_1 + \tau) - E[\text{val}(t_1 + \tau)]\rangle\}. \tag{10.42}$$

A positive valley autocovariance means that it is more likely that the electron is in equivalent valleys at the two times than the unconditional probability of finding the electron in the original valley i at time $t + \tau$.

The velocity fluctuation spectrum is given by the Fourier transform of the velocity fluctuation autocovariance with respect to the lag time τ (see, e.g., Ref. 8); that is,

$$S_v(f) = \int_{-\infty}^{\infty} K_v(t_1, t_1 + \tau) \, e^{-j2\pi f\tau} \, d\tau. \tag{10.43}$$

If the spectrum is to be well defined, the dependence of K_v on the absolute time t_1 must be known. The usual assumption of wide-sense stationarity adopted here (but modified in our discussion of time-periodic systems) requires K_v to be independent of t_1. At first glance there would seem to be a factor of 2 discrepancy between (10.43) and (10.1). That is because here we do not introduce the factor of 2 used in (10.1) to convert a double-sided Fourier transform into a single-sided transform.

10.3.3. Statistical Procedures

The basic techniques for using time series data for the estimation of autocovariances and spectra are well understood and described in a variety of texts (e.g., Ref. 8), and computer programs for their evaluation are available publicly (e.g., Ref. 9). The discussion in this section therefore concentrates mainly on the particular methods and problems associated with time series analysis of Monte Carlo generated hot-carrier transport data. The discussion is divided into three parts: a general discussion of time series analysis, then of window function properties, and finally of the selection of time series length and sampling rate.

Background to Time Series Analysis. A crude Monte Carlo estimation of an autocovariance may be obtained by first generating k independent realizations or time series of the basic Monte Carlo experiment and then averaging the quantities $\Delta v(t_1) \, \Delta v(t_1 + \tau)$ over each realization. Under fairly weak conditions the ensemble average of these results (over the ensemble of realizations) converges (almost) surely to the actual autocovariance of the underlying random process. Alternatively, a single time series can be used with the additional assumption of ergodicity to obtain an estimate of the same autocovariance. In particular, if the single time series is observed at intervals Δt with $r\,\Delta t = \tau$ then[8]

$$K_{\Delta V}(\tau) = \lim_{L \to \infty} \frac{1}{L} \sum_{j=M}^{M+L-r} \Delta v(j) \, \Delta v(j + r), \tag{10.44}$$

where M is some starting value, L is the number of time points for which data are available, and equality means convergence with probability 1. The corresponding estimate of the spectrum from the same single time series is the periodogram, defined by[8]

$$I_v(f) = \frac{\Delta t}{N} \left| \sum_{n=-N/2}^{n=N/2-1} v(t) \exp(-j2\pi f \, \Delta t) \right|^2, \tag{10.45}$$

where

$$-\frac{1}{2\,\Delta t} \le f \le \frac{1}{2\,\Delta t}$$

and $v(t)$ is sampled at N data points, spaced Δt apart, over the interval $(-N\,\Delta t/2, [N/2-1]\,\Delta t)$. The inverse transform of $I_v(f)$ is the time-average autocovariance of the data set. (In practice, it is computationally more efficient to use fast Fourier transforms (FFTs) to calculate and invert $I_v(f)$ than it is to calculate $K_v(\tau)$ directly from the time average.)

It is important to note that even for an ergodic process

$$\lim_{L \to \infty} I_v(f) \ne S_v(f); \tag{10.46}$$

i.e., the periodogram does not limit the spectrum, as is sometimes assumed intuitively. As Jenkins and Watt[8] describe, the left-hand side of (10.45) is actually a random variable whose variance is proportional to $S_v^2(f)$, the spectrum of the random process which is generating the time series. Since the properties of the underlying process are independent of the time series length, increasing the time series length has no effect on the variance or mean square error of the periodogram estimate of $S_v(f)$. Increasing the time series length improves the resolution (it makes each periodogram point an estimate of the power in a narrower frequency band) without improving the accuracy of the estimate.

Having noticed that the periodogram is a very noisy estimate of $S_v(f)$, the natural question to ask is, how does one estimate a noise spectra? While there is a rich literature on this topic, only a few methods have been applied in the context of our problem here. In particular, two approaches have previously been used to reduce the variance in estimates of $I_v(f)$ obtained from Monte Carlo generated time series. One approach is to average many periodograms,[10] while the other is to Fourier-transform a truncated velocity fluctuation autocorrelation function.[11,12] Actually, both of these procedures are particular cases of a more general smoothing procedure in which the periodogram is smoothed by the use of a "lag window."[13] In the window smoothing procedure the spectral estimate is obtained by the Fourier transform[8]

$$\int_{-\infty}^{\infty} K_v(\tau) w(\tau) e^{-j2\pi f\tau} \, d\tau, \tag{10.47}$$

where $w(\tau)$ is the lag window function. Averaging many periodograms is equivalent to using a Bartlett window (see Table 10.1), while truncating the autocorrelation is equivalent to imposing a rectangular window. Since a window smoothing is implicitly involved, it is necessary to understand the effect of the window on the estimate. It turns out that this window-smoothing approach is very similar to a convolution used by Rees[14] to eliminate unphysically long time constraints in a distribution function based calculation of transferred electron device noise.

TABLE 10.1. Lag and Spectral Window Functions

Window name	Lag Window Function	Spectral window function
Rectangular	$w_R(u) = \begin{cases} 1 & \lvert u \rvert \leq M \\ 0 & \lvert u \rvert > M \end{cases}$	$W_R(f) = 2M \sin(2\pi f M)/2\pi f M$
Bartlett	$w_B(u) = \begin{cases} 1 - \lvert u \rvert / M & \lvert u \rvert \leq M \\ 0 & \lvert u \rvert > M \end{cases}$	$W_B(f) = M[(\sin \pi f M)/\pi f M]^2$
Parzen	$w_P(u) = \begin{cases} 1 - 6(u/M)^2 + 6(\lvert u \rvert / M)^3 & \lvert u \rvert \leq M/2 \\ 2[1 - (\lvert u \rvert / M)]^3 & M/2 < \lvert u \rvert \leq M \\ 0 & \lvert u \rvert > M \end{cases}$	$W_P(f) = \tfrac{3}{4} M[(\sin \pi f M/2)/(\pi f M/2)]^4$

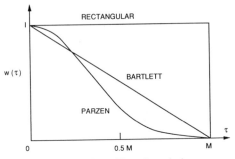

FIGURE 10.1.   Three lag windows.

Window Function Properties.   The functional forms of the rectangular, Bartlett, and Parzen windows are given in Table 10.1 and plotted in Figure 10.1. Details of other windows are given in Ref. 8. Since time domain multiplication corresponds to frequency domain convolution, the Fourier transforms of the three windows, shown in Figure 10.2, are basic to understanding their effect on the estimate. The Parzen window has the particular advantage of having reduced sidelobes, thereby decreasing one mechanism (leakage) by which spurious artifacts can be introduced into the spectrum. The Parzen window also has the property that it will not provide negative estimates. The Bartlett and rectangular windows also have this property, but some windows do not.

In Table 10.2 several figures of merit for the various windows are provided. The bandwidth represents the minimum size spectral structure which can be

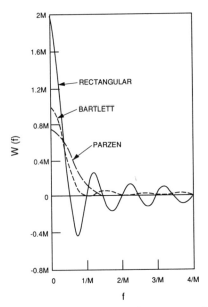

FIGURE 10.2.   Three spectral windows that correspond to Figure 10.1.

TABLE 10.2. Properties of Spectral Window Estimates

Window name	Bandwidth	Bias[a]	Variance ratio
Rectangular	0.5/M	[b]	2M/T
Bartlett	1.5/M	Order 1/M	0.667M/T
Parzen	1.86/M	$(0.152/M^2)S^{(2)}(f)^c$	0.539M/T

[a] The bias of all the windows is proportional to the second derivative of the real spectrum with respect to frequency.
[b] A convolution integral expression is provided by Jenkins and Watts.[8]
[c] $S^{(2)}(f)$ is the second derivative of $S(f)$ with respect to frequency.

resolved by the estimate. The bias and the variance are figures of merit for any statistical estimation procedure. Since any statistical estimate will yield different values for different data sets generated by the same underlying random processes, a statistical estimate can be viewed as a random variable which will have an expected or average result and a variance describing the mean square or standard deviation of the estimate from its expected value. The estimate bias is the difference between the actual quantity and the expected or average estimation result and is therefore a measure of the average or expected error. Hopefully, the bias of the estimation procedure is zero. The variance describes the expected spread or "noise" of the estimate. The variance ratio shown in Table 10.2 is the ratio of the smoothed spectral estimate variance to the variance of the unsmoothed periodogram (which is proportional to S_v). The bias is proportional to the second derivative of $S_v(f)$ with respect to frequency, so the spectrum tends to be underestimated around peaks and overestimated around troughs. While ideally both the bias and variance would be zero, in practice there is a trade-off between the two. Table 10.2 shows that increasing the window width M decreases the bias but increases the variance. The optimum M is found by performing a window-closing experiment in which the time series is analyzed using increasing values of M. The window length is increased until the resulting spectrum contains all the structures which are known to be present and no additional structures which cannot be explained. A more useful criterion does not exist.

Choice of Time Series Length and Sampling Rate. There are two considerations involved in choosing a lower bound for the time series length T. The first is that the finite length of the time series itself imposes a window, the data window, implying a window convolution of the form

$$S_{obs}(f) = \int_{-\infty}^{\infty} S_v(g) \frac{T \sin \pi (f - g) T}{\pi (f - g) T} dg. \qquad (10.48)$$

Therefore, T should be chosen such that $S_v(f)$ is constant over the main sine function lobe, e.g.,

$$\frac{0.5}{T} \leq \Delta f_{flat}, \qquad (10.49)$$

where Δf_{flat} is the smallest frequency band across which the spectrum can be approximated as flat. A good choice for Δf_{flat} for hot-electron transport is 20 to 30 GHz. The corresponding time series length is 20 ps or longer. The other factor is the desirability of ensuring that the behavior with the longest associated time constant repeats often enough to be adequately modeled statistically.

The time step chosen (or, equivalently, sampling rate) can play a critical role. The process of sampling the data at a fixed rate, every Δt, alters the spectrum to

$$S_{\text{alias}}(f) = \frac{1}{\Delta t} \sum_{n=-\infty}^{\infty} S_v\left(f - \frac{n}{\Delta t}\right). \tag{10.50}$$

If Δt is not small enough to ensure that

$$S_v(|f|) \ll S_v\left[\left|f - \frac{1}{2\Delta t}\right|\right] \tag{10.51}$$

for all $f \geq (2\Delta t)^{-1}$, then significant spectral power will be transferred from high frequencies to low frequencies by the sampling process. This source of error is called aliasing. For hot-electron transport, as we indicated in our earlier analysis, the spectral power is expected to become small only at frequencies in excess of the most rapid scattering rates. There is significant spectral power out to several terahertz, and time steps in excess of a few hundreths of a picosecond could lead to significant aliasing problems.

10.3.4. Discussion of Monte Carlo Results for Velocity Fluctuation Spectra

A variety of workers[10,11,13] have estimated temporal correlations and power spectra for hot electrons. InP, GaAs, and Si have all been investigated and found to exhibit the same fundamental behaviors. The most important observation is the split between fluctuations of components parallel to the field and components perpendicular to the field. In Figure 10.3 we show the type of behavior seen for the perpendicular component, which is a simple decay behavior along the lines of (10.11). The spectral density that corresponds to this behavior is shown in Figure 10.4. It indeed does resemble the form predicted by the analysis of Section 10.3.1.

In most cases, however, the velocity component which is parallel to the field is more important. For it a very different behavior is seen. The autocorrelation function for velocity fluctuations parallel to the field behaves in a fashion similar to that shown in Figure 10.5. There is strong swing into a negative-valued region during the second half of the first picosecond. This is easily understood when one pictures a single electron, being accelerated by a field between scattering events, with a well-established average velocity. Since the electron is fluctuating about this average, there is a tendency for positive velocity fluctuations to be transformed into negative-valued fluctuations (by scattering) and for negative

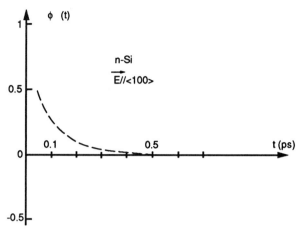

FIGURE 10.3. Transverse velocity fluctuation correlation function for electrons in silicon. A field of 20 kV/cm was applied along the ⟨100⟩ direction. After Fauquembergue *et al.*[11]

fluctuations to be transformed into positive-valued fluctuations (by both scattering and field acceleration). This change in the sign of the fluctuation is illustrated by this negative-valued portion of the autocorrelation function. It produces the strong peak seen in the low-terahertz range in the spectral density shown in Figure 10.6.

Several of the estimates have considered materials with nonequivalent valleys. Only one group however[13] has reported seeing the partition noise component expected from the analysis of Shockley *et al.* (reviewed in Section 10.2.3). They saw a bump in the "low-frequency" portion of the spectrum (around several hundred gigahertz), which they attributed to an intervalley cycling time shown by their valley autocovariance function.

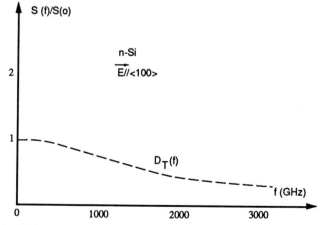

FIGURE 10.4. Noise power spectral density that corresponds to Figure 10.3. After Fauquembergue *et al.*[11]

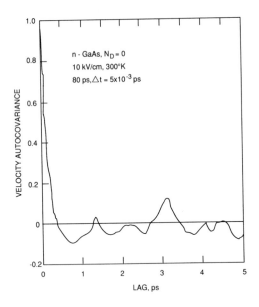

FIGURE 10.5. Parallel velocity fluctuation autocorrelation (or velocity autocovariance) for electrons in GaAs. A two-conduction-band valley model was used. After Grondin et al.[13]

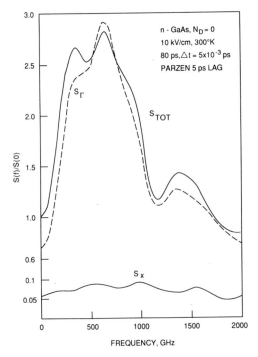

FIGURE 10.6. Noise spectral density function that corresponds to Figure 10.5. After Grondin et al.[13]

10.3.5. The Quantum Correction Factor

At sufficiently high frequencies, quantum mechanical effects appear in our statistical descriptions via the Planck distribution function.[2,15] Under these circumstances the spectral density of both thermal and shot-noise is multiplied by a quantum correction factor $p(f)$, defined as

$$p(f) = \frac{hf}{kT}\left[\frac{1}{2} + \frac{1}{\exp[hf/kT] - 1}\right], \tag{10.52}$$

where T is the absolute temperature, k is Boltzmann's constant, and h is Planck's constant. This correction factor includes both the Planck distribution function and zero-point fluctuation. The zero-point fluctuations dominate in the infrared $(f > 10\text{ THz})$, and the Planck distribution is dominant at microwave and millimeter-wave frequencies. The quantum correction factor is unity out to frequencies in the terahertz range at room temperature.

10.3.6. Effective Noise Temperatures

As the above discussion clearly shows, the process of modeling the noise sources can be extremely complex. Therefore, quite commonly this is not done. A common approximation is to model the velocity fluctuation noise as being a white thermal noise with an effective noise temperature. In this section we review some of the forms that have been used in GaAs FET noise modeling for the field dependence of this noise temperature. Our starting point is the noise figure (NF), which is related to equivalent noise temperature T_n by

$$\text{NF} = 10\log_{10}\left(1 + \frac{T_n}{290}\right), \tag{10.53}$$

where T_n depends on the electric field and the lattice temperature T_o. Baechtold[16] described the enhancement of noise temperature by an electric field as

$$T_n = 1 - \delta\left(\frac{E}{E_s}\right)^n \tag{10.54}$$

where E is the electric field, E_s is the velocity saturation field and n is an empirical constant.

Other workers have used both different parameters and different field dependencies. Pucel, Haus, and Statz[17] used the form of (10.54) with the parameters are listed in Table 10.3. J. Frey[18] used Monte Carlo techniques to investigate this noise using the Baechtold values (see Table 10.3). J. Graffeuil et al.[19] suggested the form

$$\frac{T_e}{T_o} = 1 + \alpha\,\exp\left(\beta + \frac{E}{E_{\text{sat}}}\right), \tag{10.55}$$

TABLE 10.3. Noise Temperature Parameters
at 300 K for $n = 3$

Reference	δ	E_s (kV/cm)
Pucel et al.[17]	1.2	5
Frey[18]	6	5
Graffeuil et al.[19]	0.7–2.5 (varies inversely with T)	not used
Golio and Trew[20]	3.09	5
Weinreb[22]	0.64	2.96
	1.2	2.9
	1.2	3.8

which reduces to (10.54) for $E/E_{sat} \ll 1$, with $\delta = \alpha \exp \beta$. While Graffeuil found that the value of $\beta = 3$ was appropriate for a large collection of GaAs devices, he noted that even small variations in β would lead to large variations in T_e, and values ranging from 0.7 to 2.5 would be obtained. J. M. Golio and R. J. Trew[20] also used Monte Carlo techniques; they obtained the values shown in Table 10.3. They noted that the variations in δ account for changes in noise figure in the range of 0.15 dB. While Frey,[18] Graffeuil,[19] and Takagi and van der Ziel[21] all found δ to be a strong function of temperature, S. Weinreb[22] suggests that this temperature dependence is partly a result of a normalization scheme in which both thermal and nonthermal components have been normalized to the ambient temperature. He suggests the form

$$T_e = 290 \left[\frac{T_o}{290} + f(E) \right], \tag{10.56}$$

where $f(E)$ is a term that is strongly dependent on field but only weakly dependent on temperature. His $f(E)$ can be related to the more common form by $f(E) = (E/E_s)^3$. The parameters used by Weinreb are also given in Table 10.3.

10.4. SPATIAL CORRELATIONS AND NONSTATIONARY TRANSPORT

It seems strange, but the realization that velocity fluctuations have a spatial correlation occurred rather late. The arguments for the existence of these spatial correlations are identical with those made for the existence of velocity overshoot or quasi-ballistic transport in submicron-gate GaAs FETs, and a proper noise theory for these devices therefore should include this correlation. Yet while there is an occasional hint of earlier awareness (e.g., Ref. 13), these correlations were not actually considered until Lugli and the present authors estimated them in 1982[23] and Nougier *et al.* demonstrated their existence analytically in 1983.[24] Yet, all of the terminal noise calculations either explicitly or implicitly assumed that the local velocity fluctuations at two points x and x' are uncorrelated.

10.4.1. Energy and Velocity Fluctuation Correlation Functions

As discussed in Chapter 5, ensemble Monte Carlo (EMC) techniques are particularly suited for the calculation of transient dynamic response. As developed by Lebwohl and Price[25] and subsequently used by one of the present authors[26,27] the EMC technique is a hybrid method in which an ensemble of electrons is adopted. The EMC method has advantages over the normal MC technique in that an ensemble distribution function exists at each time step. Lugli *et al.*[23] extended an EMC model of electron transport in silicon to include one spatial dimension. By solving Poisson's equation simultaneously with the EMC transport model, one can model a submicron device. They used this method to investigate the correlations of energy and momentum fluctuations. Both spatial and temporal and auto- and cross-correlations were estimated. The temporal correlations are studied in a spatially homogeneous system, while the spatial correlations are evaluated inside a generic Si submicron device at 300 K.

In Figures 10.7 and 10.8, the temporal correlation of the velocity and energy fluctuations are shown. In these steady-state results the homogeneous electric field is 50 kV/cm, and the results are consistent with earlier results.[26,27] These fluctuations are found to decay in a time of the order of a few tenths of picoseconds. Of particular interest are the cross-correlations of energy and momentum shown in Figure 10.8. These cross-correlations are not negligible and are asymmetric in the choice of reference variable. This type of behavior is suggested by the argument that energy fluctuations provide a source of velocity fluctuations.[26,28,29] In particular, the nonnegligible nature of these correlations are characteristic of a non-Maxwellian distribution function, as these correlations are usually ignored in generalized moment equation approaches.[30,31]

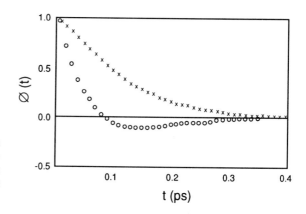

FIGURE 10.7. Autocorrelation functions for fluctuations in velocity (o) and energy (x) for electrons in silicon. A field of 50 kV/cm was applied. After Lugli *et al.*[23]

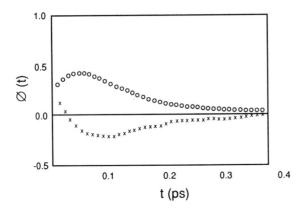

FIGURE 10.8. Cross-correlation of energy and velocity fluctuations. This plot corresponds to Figure 10.7. Circles are for the case $\langle \Delta v(t) \Delta E(t + \tau) \rangle$ while crosses are for the case $\langle \Delta E(t) \Delta v(t + \tau) \rangle$. After Lugli et al.[23]

The spatial correlation functions were calculated in a $n^+ - n - n^+$ Si structure with an active region length of 0.2 μm. In Figure 10.9 we show the electric field and carrier concentration profiles through the device for a bias current of 1.5×10^5 A/cm^2 and a voltage drop of 1.16 V across the device. In analyzing spatial correlation functions in such a device, care must be taken to include accurately the possibility that an individual electron may cycle through a single spatial cell several times (in opposite directions) as it is scattered. The procedure used has started with an ensemble in the cathode. Each electron was simulated until it left the device. For each electron, the average values of the state descriptors in each cell are found. This is the average over the various Monte Carlo time steps in which the electron was located inside the cell of interest. The spatial correlations were then evaluated over these values. During this study, the field profile was fixed to that given by the steady-state device simulation.

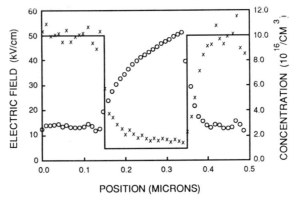

FIGURE 10.9. The electric field (o) and carrier density (x) profiles for a silicon diode. A voltage of 1.16V was applied across the device. After Lugli et al.[23]

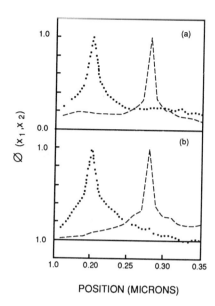

FIGURE 10.10. Normalized spatial autocorrelation functions for velocity (a) and energy (b) fluctuations inside the structure of Figure 10.9. The peaks represent different initial reference locations. After Lugli *et al.*[23]

In Figures 10.10 and 10.11, typical spatial correlation functions are shown. It is obvious that the fluctuations remain correlated over lengths of the order of several hundred angstroms and that the cross-correlation of the fluctuations in energy and momentum is not negligible. These results have significant implications in noise modeling. To model noise using an energy-momentum balance approach

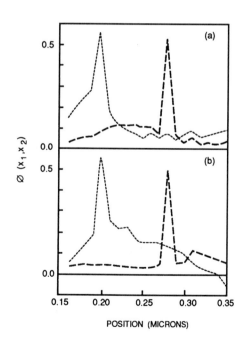

FIGURE 10.11. Spatial cross-correlations that correspond to Figure 10.9. (a) is for $\langle \Delta E(x) \Delta v(x + \delta x) \rangle$. (b) is for $\langle \Delta v(x) \Delta E(x + \delta x) \rangle$. The peaks represent the initial reference locations. After Lugli *et al.*[23]

both velocity and energy fluctuations must be included. Furthermore, the spatial correlation must also be included in the model, either in the source terms or in the equations which describe the response of the system to the fluctuations.

It is worth remembering that noise theory has previously gone astray as a result of an improper neglect of spatial correlations. In the early salami method of noise calculation, as we mentioned at the very beginning of the chapter, one neglected the correlation between the noise voltages measured across two different regions of the device. This is now known to be incorrect.[32] Yet, in its first application[33] it yielded exactly the correct answer. We expect that the assumption that the spatial correlation of the velocity and energy fluctuations is negligible will hold in devices which are much longer than a mean free path. However, practical application of these models[17] involves the subdivision of the device into a grid. We expect errors to occur whenever the length of the individual grid element becomes small when compared with a mean free path. This will be the case for short-gate FETs.

10.4.2. Nonstationary Transport of Average Values

Any noise calculation deals with fluctuations about a set of average values. The accuracy of the noise model therefore cannot be separated from the accuracy with which these average values are predicted. These average values furthermore are intimately related to the Green's functions that translate the fluctuations in the interior into fluctuating terminal voltages and currents. This leads to a second difficulty in the development of a noise model for submicron devices. The same conditions which lead to the existence of velocity overshooting and spatial correlation of the noise sources invalidate the use of a drift-diffusion model for carrier transport. This point to date has received relatively little investigation, most of which has focused on short-gate GaAs FETs.

The initial work in this direction was reported by Zimmermann et al.,[34] who found that the magnitude of the velocity fluctuations in a short-gate FET is a strong function of position (this is a noise analog of the velocity overshoot effect). Recently, Zimmermann and Constant[35] extended this approach to the calculation of terminal noise voltages or currents in unipolar one-dimensional structures. This has several consequences. First, all of the common techniques for noise modeling are based on the application of a drift-diffusion law for the production of the prerequisite Green's functions. Second, as we have seen, the magnitude of the velocity fluctuations at low frequencies is commonly equated with the diffusion coefficient. Under the conditions expected in a short-gate FET, a diffusion coefficient can no longer be properly defined.[11,24,26,36] One certainly cannot use the normal field-dependent diffusion coefficient and have confidence in the results. Lastly, we expect to see strong variations in the mean carrier energy in a short-gate FET, and, as we shall soon note, some attempts have been made to model noise in these devices with a set of balance equations which describe energy and momentum conservation. The set of equations used in this balance approach directly couple the evolution of the mean velocity (and therefore the decay of a velocity fluctuation) with the mean carrier energy. We therefore expect that a proper theory for noise in short-gate FETs must incorporate energy fluctuations as well as velocity fluctuations, and include the strong cross-correlation of energy and velocity fluctuations.

Carnez et al.[37] suggested a modification of the Shur energy momentum balance model[38] for a FET for use in FET noise calculations. They used the simple equations of Shur as a basis for the interior or intrinsic FET model. As discussed in Chapter 4, these equations are somewhat limited in their scope and they are not strictly true in a spatially inhomogenous region, such as the channel of a FET. Carnez et al. used the solution which they obtained for the mean carrier energy to produce the velocity fluctuations in a given section of the FET. This was done by replacing the normal field dependency of the diffusion coefficient with the equivalent steady-state energy dependency. However, they neglected the correlation of the fluctuations in adjoining sections and did not include energy fluctuations. This ad hoc model, in spite of its flaws, is the best attempt to date at modeling the physical noise arising in the FET channel. The results of Carnez et al. indicate that the neglect of relaxation effects can produce errors of 4 to 5 dB in the prediction of the intrinsic FET noise figure.

10.5. SHOT NOISE

A semiconductor device is generally connected to an external circuit. Carriers continually flow into the device through its terminal contacts. While we model current flow in a circuit as a continuous parameter, the actual flow of particles across the ends of the device is, in a semiclassical picture, discrete. This appears in circuits as shot noise, a noise associated fundamentally with the statistical distribution of times of entry of a carrier into the device.

10.5.1. Transit-Time Effects in Shot Noise

Typically, shot noise is treated as having a white-noise spectral density of magnitude $2eI_{dc}$, where e is the electronic charge and I_{dc} is the dc current density. This expression, though, assumes that the time needed for the carriers to transit the barrier is negligible. Here transit-time effects will be included by a simple model. It is assumed that carriers transit the barrier at a constant velocity and that the carrier transit time is τ_b. We will use the general Poisson point process shot-noise model[39] (a more detailed discussion of such models is found in Section 10.6)

$$i_e = \sum_{n=1}^{N_t} h(t - t_n), \tag{10.57}$$

where i_e is the total instantaneous current of the electrons, N_t is the total number of electrons that have entered the barrier region since the device was turned on, t_n is the time at which the nth electron enters the barrier, and $h(t - t_n)$ is the time-varying terminal or induced current contribution of the nth electron. This quantity, when integrated, must yield a single electronic charge and is the vehicle by which transport effects enter such models. For our assumptions

$$h(t - t_n) = \begin{cases} e/\tau_b, & t_n \le t \le t_n + \tau_b, \\ 0, & \text{elsewhere.} \end{cases} \tag{10.58}$$

When the injection process is not space-charge limited, N_t is statistically described by a Poisson counting process. As an example, Snyder[39] develops the following expression for the spectral density of shot noise:

$$S_i(f) = \lambda |H(f)|^2, \tag{10.59}$$

where $\lambda = I_{dc}/e$ and $h(f)$ is the Fourier transform of $h(t)$. If transit-time effects are negligible, then $h(t - t_n)$ is a delta function and $H(f)$ is white. Then we obtain the classic Schottky formula for shot noise,

$$S_i(f) = \lambda |H(0)|^2 = eI_{dc}. \tag{10.60}$$

Here we use (10.58) and (10.59) to develop a simple transit-time shot-noise model. We follow the line of thought:

$$H(f) = \int_0^\infty h(t) \exp[-j2\pi ft] \, dt \tag{10.61}$$

$$= \int_0^{\tau_b} \frac{e}{\tau_b} \exp[-j2\pi ft] \, dt \tag{10.62}$$

$$= \frac{e}{j2\pi f\tau_b} \{\exp(-j2\pi f\tau_b) - 1\} \tag{10.63}$$

$$= j\frac{e}{\theta_b} (1 - \cos\theta_b + j\sin\theta_b), \tag{10.64}$$

where $\theta_b = 2\pi f\tau_b$. This expression, not unsurprisingly, is very similar to the Gilden–Hines transit factor for the drift region of an IMPATT.[40,41] The spectral density function then is

$$S_i(f) = eI_{dc}|\Gamma|^2, \tag{10.65}$$

where

$$\Gamma = \frac{\sin\theta_b}{\theta_b} - j\frac{\cos\theta_b - 1}{\theta_b}. \tag{10.66}$$

As can be seen, the transit-time factor causes the shot noise to significantly deviate from the classic white spectral density predicted by the Schottky formula in the millimeter-wave region. While here we assume constant velocity, another interesting case, that of constant acceleration or ballistic transport, is example 10.13 in Papoulis.[42]

10.5.2. Barrier Transit-Time Diffusion Case

If diffusion is the main transport mechanism[2] then an estimate of the transit time can be obtained by using the Einstein formula (10.9). For a 0.3-μm region in silicon, the appropriate diffusion time t obtained using (10.9) is approximately 10 ps. As the limitation to this formula is that t must be much longer than the mean free time between phonon scattering events, its use is not inappropriate here.

10.5.3. Noise Sideband Correlation in Time-Periodic Systems

A common element in the noise theory for two-terminal junction devices is noise correlation. The presence of correlation effects in mixers was first noted by Strutt[43] in 1946. Simple analyses were presented later for mixer tubes by van der Ziel and Watters[44] and for tunnel diode mixers by Kim.[45] In 1958, Uhlir[46] presented a general analysis of correlated shot noise in p-n junction frequency converters, which was later extended by Dragone.[47] Correlation effects were included in the BARITT mixer analysis of McCleer[48] and in theory of Josephson junction mixers by Taur.[49] Standard Schottky barrier mixer analysis (e.g., Held and Kerr[50]) include such effects as they are both experimentally observable[51] and contradict several common misconceptions concerning the interrelationship of conversion loss and noise figure.[52]

The existence of similar correlation effects in IMPATT oscillators was first noted by Sjolund.[53] The prominent role played by these effects in IMPATT oscillators and amplifiers was clearly demonstrated by Kuvas.[54] Goedbloed and Vlaardingerbroek[55] included these correlation effects in an IMPATT model by use of a correlation matrix. Such correlation matrices were first introduced into mixer noise analysis by Dragone[47] and substantially ease the analytic formulation of the noise problem. The basic identity of noise correlation in mixers and noise correlation in oscillators makes them useful in both situations.

Here a unified theory of noise correlation in two-terminal devices is developed. First, a random process is used to demonstrate that noise correlation is expected in any device when it is driven by a periodic signal or pump. The correlation matrix elements are then formulated. It is shown that, although correlation effects in velocity fluctuation noise are conventionally neglected, such correlations are expected.

10.5.4. Mathematical Introduction

In this section the mathematical notation and operations used are introduced. The statistical average, mean, or expectation of a function g is denoted by the expectation operator $E[g]$. The autocovariance function of a random process $x(t)$ is defined as

$$K_x(t_1, t_2) = E[\langle x(t_1) - E[x(t_1)]\rangle\langle x(t_2) - E[x(t_2)]\rangle]. \qquad (10.67)$$

The autocorrelation function is defined as

$$R_x(t_1, t_2) = E[x(t_1)x(t_2)]. \qquad (10.68)$$

By replacing t_2 by $t_1 + \tau$ the autocovariance can be expressed as $K_x(t_1, t_1 + \tau)$. The Fourier transform of $K_x(t_1, t_1, +\tau)$ with respect to the *lag* time τ defines the power spectral density of the process as

$$S_x(f) = \frac{1}{2\pi} \int_{-\infty}^{\infty} K_x(t_1, t_1 + \tau) \, e^{-j2\pi f\tau} \, dt. \qquad (10.69)$$

Typically, a wide-sense stationary process is assumed in which K_x does not depend on the real time t_1. The crucial point in understanding the correlation effects is that they arise from a *periodic* dependency of K_x on t_1. Then the power spectral density as defined in equation (10.69) must also have a periodic dependency on t_1.

10.5.5. Origin of the Noise Correlation.

In this section a periodic dependence of $K_x(t_1, t_1 + \tau)$ on t_1 is shown to lead to noise correlation. A similar but abbreviated discussion is found in an appendix of Kuvas[54] and also in the paper of Sjolund.[53] The quantity needed in noise calculations is the autocorrelation of a noise variable Y. Since the noise variable has a zero mean,

$$R_Y(t_1, t_1 + \tau) = K_Y(t_1, t_1 + \tau). \qquad (10.70)$$

The goal of the discussion is to establish the correlation between the spectral power density at two frequencies ω_1 and ω_2. This quantity is

$$E[Y(\omega_1) Y(\omega_2)]$$
$$= \int_{-\infty}^{\infty} dt_1 \int_{-\infty}^{\infty} dt_2 \exp[-j(\omega_1 t_1 + \omega_2 t_2)]E[Y(t_1) Y(t_2)], \qquad (10.71)$$

where the factors $1/2\pi$ are temporarily dropped. They are reinserted at the end of this section. It is assumed that the autocovariance of Y is periodic in absolute time; i.e.,

$$E[Y(t)Y(t+\tau)] = K_Y(t, t+\tau) = \sum_{m=-\infty}^{\infty} K_m(\tau) \exp[jm\omega_{LO}], \quad (10.72)$$

where ω_{LO} is the radian frequency of the periodicity and

$$K_m(\tau) = \int_{-\infty}^{\infty} K_Y(t, t+\tau) \exp(-jm\omega_{LO}t)\, dt. \quad (10.73)$$

The first step in the discussion is to rewrite (10.71). The variables of integration are changed to

$$t_s = \tfrac{1}{2}(t_1 + t_2) \quad (10.74)$$

and

$$\tau = t_2 - t_1. \quad (10.75)$$

The arguments of the exponentials in (10.71) are rewritten as

$$(\omega_1 + \omega_2)t_s - (\omega_1 - \omega_2)\frac{\tau}{2} = (\omega_1 + \omega_2)\tfrac{1}{2}(t_1 + t_2) - (\omega_1 - \omega_2)\frac{t_2 - t_1}{2}$$

$$= \tfrac{1}{2}\omega_1 t_1 + \tfrac{1}{2}\omega_2 t_2 + \tfrac{1}{2}\omega_2 t_1 + \tfrac{1}{2}\omega_1 t_2$$

$$- \tfrac{1}{2}\omega_1 t_2 - \tfrac{1}{2}\omega_2 t_1 + \tfrac{1}{2}\omega_1 t_1 + \tfrac{1}{2}\omega_2 t_2$$

$$= \omega_1 t_1 + \omega_2 t_2 \quad (10.76)$$

The covariance term in the integrand becomes

$$E[Y(t_1)Y(t_2)] = E\left[Y\left(t_s - \frac{\tau}{2}\right)Y\left(t_s + \frac{\tau}{2}\right)\right]$$

$$= \sum_{m=-\infty}^{\infty} K_{ms}(\tau) \exp(jm\omega_{LO}t_s), \quad (10.77)$$

where

$$K_{ms}(\tau) = \int_{-\infty}^{\infty} K_Y\left(t_s - \frac{\tau}{2}, t_s + \frac{\tau}{2}\right) \exp(-jm\omega_{LO}t_s)\, dt_s. \quad (10.78)$$

Note that while we ordinarily think of a correlation in terms of the time differential between the two arguments, here we are expressing each argument in terms of a differential between some common central point in time, the time t_s. We shift the periodicity to this central time. Now (10.71) can be rewritten as

$$E[Y(\omega_1)Y(\omega_2)] = \int_{-\infty}^{\infty} dt_s \int_{-\infty}^{\infty} d\tau \exp\left[-j(\omega_1 + \omega_2)t_s + j(\omega_1 - \omega_2)\left(\frac{\tau}{2}\right)\right]$$

$$\cdot \sum_{m=-\infty}^{\infty} K_{ms}(\tau) \exp(jm\omega_{LO}t_s). \tag{10.79}$$

The next step is to evaluate (10.79). First, the summation is brought out of the integrals and the order of integration is interchanged:

$$E[Y(\omega_1)Y(\omega_2)] = \sum_{m=-\infty}^{\infty} \int_{-\infty}^{\infty} d\tau\, K_{ms}(\tau) \exp\left[j(\omega_1 - \omega_2)\left(\frac{\tau}{2}\right)\right]$$

$$\cdot \int_{-\infty}^{\infty} dt_s \exp[j(m\omega_{LO} - \omega_1 - \omega_2)t_s]. \tag{10.80}$$

The t_s integral is $2\pi\delta(\omega_1 + \omega_2 - m\omega_{LO})$. The τ integral is

$$\int_{-\infty}^{\infty} d\tau\, K_{ms}(\tau) \exp\left[j(\omega_1 - \omega_2)\left(\frac{\tau}{2}\right)\right]$$

$$= \int_{-\infty}^{\infty} ds \exp(-j\omega_{LO}s)$$

$$\cdot \int_{-\infty}^{\infty} d\tau\, K_Y\left(s - \frac{\tau}{2}, s + \frac{\tau}{2}\right) \exp\left[j(\omega_1 - \omega_2)\left(\frac{\tau}{2}\right)\right]. \tag{10.81}$$

Equation (10.81) defines a quantity which can be viewed as an instantaneous power spectral density

$$S(\omega, s) = \int_{-\infty}^{\infty} d\tau\, K_Y\left(s - \frac{\tau}{2}, s + \frac{\tau}{2}\right) \exp(-j\omega\tau)$$

$$= \sum_{m=-\infty}^{\infty} S_m(\omega) \exp(jm\omega_{LO}s), \tag{10.82}$$

where

$$S_m(\omega) = \int_{-\infty}^{\infty} ds \exp(-jm\omega_{LO}s)S(\omega, s). \tag{10.83}$$

Therefore, with the 2π factor that was dropped in (10.71) reinserted,

$$E[Y(\omega_1)Y(\omega_2)] = \sum_{m=-\infty}^{\infty} \delta(\omega_1 + \omega_2 - m\omega_{LO})S_m\left(\frac{\omega_2 - \omega_1}{2}\right). \quad (10.84)$$

Equation (10.84) shows that the noise power at the frequencies ω_1 and ω_2 is correlated when ω_1 and ω_2 are separated by a harmonic of the basic periodic frequency ω_{LO} of $K_Y(t_1, t_1 + \tau)$. For any other spacing of ω_1 and ω_2 they are not correlated even if $K_Y(t_1, t_1 + \tau)$ is periodically varying. If the process is stationary, then $S_m(\omega)$ is zero for all nonzero m.

10.6. POISSON POINT PROCESS MODEL OF CARRIER INJECTION

10.6.1. Poisson Point Processes

The current is a summation of the contributions of each individual charge carrier as the carrier transits the device. The noise results from statistical variations in the rate at which carriers transit the injection region. An appropriate model of this process is the filtered Poisson point process[39]

$$I(t) = \sum_{n=1}^{N(t)} h(t - \tau_n). \quad (10.85)$$

In (10.85), $I(t)$ is the current and $N(t)$ is the total number of carriers that have been injected into the device since t_0, the point in time at which the device was turned on. The quantity $N(t)$ is a random variable called the counting variable. For a Poisson point process, $N(t)$ satisfies

1. $\Pr[N(t_0) = 0] = 1.$ $\quad (10.86)$

2. $\Pr[N(t) - N(s) = n] = \dfrac{1}{n!}[\Lambda(t) - \Lambda(s)]^{\Lambda} \exp\{-[\Lambda(t) - \Lambda(s)]\}.$ $\quad (10.87)$

3. $N(t)$ has independent increments; i.e., the number of points in any set of nonoverlapping intervals is statistically independent.

The points counted by the counting variable $N(t)$ are carriers entering the device. Property 2 states that the number of points in any interval is Poisson distributed and therefore is responsible for the nomenclature "Poisson point process." The parameter function $\Lambda(t)$ in (10.87) need only be nonnegative and nondecreasing. It need not be either continuous or differentiable. Here, however, it is useful to assume that

$$\Lambda(t) = \int_{t_0}^{t} \lambda(s) \, ds, \tag{10.88}$$

where $\lambda(s)$ is a nonnegative function of time called the intensity function. The intensity function is the instantaneous average rate at which points occur; i.e., carriers enter the device (i.e., the particle current).

The only parameter not yet defined in (10.85) is $h(t - \tau_n)$; τ_n is the time at which the nth carrier enters the device, and $h(t - \tau_n)$ is the filter impulse-response function. It satisfies all the properties of a causal linear time-invariant filter-response function. In particular,

$$h(t - \tau_n) = 0, \qquad t < \tau_n, \tag{10.89}$$

and

$$\int_{0}^{\infty} h(s) \, ds = e, \tag{10.90}$$

where e is the charge of the carrier. Although here $h(t - \tau_n)$ is assumed to be a delta function, the effects of a nonzero injection region transit time can be included by using a nondelta function form for $h(s)$.

Snyder[39] shows that the autocovariance of $I(t)$ is

$$K_I(t_1, t_2) = \int_{0}^{\min(t_1, t_2)} \lambda(\tau) E[h(t_1 - \tau)h(t_2 - \tau)] \, d\tau. \tag{10.91}$$

For the deterministic delta-type h functions used here

$$K_I(t_1, t_2) = \int_{t_0}^{\min(t_1, t_2)} \lambda(\tau) e^2 \delta(t_1 - \tau) \, d\tau = e^2 \lambda[\min(t_1, t_2)]. \tag{10.92}$$

If $t_2 = t_1 + \tau$, then

$$K_I(t_1, t_1 + \tau) = \begin{cases} e^2 \lambda(t_1), & \tau \geq 0, \\ e^2 \lambda(t_1 + \tau), & \tau < 0. \end{cases} \tag{10.93}$$

If $\lambda(s)$ is periodic, then $K_I(t_1, t_1 + \tau)$ will also be periodic. Therefore, if the instantaneous rate at which carriers enter the device is periodic (i.e., the current is periodic), $K_I(t_1, t_1 + \tau)$ is also periodic in the absolute time t_1 and noise correlation is expected.

10.6.2. Connection with Deterministic Models

The expected value of $I(t)$ is[39]

$$E[I(t)] = \int_{t_0}^{t} \lambda(s)E[h(t-s)]\,ds. \qquad (10.94)$$

For deterministic delta h functions

$$E[I(t)] = e\lambda(t). \qquad (10.95)$$

Since the deterministic models of $I(t)$ typically used to determine the injection behavior actually predict $I(t)$, they can be used in conjunction with (10.95) to determine $\lambda(t)$.

When the cases of periodic $\lambda(t)$ are considered, (10.92) and (10.95) lead to the result

$$K_I(t_1, t_1 + \tau) = \begin{cases} eI(t_1), & \tau \geq 0, \\ eI(t_1 + \tau), & \tau < 0, \end{cases} \qquad (10.96)$$

and therefore, the Fourier expansion of K_I with respect to t_1 is e times the Fourier expansion of $I(t)$. Equation (10.84) becomes

$$E[I(\omega_1)I(\omega_2)] = e \sum \delta(\omega_1 + \omega_2 - m\omega_{LO})I_m[\tfrac{1}{2}(\omega_2 - \omega_1)], \qquad (10.97)$$

where I_m is the mth Fourier coefficient of $I(t)$.

10.6.3. Correlated Velocity Fluctuation Noise

Generally, velocity fluctuation noise is treated as being uncorrelated. Under these conditions it, like thermal noise, contributes only to the main diagonal of the correlation matrix. It is not difficult, however, to show that in a periodic field, velocity fluctuation noise is expected to have correlation structure. The following random flight process is considered:

$$V(t) = V_0 + \sum_{i=0}^{N(t)} \Delta v_i. \qquad (10.98)$$

Here V_0 is the initial carrier velocity at time t_0, $N(t)$ is some counting random variable, and ΔV_i is the total change in velocity as a result of the ith flight. Now, ΔV_i has two components. One is the random change due to scattering at the end of the flight. Since the scattering rates are determined by the instantaneous velocity (momentum), the scattering component of ΔV_i will be periodic when the instantaneous average velocity is periodic. This happens when the field is periodic. The second component of ΔV_1 is the field acceleration during the flight. When the field is periodic, this component will be periodic also.

If ΔV_i is periodic and if $N(t)$ is Poisson process with a constant rate parameter, (10.91) clearly shows that the autocovariance of $V(t)$ is periodic and therefore correlation structure will be present. In fact, in all Monte Carlo programs the model actually implemented is an expanded version of (10.98) (V_0 and ΔV_1 become vectors with additional components representing energy and valley) with a Poisson distributed $N(t)$ that has a constant rate parameter. This results from the use of the self-scattering mechanism of Rees.[56] Self-scattering, as discussed in Chapter 4, is a fictitious scattering mechanism in which the carrier state is left totally unchanged. It, therefore, has no effect on carrier transport. However, by properly specifying its rate as a function of an electronic wave vector the "total" scattering rate (real scattering plus self-scattering) becomes constant and independent of wave vector. When the total scattering rate is constant, the flight duration between scatterings is an exponentially distributed random variable. Self-scattering is used in Monte Carlo programs because it is less expensive to compute this exponentially distributed flight duration than the real flight duration between real scattering events. In spite of this great simplification, self-scattering is rarely used in analytical transport models. Here, since any point process in which the distance between points (the arrival time) is exponentially distributed is a Poisson process, employing self-scattering in the analytic model allows a well-understood random process to be used as the basic model. One obvious conclusion of such an analysis is that velocity fluctuation noise in the presence of a periodically varying field is expected to have correlated sideband structure.

10.7. NOISE IN BALLISTIC DIODES

Recently attention has been focused on noise in short "ballistic" diodes.[57-59] These short diodes are not envisioned as useful devices, but instead have been used in attempts at experimentally observing ballistic transport. Initially it was argued that this could be done by observing the dc current–voltage characteristics of these diodes.[60] Later work, however,[61-63] showed that the dc i-v curves by themselves cannot prove that the current is transported ballistically, and interest in alternative methods, such as noise measurements, of observing ballistic transport arose. Van der Ziel and Bosman[57,58] have developed a noise theory for the pure ballistic case which does not agree with the experimental work of Schmidt et al.,[59] who argue that this can be attributed to carrier drag effects. This appears to be in good agreement with a wide variety of Monte Carlo studies, which can be interpreted as showing drag effects at very short time scales.[64] Schmidt et al. found that in short n^+nn^+ devices there is little $1/f$ noise and argued that this illustrates a transport or lattice scattering basis for the existence of $1/f$ noise. There has been a controversy over $1/f$ noise virtually forever. The most recent version of a universal (and controversial) theory for $1/f$ noise is discussed by van der Ziel in a recent tutorial article.[65]

REFERENCES

1. K. M. van Vliet, A. Friedman, R. J. J. Zijlstra, A. Gisolf, and A. van der Ziel, *J. Appl. Phys.* **46**, 1804 (1975).
2. A. van der Ziel and E. R. Chenette, *Advances in Electronics and Electron Physics* (L. Marton, ed.), vol. 46, pp. 313–383, Academic Press, New York (1978).
3. W. Shockley, J. A. Copeland, and R. P. James, in: *Quantum Theory of Atoms, Molecules and the Solid State* (P. O. Lowdin, ed.), Academic Press, New York (1966).
4. P. J. Price, *J. Appl. Phys.* **31**, 949 1960.
5. J. P. Nougier, in: *Physics of Nonlinear Transport in Semiconductors* (D. K. Ferry, J. R. Barker, and C. Jacoboni, eds.), Plenum, New York (1980).
6. K. M. van Vliet and A. van der Ziel, *Solid-State Electron.* **20**, 931 (1977).
7. J. L. Moll, *Physics of Semiconductors*, McGraw-Hill, New York (1964).
8. G. M. Jenkins and D. G. Watts, *Spectral Analysis and Its Applications*, Holden-Day, San Francisco (1968).
9. *Programs for Digital Signal Processing* (Digital Signal Processing Committee, ed.), IEEE, New York (1980).
10. G. Hill, P. N. Robson, and W. Fawcett, *J. Appl. Phys.* **50**, 356 (1979).
11. R. Fauquembergue, J. Zimmermann, A. Kaszynski, E. Constant, and G. Microondes, *J. Appl. Phys.* **52**, 1065 (1980).
12. A. Kaszynski, *Etude des phenomenes de transport dans les materiaux semiconducteurs par les methodes de Monte Carlo: Application a l'arseniure de gallium de type N*, Theses Docteur Ing., L'Universite des Sciences et Techniques de Lille (1979).
13. R. O. Grondin, P. A. Blakey, J. R. East, and E. D. Rothman, *IEEE Trans. Electron Dev.* **ED-28**, 914 (1981).
14. H. D. Rees, *IEEE Solid St. Electron Dev.* **1**, 165 (1977).
15. B. M. Oliver, *Proc. IEEE* **53**, 436 (1965).
16. Baechtold, W., *IEEE Trans. Electron Dev.* **ED-19**, 674 (1972).
17. R. A. Pucel, H. A. Haus, and H. Statz, in: *Advances in Electronics and Electron Physics* (L. Marton, ed.), vol. 38, Academic Press, New York (1975).
18. J. Frey, *IEEE Trans. Electron Dev.* **ED-23**, 1298 (1976).
19. J. Graffeuil, J-F. Sautereau, G. Blasquez, and P. Rossel, *IEEE Trans. Electron Dev.* **ED-25**, 596 (1978).
20. J. M. Golio and R. J. Trew, *IEEE Trans. Electron Dev.* **ED-27**, 1256 (1980).
21. K. Takagi and A. van der Ziel, *Solid-State Electron.* **22**, 285 (1979).
22. S. Weinreb, *IEEE Trans. Microwave Theory Tech.* **MTT-28**, 1041 (1980).
23. P. Lugli, R. O. Grondin, and D. K. Ferry, in: *The Physics of Submicron Structures* (H. L. Grubin, K. Hess, G. J. Iafrate, and D. K. Ferry, eds.), Plenum Press, New York (1984).
24. J. P. Nougier, J. C. Vaissiere, and C. Gontand, *Phys. Rev. Lett.* **51**, 513 (1983).
25. P. A. Lebwohl and P. J. Price, *Appl. Phys. Lett.* **19**, 530 (1971).
26. D. K. Ferry and J. R. Barker, *J. Appl. Phys.* **52**, 818 (1981).
27. P. Lugli, J. Zimmermann, and D. K. Ferry, *J. Phys.* **42** (Suppl. 10), C7-103 (1981).
28. P. J. Price, in: *Fluctuation Phenomena in Solids* (R. E. Burgess, ed.), p. 355, Academic Press, New York (1965).
29. V. L. Gurevich, *Sov. Phys.-JETP* **17**, 1252 (1963).
30. V. P. Kalashnikov, *Physica* **48**, 93 (1970).
31. D. K. Ferry, *J. Phys.* **42** (Suppl. 10), C7-253 (1981).
32. K. K. Thornber, *Solid-State Electron.* **17**, 95 (1973).
33. A. van der Ziel, *Solid-State Electron.* **9**, 899 (1966).

34. J. Zimmermann, Y. Leroy, A. Kaszynski, and B. Carnez, Monte Carlo Calculation of Nonsteady State Hot Electron Noise in very short Channel n-Si and n-GaAs devices, in: *Proc. 5th Int. Conf. Noise in Physical Systems* (D. Wolf, ed.), Springer Series on Electrophysics, vol. 2, Springer, New York (1978).

35. J. Zimmermann and E. Constant, Application of Monte Carlo techniques to hot carrier diffusion noise calculation in unipolar semiconducting components, *Solid-State Electron.* **23**, 915–925 (1980).

36. C. Jacoboni, *Phys. Stat. Solids* (*b*) **65**, 61 (1974).

37. B. Carnez, A. Cappy, R. Fauquembergue, E. Constant, and G. Salmer, *IEEE Trans. Electron Dev.* **ED-28**, 784 (1981).

38. M. Shur, *Electron. Lett.* **12**, 615 (1976).

39. D. L. Snyder, *Random Point Processes*, Wiley, New York (1975).

40. P. J. McCleer, D. Snyder, R. O. Grondin, and G. I. Haddad, *Solid-State Electron.* **24**, 37 (1981).

41. M. Gilden and M. E. Hines, *IEEE Trans. Electron Dev.* **ED-13**, 169 (1966).

42. Atharosios Papoulis, *Probability, Random Variables, and Stochastic Processes*, McGraw-Hill, New York (1965).

43. M. J. O. Strutt, *Proc. IRE* **34**, 942 (1946).

44. A. van der Ziel and R. L. Watters, *Proc. IRE* **46**, 1426 (1958).

45. C. S. Kim, *IEEE Trans. Electron Dev.* **ED-8**, 394, 1961.

46. A. Uhlir, Jr., *Bell System Tech. J.* **37**, 951 (1958).

47. C. Dragone, *Bell System Tech. J.* **47**, 1883 (1968).

48. P. J. McCleer, *Frequency Conversion in Punch-Through-Semi-conductor Devices*, Technical Report No. 143, Electron Physics Laboratory, Doctoral Dissertation, Department of Electrical and Computer Engineering, The University of Michigan, Ann Arbor, August (1978).

49. Y. Taur, *IEEE Trans. Electron Dev.* **ED-27**, 1921 (1980).

50. D. N. Held and A. R. Kerr, *IEEE Trans. Microwave Theory Techn.* **MTT-26**, 49 (1978).

51. N. J. Keen, *Electron. Lett.* **13**, 282 (1977).

52. A. R. Kerr, *IEEE Trans. Microwave Theory Tech.* **MTT-27**, 135 (1979).

53. A. Sjolund, *Int. J. Elec.* **34**, 551 (1973).

54. R. L. Kuvas, *IEEE Trans. Electron Dev.* **ED-23**, 395 (1976).

55. J. J. Goedbloed and M. T. Vlaardingerbroek, *IEEE Trans. Microwave Theory Tech.* **MTT-25**, 324 (1977).

56. H. D. Rees, *J. Phys. Chem. Solids* **30**, 643 (1969).

57. A. van der Ziel and G. Bosman, *Phys. Stat. Solids* (*a*) **73**, K87 (1982).

58. A. van der Ziel and G. Bosman, *Phys. Stat. Solids* (*a*) **73**, K93 (1983).

59. R. R. Schmidt, G. Bosman, C. M. van Vliet, L. F. Eastman, and M. Hollis, *Solid-State Electron.* **26**, 437 (1983).

60. M. S. Shur and L. F. Eastman, *Electron. Lett.* **16**, 522 (1980), L. F. Eastman, R. Stall, D. Woodward, N. Dandekar, C. E. C. Wood, M. S. Shur, and K. Board, *Electron. Lett.* **16**, 524 (1980).

61. A. J. Holden and B. T. Debney, *Electron. Lett.* **18**, 558 (1982).

62. J. R. Barker, D. K. Ferry, and H. L. Grubin, *IEEE Electron Dev. Lett.* **EDL-1**, 209 (1980).

63. D. K. Ferry, J. Zimmermann, P. Lugli, and H. Grubin, *IEEE Electron Dev. Lett.* **EDL-2**, 228 (1981).

64. R. O. Grondin, P. Lugli, and D. K. Ferry, *IEEE Electron Dev. Lett.* **EDL-3**, 373 (1982).

65. A. van der Ziel, *Proc. IEEE* **76**, 233 (1988).

Index